VOLUME FIFTY SIX

ANNUAL REPORTS IN MEDICINAL CHEMISTRY

The Design of Covalent-Based Inhibitors

ANNUAL REPORTS IN MEDICINAL CHEMISTRY

VOLUME 56

Editor-in-Chief

STEPHEN NEIDLE

School of Pharmacy, University College London, London, United Kingdom

VOLUME FIFTY SIX

Annual Reports in MEDICINAL CHEMISTRY

The Design of Covalent-Based Inhibitors

Volume Editors

RICHARD A. WARD
Oncology R&D Business Development, AstraZeneca, Cambridge, United Kingdom

NEIL P. GRIMSTER
Oncology R&D, AstraZeneca, Waltham, MA, United States

ELSEVIER

ACADEMIC PRESS
An imprint of Elsevier

Academic Press is an imprint of Elsevier
50 Hampshire Street, 5th Floor, Cambridge, MA 02139, United States
525 B Street, Suite 1650, San Diego, CA 92101, United States
The Boulevard, Langford Lane, Kidlington, Oxford OX5 1GB, United Kingdom
125 London Wall, London, EC2Y 5AS, United Kingdom

First edition 2021

ISBN: 978-0-12-821689-7
ISSN: 0065-7743

For information on all Academic Press publications
visit our website at https://www.elsevier.com/books-and-journals

Publisher: Zoe Kruze
Acquisitions Editor: Jason Mitchell
Developmental Editor: Federico Paulo S. Mendoza
Production Project Manager: Denny Mansingh
Cover Designer: Miles Hitchen

Typeset by SPi Global, India

Contents

Contributors

Preface

Richard A. Ward[a,*] and Neil P. Grimster[b,*]
[a]Oncology R&D Business Development, AstraZeneca, Cambridge, United Kingdom
[b]Oncology R&D, AstraZeneca, Waltham, MA, United States
[*]Corresponding authors: e-mail address: richard.a.ward@astrazeneca.com; neil.grimster@astrazeneca.com

Targeted covalent inhibition (TCI) has emerged as a powerful strategy in medicinal chemistry over the preceding decade and a half. This approach relies on the reaction of an electrophilic warhead, often an acrylamide, with a nucleophilic amino acid sidechain residue, e.g., cysteine, to form a site-specific covalent linkage to a target protein. In so doing, TCI can provide a number of distinct advantages over reversible inhibitors, including increased potency, selectivity, and prolonged duration of action. However, TCI is not a panacea, and knowing when and where to deploy it as a strategy is key. In this book, we have brought together leading industrial and academic medicinal chemists in order to provide a road map to help the reader better understand the intricacies involved in the development of covalent inhibitors.

It has previously been stated that TCI has seen a "resurgence"[1] in popularity, due to the large number of compounds now entering the market that utilize a covalent mechanism of action. Many of the earliest and most well-known therapeutic agents also elicit their biological effects through TCI, e.g., aspirin, but these compounds were often discovered phenotypically, and their mode of action was only revealed decades later. In fact, until recently medicinal chemists had eschewed covalent inhibitors and diligently removed any suspected reactive compounds from high-throughput screening decks. This rejection of TCI as a viable strategy was invariably driven by a fear of idiosyncratic toxicities, which it was believed would arise from indiscriminate reaction of the covalent warheads with off-target proteins, leading to unwanted biological effects and/or haptenization. What has evolved over the preceding decades is our understanding of the factors that go into a successful covalent inhibitor program. First, we now possess a more complete understanding of the reactivity profile that is required of a covalent warhead. To this end, covalent inhibitor-based medicinal chemistry programs will often utilize glutathione counter screens to evaluate the reactivity of warheads to help prevent both indiscriminate reactivity and rapid metabolism. Second, through computational and structural evaluations, it is now possible to determine whether a nucleophilic residue of interest is predisposed to covalent modification. These approaches examine the chemical environment of the residue within the binding site to determine if its pK_a has been perturbed to positively or negatively modulate its reactivity toward covalent modification. The combination of these two factors now allows rapid determination of whether a covalent approach should be utilized in the prosecution of a desired target.

It should be noted that covalent drug discovery is still in its infancy, with large areas still in need of further exploration. In this book, we have tried to cover all aspects of TCI, to give the reader a complete understanding of the factors that influence a covalent-based drug discovery approach. Although each of these chapters can be read as an individual standalone review of the topic being covered, there are themes which are picked up in multiple chapters which allow the readers to review key areas from the perspective of multiple authors.

In Chapter 1 McWhirter describes the physical chemistry underpinnings and the mathematical models used to fully characterize covalent inhibitors, and provides a description of the types of PK/PD models that have been used to evaluate them in an in vivo setting. While TCI has shown great promise in oncology, other disease areas have been less willing to undertake a covalent-based approach, but in Chapter 2 Sung details some of the latest efforts to leverage covalent inhibition in the area of immune-mediated diseases.

In Chapter 3, Gao details some of the latest research looking at the use of reversible-covalent warheads in TCI, and how this strategy can provide additional benefits beyond traditional approaches. One limitation of TCI, when targeting non-activated side-chain residues, is its reliance on the placement of a cysteine proximal to a small molecule binding site. Jones details in Chapter 4 novel methods that target side-chain residues other than cysteine for covalent modification, and further expanding the utility of TCI. One novel warhead that has been clinically validated is boronic acids, and in Chapter 5 Leach provides examples of the boron-based reversible covalent warheads. Chapter 6 by Rowley gives an in-depth review of the computational tools that have been developed to predict the reactivity of both warheads and residues, and how this can be utilized to identify opportunities to deploy TCI, while in Chapter 7, Abel examines the cutting-edge molecular modeling of covalent inhibitors, detailing approaches such as docking, structure-based drug design (SBDD), virtual screening, molecular dynamics (MD), and free energy perturbation

(FEP). Chapter 8 by London provides a description of some of the novel screening approaches used to identify covalent inhibitors' starting points, and these include fragment-based screening, disulfide screening, and chemo-proteomics. Finally, Keserű in Chapter 9 provides a prospective of how the area of TCI will evolve over the coming decade.

We would like to thank all of the authors and coauthors for their hard work in contributing to the chapters for this book. These authors have brought a huge wealth of experience in drug discovery research, across many research organizations in the pharmaceutical industry and academia.

Reference

1. Singh, J.; Petter, R. C.; Baillie, T. A.; Whitty, A. The Resurgence of Covalent Drugs. *Nat. Rev. Drug Discov.* **2011**, *10*, 307–317.

CHAPTER ONE

Kinetic mechanisms of covalent inhibition

Claire McWhirter*
Artios Pharma Limited, Babraham Research Campus, Cambridge, United Kingdom
*Corresponding author: e-mail address: claire.mcwhirter@artiospharma.com

Contents

1. Introduction

The resurgence of covalent inhibitors over the past decade has been brought about by multiple factors including the clinical success of several covalent inhibitors and a better understanding of the safety risks associated with covalent inhibition.[1,2] In addition, there has been a realization that for

Annual Reports in Medicinal Chemistry, Volume 56
ISSN 0065-7743
https://doi.org/10.1016/bs.armc.2020.11.001

many targets, extended target occupancy through a long target residence time is required for clinical efficacy.[3,4] Once an irreversible covalent inhibitor has formed a covalent complex with the target protein there is, by definition, no dissociation and so the target residence time effectively becomes infinite (although in practice other factors such as drug clearance and protein turnover may still limit target occupancy).[4] Similarly, there is an increasing understanding that in some circumstances an extended target residence time may not be clinically desirable and so reversible covalent inhibitors, where the target residence time may be tuned, have received increasing attention as a therapeutic strategy.[3,5]

Due to the increased understanding of both the benefits and risks of covalent inhibition modern covalent drug discovery programs involve the careful balancing of nucleophile reactivity with a high affinity scaffold. As a result, covalent inhibitors require detailed mechanistic characterization throughout the drug discovery process.

Covalent inhibition is time dependent and is intrinsically related to both the affinity and reactivity of the covalent inhibitor. Therefore, the use of biochemical IC_{50} to rank covalent inhibitors is discouraged as the ranking may vary depending on the length of any preincubation step (or assay duration).[6] In the extreme, if left long enough before measuring, the IC_{50} of a covalent inhibitor will reflect the concentration of the target protein in the assay rather than the affinity and reactivity of the individual inhibitor.[6] Therefore, detailed kinetic characterization is essential to properly rank and interpret structure-activity relationships of covalent inhibitors by taking into account both the affinity and reactivity of the inhibitor and generation of k_{inact}/K_I values.[6,7]

2. Mechanisms

Covalent inhibitors are characterized by the formation of a chemical bond between an electrophilic functional group on the inhibitor (often referred to as a covalent warhead) and a nucleophilic residue on the target protein.

It is usually assumed that the formation of a covalent enzyme-inhibitor (*E*-I) complex is irreversible. However, there are several examples of reversible covalent inhibitors including Avibactam which covalently modifies the active-site Serine residue of TEM-1 and cyanoacrylamide inhibitors of several kinases including RSK1 and JAK3.[5,8–10]

There are several mechanisms by which a covalent inhibitor may react to form a covalent *E*-I complex.[11] These mechanisms vary in the requirement for prior formation of a reversible EI complex and whether they utilize the catalytic machinery of an enzyme's active site. The covalent mechanisms most commonly encountered in drug discovery are described below.

2.1 One-step covalent inhibition

In a one-step mechanism of covalent inhibition (which may also be referred to as non-specific or residue affinity labeling), the inhibitor reacts directly with the protein without prior formation of a reversible EI complex (Fig. 1A). The lack of formation of an initial reversible EI complex means that any selectivity is due to the relative nucleophilicity and steric availability of the target residues along with chemoselectivity of the covalent warhead for specific nucleophilic residues.[11] One-step inhibitors are often highly reactive and relatively promiscuous inhibitors such as iodoacetate or N-ethylmalemide, which are assumed to have negligibly low initial binding affinity.[7] As they have high intrinsic reactivity, there are few examples of drugs with one-step mechanisms of inhibition. DNA alkylating agents, which are used in the treatment of some forms of cancer, are a notable exception.

2.2 Two-step covalent inhibition

Two-step covalent inhibition requires an initial reversible binding event where the enzyme and inhibitor form a non-covalent EI complex, which may then either dissociate or react to form the covalent *E*-I complex. Fig. 1B depicts a two-step mechanism which is in rapid equilibrium and it is assumed that the inactivation rate constant (k_2) is negligibly small compared to the dissociation constant of the reversible EI complex. As the reversible binding is in rapid equilibrium, only the macroscopic equilibrium constant (K_i) can be measured for the initial encounter complex.

While most literature treatments of two-step covalent inhibition assume rapid equilibrium for the initial reversible binding event, there are numerous examples of reversible inhibitors with slow binding kinetics and so the assumption of rapid equilibrium for the initial binding event may not always be valid.[12–14] The steady-state model shown in Fig. 1C is a more general description of two-step covalent inhibition in which the magnitudes of the individual rate constants for the initial reversible binding event (k_{on} and k_{-off}) may be any arbitrary value.

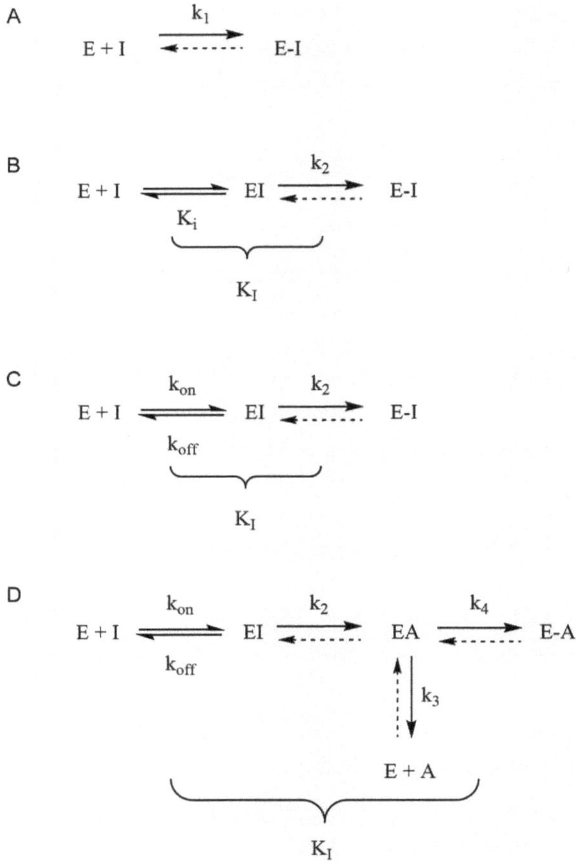

Fig. 1 Mechanisms of covalent inhibition. (A) One-step inhibition, (B) two-step rabid equilibrium, (C) two-step steady state, (D) mechanism-based covalent inhibition. E-I (or in the case of mechanism-based inhibition E-A) denotes the formation of a covalent bond. For irreversible covalent inhibition, there is no breakdown of the covalent E-I complex.

As two-step targeted covalent inhibitors are usually derived from reversible scaffolds with relatively high affinity, the intrinsic reactivity of the covalent warhead does not generally need to be as high as for one-step inhibitors, which may therefore reduce the likelihood of off-target covalent modification and compound toxicity. Examples of clinically relevant two-step inhibitors include the kinase inhibitors afatinib, osimertinib, ibrutinib and acalabrutinib, along with the kRas G12C inhibitor AMG 510.[15]

2.3 Mechanism-based inactivation

In addition to the mechanisms described above, it is also possible to have more complex mechanisms of covalent inhibition where a substrate analogue is converted to a chemically reactive intermediate by the enzyme's catalytic machinery. This intermediate is then able to form a covalent adduct with the target protein (Fig. 1D).[7,16] If the reactive intermediate dissociates and inactivates another target, it is considered to be a prodrug.[11] The ratio of dissociation of the reactive intermediate (k_3 in Fig. 1D) to reaction with the target enzyme (k_4 in Fig. 1D) is known as the partition ratio.[16] The magnitude of the partition ratio depends on the reactivity of the intermediate, the proximity of a nucleophilic target residue, and the rate of diffusion of the reactive intermediate from the active site.[16] An inhibitor with a partition ratio of zero generates an inactivated enzyme with every turnover.

The perceived advantage of mechanism-based inhibitors is that as they target active site nucleophiles, which are required for the enzyme to function, they are less likely to develop resistance than targeted covalent inhibitors. In addition, the reactive species is generated through catalysis by the target enzyme rather than inherent reactivity of the inhibitor and so may be less prone to off-target covalent modification than one- or two-step inhibitors.[11] There are numerous clinically relevant mechanism-based covalent inhibitors, many of which have been in use for decades such as aspirin, penicillin, vigabatrin and warfarin.[1] In spite of these potential advantages and historical precedent mechanistic covalent programs have often focused on the development of targeted covalent inhibitors rather than mechanism-based inhibitors. Which is in part due to the perceived difficulty in rational design of mechanism-based covalent inhibitors.[1,11]

2.4 Kinetic constants

k_{inact} (k_2 in Fig. 1B and C) is the first-order rate constant that defines the maximum rate of enzyme inactivation. For mechanism-based inhibition (Fig. 1D), k_{inact} is described by Eq. (1).[16]

$$k_{\text{inact}} = \frac{k_2 k_4}{(k_2 + k_3 + k_4)} \tag{1}$$

K_i is the dissociation constant of the initial reversible EI complex (Eq. 2).

$$K_i = \frac{k_{off}}{k_{on}} \tag{2}$$

K_I is the concentration of inhibitor required to achieve a rate of inactivation that is equivalent to half of k_{inact}. Under certain circumstances, K_I approximates to K_i and so they are often treated as interchangeable. However, the equation describing K_I varies depending on the mechanism of inhibition and rate-limiting step of the reaction and in many cases, the measured K_I will be larger than the affinity of the initial reversible binding.[7,17] For two-step steady-state inhibitors, K_I is described by Eq. (3).[17,18]

$$K_I = \frac{(k_{off} + k_2)}{k_{on}} \tag{3}$$

Under the rapid equilibrium approximation for two-step inhibitors where the chemical bond formation step is rate-limiting, the contribution from k_2 is negligible and so K_I approximates to K_i.[17,19] This is analogous to the Briggs–Haldane treatment of enzyme-substrate kinetics where K_M approximates to K_S for systems where the catalytic step is rate-limiting but is larger than K_S when the catalytic step is not rate-limiting.[17,20] For low affinity weakly reactive two-step covalent inhibitors the approximation of $K_I \approx K_i$ is often valid. However, for highly optimized clinical candidates which comprise of a moderately reactive warhead and a high affinity binding scaffold the value of k_2 may be similar to, or larger than, k_{off}. Under these conditions the measured K_I value will be larger than the initial reversible binding affinity.

For mechanism-based inhibitors, K_I is described by Eq. (4).[16]

$$K_I = \left(\frac{(k_{off} + k_2)}{k_{on}}\right)\left(\frac{(k_3 + k_4)}{k_2 + k_3 + k_4}\right) \tag{4}$$

k_{inact}/K_I (sometimes referred to as the covalent efficiency constant) is a second-order rate constant which accounts for both the affinity of the initial reversible encounter complex and the maximal rate of covalent bond formation. It is analogous to the enzyme specificity constant k_{cat}/K_M in enzyme-substrate kinetics.[20] As it is a time independent parameter, which reflects both the affinity and reactivity of an inhibitor, k_{inact}/K_I is the key metric for ranking and SAR determination of irreversible inhibitors.[6,7] For one-step inhibitors, k_1 (Fig. 1A) is equivalent to k_{inact}/K_I. This may be rationalized as the initial binding affinity being too weak to be measurable in the assay system and is similar conceptually to the "hit-and-run" approximation in enzyme substrate kinetics literature.[6,7,21]

Historically, covalent inhibitors have tended to be mechanism-based inhibitors, although many of these inhibitors were discovered serendipitously and the mechanism of action determined post hoc rather than being rationally designed.[1] Examples of mechanism-based covalent inhibitors include beta lactam inhibitors (such as penicillin) which bind covalently to the active site serine of penicillin binding protein 1b, covalent monoamine oxidase inhibitors which form irreversible covalent adducts with flavin, and protease inhibitors which acylate the active site nucleophile.[22–25]

While mechanism-based inhibitors have been a successful strategy for covalent inhibition, many enzymes do not form intermediates during catalysis, which limits the applicability of mechanism-based inhibition. An alternative approach is to target non-catalytic nucleophiles. Such inhibitors, which may be referred to as targeted covalent inhibitors, are often developed from an existing reversible scaffold to which a covalent warhead is attached via a linker.[1] While the targeted nucleophile is frequently located within the protein's active site, the strategy has also been successfully utilized for inhibitors which bind in allosteric sites.[26–28] This approach has greatly expanded the number of proteins that may be targeted by a covalent inhibition strategy.[29,30] For example, while kinases do not form a covalent bond with their substrate and are therefore not targetable using mechanism-based covalent inhibitors, it is estimated that over two hundred kinases have a cystine available for a targeted covalent approach.[31] Notable successes of a targeted covalent approach include irreversible inhibitors of the kinases EGFR and BTK, for which there are several targeted covalent inhibitors on the market and the kRas G12C inhibitors AMG 510 and MRTX849 which are currently in clinical development.[32–36]

3. Data analysis techniques

One of the main challenges with covalent inhibitor design is accurate assessment of inhibitor potency as it depends on both the affinity and chemical reactivity of the molecule. Unlike rapid-equilibrium reversible inhibitors, covalent inhibitors have slow kinetics and the progress curves of irreversible inhibition show curvature over time (Fig. 2A). Several analytical methodologies have been developed to characterize covalent inhibitors which differ in both their suitability to rank covalent inhibitors, and their ease of use.

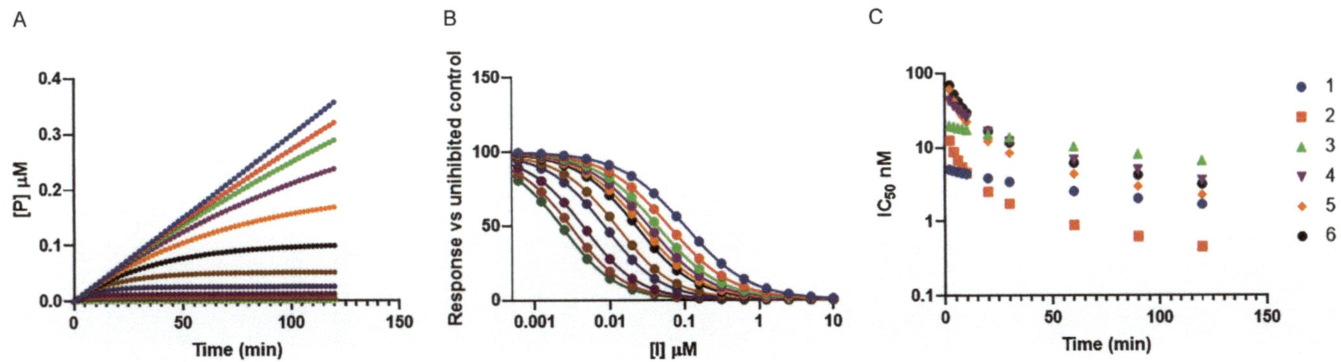

Fig. 2 (A) Simulated time-course data for a rapid-equilibrium covalent inhibitor with a K_I of 50 nM and k_{inact} 0.01 s^{-1}. Simulated assay conditions [S] = 10 µM (K_M), [E] = 100 pM, k_{cat} = 1 µM s^{-1}. 16 × 2-fold dilutions of inhibitor (10 µM top concentration) and a no inhibitor control were simulated. (B) IC_{50}s of covalent inhibitor with a K_I of 50 nM and k_{inact} of 0.001 s^{-1}. t = 2, 4, 6, 8, 10, 20, 30, 60, 90 and 120 min. A reversible inhibitor with a K_I of 50 nM is used as a reference. (C) IC_{50} vs time for a range of inhibitors with clinically relevant values of k_{inact} and K_I. 1 = 2.5 nM K_I, 0.0005 s^{-1} k_{inact}, 2 = 10 nM K_I, 0.01 s^{-1} k_{inact}, 3 = 10 nM K_I, 0.0005 s^{-1} k_{inact}, 4 = 25 nM K_I 0.003 s^{-1} k_{inact}, 5 = 50 nM K_I 0.01 s^{-1} k_{inact}, 6 = 50 nM K_I 0.007 s^{-1} k_{inact}.

3.1 IC_{50} determination

The most common method of ranking inhibitor potency in drug discovery is the IC_{50} which is the concentration of compound required to give 50% inhibition of the target protein at a fixed time point (Eq. 5).[7,37]

$$v_i = \frac{v_o}{\left(1 + \frac{IC_{50}}{[I]}\right)} \tag{5}$$

However, in the case of irreversible inhibitors, as the inhibition is time dependent the IC_{50} decreases over time (Fig. 2B). If left long enough the IC_{50} of all covalent inhibitors approximate to half the enzyme concentration in the assay.[38]

As the IC_{50} of a covalent inhibitor depends on the affinity, reactivity and assay duration, the ranking of inhibitors will vary depending on the time point chosen for the assay (Fig. 2C, Table 1) and thus may give misleading compound SAR. A further complication with using IC_{50}s determined from a single time point is that for two compounds with the same IC_{50}, it is impossible to differentiate between a compound with a relatively high initial binding affinity and relatively low chemical reactivity or vice versa.

Despite these severe limitations, the IC_{50} remains a common method of assaying covalent inhibitors.[6,10,15] This is primarily due to the ease and scale at which such assays can be deployed, especially during the initial phases of a drug discovery program where a high-throughput screening campaign may be used to identify initial starting points. As a result of this, several methods have been developed to extract more useful information from IC_{50} assay formats.

Table 1 IC_{50} ranking of compounds with various combinations of k_{inact} and K_I over time.

Time (min)	2	10	20	30	60	120
2.5 nM K_I, 0.0005 s^{-1} k_{inact}	1	1	2	2	2	2
10 nM K_I, 0.01 s^{-1} k_{inact}	2	2	1	1	1	1
10 nM K_I, 0.0005 s^{-1} k_{inact}	3	3	4	6	6	6
25 nM K_I, 0.003 s^{-1} k_{inact}	4	5	5	5	5	5
50 nM K_I, 0.01 s^{-1} k_{inact}	5	4	3	3	3	3
50 nM K_I, 0.007 s^{-1} k_{inact}	6	6	6	4	4	4

The magnitude of both k_{inact} and K_I were chosen to cover the range of values that are typically found for covalent inhibitors entering the clinic.[32–35]

Assaying compounds with and without a preincubation step can be used to triage putative covalent inhibitors by identifying compounds which show a decrease in IC_{50} upon preincubation. Assaying with and without a preincubation step can therefore be a useful triaging step to identify inhibitors for follow-up with detailed mechanistic analysis.[6,39] While a shift in IC_{50} upon preincubation gives an indication that covalent (or at least time-dependent) inhibition is occurring, it should not be used as an assessment of inhibitor reactivity. As with the decrease in IC_{50} upon increasing assay length, the magnitude of the shift depends on the reactivity, affinity, and concentration of enzyme in the assay system as well as the preincubation time. Therefore using the magnitude of the shift on preincubation as an estimate of relative reactivity may be misleading.[6]

It is also possible to determine the affinity and reactivity of rapid-equilibrium two-step inhibitors by running the assay and stopping at numerous time points and obtaining the IC50 value at each, without having a prior preincubation step.[40] A fit of IC_{50} vs time to Eq. (6) is then used to determine k_{inact} and K_I.[40]

$$IC_{50}(t) = K_I\left(1 + \frac{[S]}{K_M}\right)\left(\frac{2 - 2\exp(\eta IC_{50}.k_{inact}.t)}{\eta IC_{50}.k_{inact}.t} - 1\right) \tag{6}$$

$$\eta IC_{50} = \frac{IC_{50}(t)}{K_I\left(1 + \frac{[S]}{K_M}\right) + IC_{50}(t)} \tag{6a}$$

A major limitation with this technique is that it is assumed that the inhibitors are irreversible and give a hyperbolic plot of k_{obs} vs [I], i.e., the inhibitors follow a two-step or mechanism-based model of irreversible covalent inhibition. As the compound's mechanism of inhibition cannot be known without prior detailed kinetic characterization, the method has limited applicability. To confirm the inhibition mechanism, data collected to determine the IC_{50}s at various time points may be re-plotted as time-course data and analysed using the algebraic or ordinary differential equation (ODE) methodologies covered in the following sections.

3.2 Algebraic analysis of covalent inhibitors

Algebraic analysis of covalent inhibitors was originally proposed by Kitz & Wilson in 1962, who showed that under a number of simplifying conditions the time-course of an irreversible inhibition assay can be described by a rising exponential curve (Eq. 7).[19]

$$[P] = \frac{v_i}{k_{obs}}(1 - \exp(-k_{obs}.t)) \quad (7)$$

By determining k_{obs} at a variety of inhibitor concentrations, secondary plots of k_{obs} vs [I] can be generated and, from those plots, the kinetic parameters determined (Fig. 3).[6,7,15,38] To carry out this form of analysis the control rate of the uninhibited reaction needs to be strictly linear for the entire duration of the assay. In practical terms, this means that initial substrate concentration must be at a far greater concentration than its K_M and/or that only a negligibly small fraction (<10%) of the substrate in the control reaction has been converted to product at the end of the assay. It is also assumed that there is no inhibitor depletion (i.e., the inhibitor is not tight binding).[41] If these assumptions are not valid for a particular assay system, then ODE analysis should be performed instead.[32,33,42]

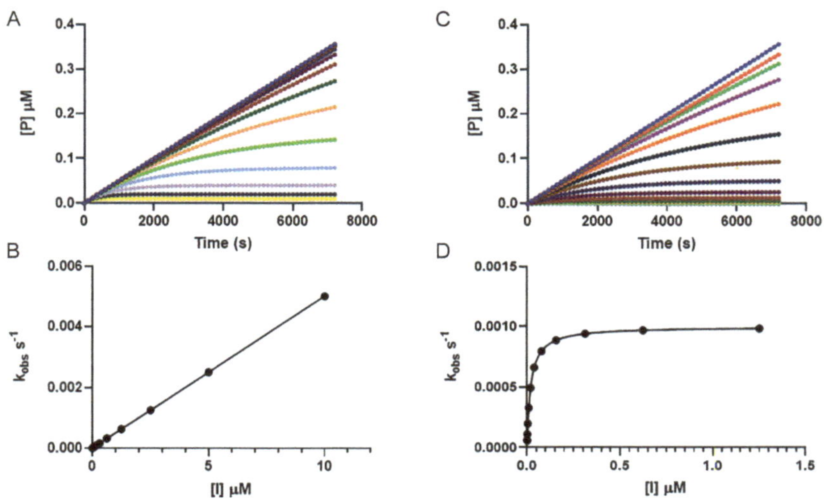

Fig. 3 Simulated time-course data and secondary k_{obs} vs [I] plots for one- and two-step irreversible covalent inhibition. Simulated assay conditions [S] = 10 μM (K_M), [E] = 100 pM, k_{cat} = 1 μM s^{-1}. 16 × 2-fold dilutions of inhibitor (10 μM top concentration) and a no inhibitor control were simulated. For both mechanism simulations, inhibitor was constrained to only bind to free E. (A) time-course data for a one-step covalent inhibitor with k_1 (k_{inact}/K_I) of 0.001 μM s^{-1}. Time-course data was fitted to Eq. 7 to determine the k_{obs} value for each concentration of inhibitor. (B) Secondary plot of k_{obs} values determined from Fig. 2A vs [I]. The data was fitted to Eq. 8 to determine the apparent k_{inact}/K_I (0.0005 μM s^{-1}). (C) A two-step covalent inhibitor with a K_I of 10 nM and k_{inact} 0.001 s^{-1}. (D) Secondary plot of k_{obs} values determined from Fig. 2C vs [I]. The data was fitted to Eq. 10 to determine k_{inact} = 0.001 s^{-1} and an apparent K_I = 20 nM, which when corrected for substrate competition gives K_I = 10 nM.

3.2.1 One-step

For one-step irreversible inhibitors (Fig. 1A), a plot of k_{obs} vs [I] is a straight line (Eq. 8) with the y intercept at the origin and a slope of k_1, i.e., k_{inact}/K_I (Fig. 3B).

$$k_{obs} = k_1[I] \tag{8}$$

With a one-step mechanism, there is no distinction between the steady-state and rapid-equilibrium approximations, because the non-covalent initial EI complex is absent.

3.2.2 Two-step

In most literature treatments of two-step covalent inhibitors, it is assumed that the initial formation of the non-covalent ES complex is under rapid equilibrium (Fig. 1B).[6,7] Under these conditions, the apparent first-order rate constant(k_{obs}) follows a hyperbolic plot vs [I] described by Eq. (9), (Fig. 3D). k_{inact} (k_2 in Fig. 1B) is the first-order inactivation rate constant and K_I is the apparent inhibition constant. These values are analogous to k_{cat} and K_M values in a Michaelis–Menten plot of substrate kinetics.[20]

$$k_{obs} = \frac{k_{inact}[I]}{KI + [I]} \tag{9}$$

For a two-step inhibitor, when the concentration of inhibitor is low (i.e., $[I] \ll KI$), the contribution of [I] in the denominator of Eq. (9) is negligible and the plot of k_{obs} vs [I] becomes linear and may be characterized as for one-step inhibition (Eq. 8).

If the covalent bond formation is rate-limiting, the initial potency reflects the affinity of the initial reversible interaction ($K_I \approx K_i$). However, if the rate of covalent bond formation is only partially rate-limiting, the measured K_I value will include a component of the covalent complex formation in addition to the initial non-covalent interaction (Eq. 3).

Many targeted covalent inhibitor programs use existing (often high affinity) reversible inhibitors as a scaffold on to which a reactive covalent warhead is incorporated.[1] For these inhibitors the assumption of rapid equilibrium for the initial formation of the non-covalent complex may not be valid, as the wide variation in both on and off rates of drug-like reversible inhibitors is well documented.[12–14] In these circumstances, a steady-state analysis of covalent inhibition as described by eqs. 5–13 in Ref. 43 or ODE analysis of the time-course data is more appropriate.[32,44]

For highly optimized two-step covalent inhibitors, where the value of k_{inact}/K_I is high, it may not be possible to accurately determine the individual parameters k_{inact} and K_I using any method of analysis.[6] This is because at high concentrations of inhibitor k_{obs} becomes too fast to accurately determine due to limitations in the speed of data collection and assay window and it becomes challenging to fully define the hyperbolic curve.[33] Under these conditions, k_{obs} values may only be defined when ($[I] \ll K_I$ and so the plot of k_{obs} vs [I] will be linear. In these circumstances the second-order rate constant k_{inact}/K_I may be determined from Eq. (8) as for one-step inhibitors.[6]

3.2.3 Mechanism-based inhibition

As with two-step inhibition, a secondary plot of k_{obs} vs [I] for a mechanism-based inhibitor will give a hyperbolic plot with k_{inact} and K_I values described by Eqs. (1) and (4), respectively. On the basis of kinetic plots alone, it is not possible to discriminate between two-step and mechanism-based inhibition. Therefore, where mechanism-based inhibition is a possibility, further experiments are required to fully characterize the inhibition.[7,16,45]

3.3 Global fitting of time-course data

With most modern data analysis packages it is possible to globally fit the progress curves for a range of inhibitor concentrations and extract the kinetic constants directly from the global fits rather than from secondary plots of k_{obs} vs [I] (Fig. 4). Substitution of the equations describing one-step (Eq. 8) or two-step irreversible inhibition (Eq. 9) into the equation describing the time-course for covalent-inhibition (Eq. 7) give rise to the global fit equations for one-step (Eq. 10) and two-step rapid-equilibrium (Eq. 11), respectively. The advantage of globally fitting the data is that it minimizes the overall sum of weighed squared deviations of all measured and calculated

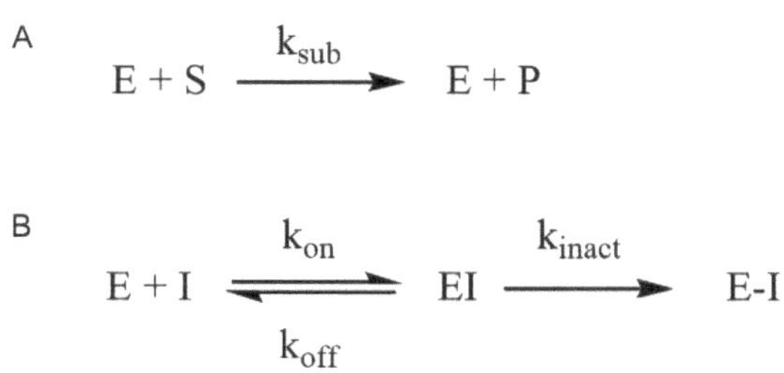

Fig. 4 Example of an input kinetic scheme for ODE analysis of covalent inhibitors. This example, as used by Schwartz et al., comprises of (A) a "hit-and-run" substrate catalysis mechanism (B) a two-step irreversible mechanism of covalent inhibition.

values for the model. The use of global analysis may therefore improve the accuracy and confidence limits of the calculated model parameters.[46]

$$[P] = \frac{v_i}{k_1[I]}(1 - \exp(-k_1[I].t)) \tag{10}$$

$$[P] = \frac{v_i(K_I + [I])}{k_{inact}[I]}\left[1 - \exp\left(-\frac{k_{inact}[I]}{K_I + [I]}.t\right)\right] \tag{11}$$

3.4 Analysis of covalent inhibition using ordinary differential equations (ODE)

While the use of ODE fitting is often viewed as a more specialized data-fitting technique, there are numerous software packages that can be used to automatically derive and fit a series of ordinary differential equations from a kinetic reaction scheme to time-course data.[47–49] In order to carry out ODE analysis all that is required is a kinetic reaction scheme (such as those in Figs. 1 and 4) which may be easily adapted to take into account additional factors such as enzyme activation or reversibility of the reaction. Taking the ODE fitting of covalent inhibitors of EGFR in Schwartz et al. as an example, the reaction scheme shown in Fig. 4, which comprises of a simple "hit and run" mechanism for substrate turnover (where $[S] \ll K_M$ and so substrate is converted directly to product with no build-up of ES complex) and a two-step model of covalent inhibition, was used to automatically compute the series of ODEs that describe the model (eqs. 8–13 in Ref. 32 supplementary materials) in the software package Dynafit. The software then globally fits the time-course data to the ODE model to generate best fit values for the microscopic rate constants. In this case, the value of k_{on} was fixed at 100 μM s^{-1} (diffusion controlled on-rate) and only the values of k_{sub}, k_{inact} and k_{off} (and from k_{off}/k_{on}, the K_i), were determined.

Although it is a less commonly utilized technique for covalent data fitting than algebraic analysis of k_{obs} vs [I] plots, there are several recent examples where ODE analysis has been used to characterize covalent inhibition.[32,33,42]

One of the major advantages of ODE fitting is that data may be fitted to kinetic schemes for which the assumptions used to derive the algebraic models are no longer valid. Such cases include non-linearity of the uninhibited reaction time-course due to substrate depletion, product inhibition, and autoactivation or degradation of the enzyme during the reaction.[32,42] ODE fitting may also be used for inhibitors where the initial binding step is not in rapid equilibrium.[33]

3.5 Model discrimination

3.5.1 Selection of data fitting methodology

Prior to fitting the data it is important to examine the uninhibited control reaction to ensure that it is linear for the duration of the experiment. Any non-linearity in the control reaction should not be ignored as it invalidates the assumptions under which the algebraic equations most commonly used for data fitting were derived. Fig. 5 shows simulated time-course data for a rapidly equilibrating two-step irreversible covalent inhibitor where there is modest loss of linearity due to substrate depletion ([S] $0.1 \times K_M$). When the time-course data is analysed to determine k_{inact} and K_I, this loss of linearity leads to an overestimation of both k_{inact} and K_I. The severity of the overestimation will vary depending on the magnitude of the substrate depletion.

It can also be seen from Fig. 5C that at low inhibitor concentration, the value of k_{obs} systematically deviates from the line of best fit and depending on the concentrations of inhibitor assayed and any noise in the measurement may give the impression that the intersect is at a positive value of k_{obs} which could then be misinterpreted as slow reversible inhibition.

Although every effort should be taken to ensure that the uninhibited control reaction is linear for the duration of the assay, this may not always be possible. This may be due to an inherent property of the enzyme system being studied such as auto activation or product inhibition. There may also be instances where for practical reasons (such as a high substrate K_M), the available assay systems have to be run at low substrate concentration relative to K_M or if the signal-to-background noise of the assay system is high, large turnover of substrate may be required to generate sufficient assay signal.[32]

In the case of loss of linearity of the uninhibited reaction due to substrate depletion an algebraic model has been derived that takes this in to account, or alternatively ODE fitting may be used to fit the more complex kinetic models required to accurately describe the data.[32,44]

3.5.2 Model discrimination

When fitting kinetic inhibition data, it is often assumed that the mechanism of covalent inhibition is known in advance, with one-step inhibitors presumed to be highly reactive non-specific affinity labels such as iodoacetamide, while lower affinity targeted covalent inhibitors will follow a two-step mechanism.

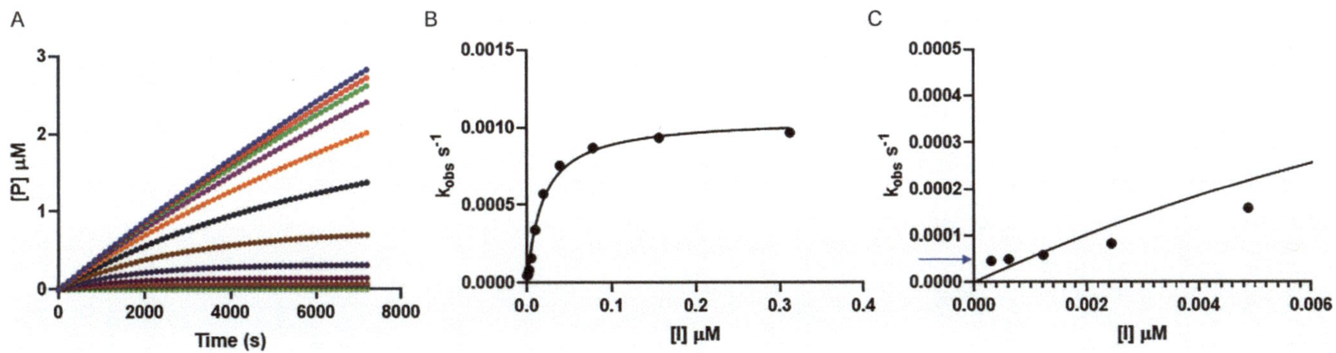

Fig. 5 (A) Simulated time-course data for a two-step irreversible covalent inhibitor ($k_{inact} = 0.001\ s^{-1}$, $K_I = 10$ nM) with modest substrate depletion. [S] = 10 µM (1/10th K_M), [E] = 5 nM, $k_{cat} = 1\ \mu M\ s^{-1}$. 16 × 2-fold dilutions of inhibitor (10 mM top concentration) and a no inhibitor control were simulated. (B) Secondary k_{obs} vs [I] plot. Fitting to Eq. 10 gives $k_{inact} = 0.0016\ s^{-1}$ which is a slight overestimation of the true k_{inact} and $K_I = 18.77$ nM, which when corrected for substrate competition gives a K_I of 17.1 nM and so overestimates the true K_I of 10 nM. (C) Truncation of the secondary k_{obs} vs [I] plot shows clear deviation away from the line of best fit for k_{obs} values at low inhibitor concentration.

However, for inhibitors in which the initial reversible complex formation is of high affinity, this assumption may no longer be valid.[17] If the inactivation rate constant k_2 (k_{inact}) for a two-step inhibitor is larger than the dissociation rate constant k_{-1}, the initial non-covalent complex will be pulled through the reaction path so rapidly that it will not make a detectable appearance in the reaction mixture. Under these conditions, the mechanism of inhibition will appear as though there is only a single irreversible step.[6]

Therefore, even if it is presumed that an inhibitor will follow a certain mechanism, it is crucial to perform model selection methods such as the F-test for nested models or Akike information criteria to accurately discriminate between models.[50] In some cases additional experiments may be required to discriminate between models (e.g., two-step vs mechanism-based inhibition).[7,16,45]

3.5.3 Estimation of K_i from initial rate data

Evaluation of the early portion of the time-course may give an indication as to whether the inhibitor is reacting via a one- or two-step mechanism. For a one-step inhibitor there is no binding event prior to covalent bond formation so during the early phase of the time-course, where the concentration of covalent *E*-I complex is low, the time-course data for each inhibitor concentration will overlay (Fig. 6). For a two-step covalent inhibitor where the initial binding event is in rapid equilibrium, even during the initial phase of

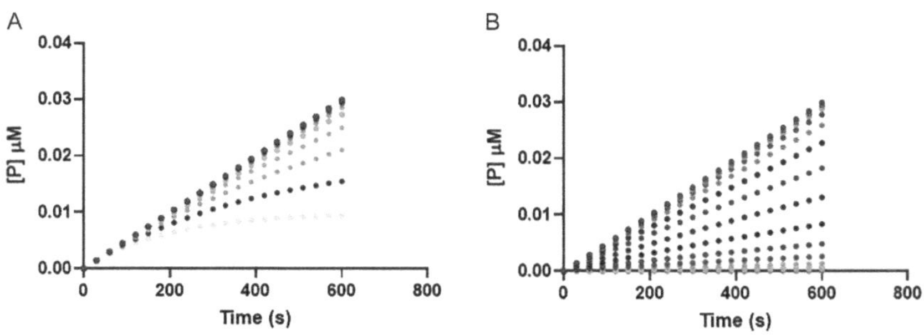

Fig. 6 Simulated time-course data showing the differences in initial rates for one- and two-step irreversible covalent inhibition. Simulated assay conditions for both one- and two-step plots are as for Fig. 3. (A) One-step inhibition. Inhibitors conforming to a one-step model of inhibition, show overlay of their time-courses are early time points. This indicated that there is no formation of a reversible EI complex prior to the formation of the covalent E-I complex. (B) Time-course data for the two-step inhibitor shows a decrease in initial rate with increasing inhibitor concentration indicating the formation of a reversible EI complex prior to covalent bond formation.

the time-course, the enzyme will be inhibited due to the rapid formation of the reversible EI complex and so the time-courses for different inhibitor concentrations will not overlay. As well as being a useful qualitative check to determine if the inhibition is via a one- or two-step inhibition mechanism, the K_i for a two-step inhibitor may also be estimated by fitting initial rate data to the early linear portion of the time-course.[32,33]

3.5.4 Reversible vs irreversible covalent inhibition

While many covalent inhibitors form an irreversible covalent adduct with the target protein, it should not be assumed that the formation of a covalent adduct alone is proof of an irreversible mechanism of inhibition, and there are numerous examples of reversible covalent inhibitors for both targeted and mechanism-based covalent inhibition.[11] Examination of the progress curves for irreversible and reversible inhibitors shows differences in the final steady-state velocity (Fig. 7).

Reversible covalent inhibitors have a non-zero final steady-state velocity. The progress curve of a reversible covalent inhibitor is described by Eq. (12) and may be characterized in the same manner as slow-binding inhibitors.[7,11,51,52] For irreversible inhibitors the progress curve has a final steady-state velocity (ν_s) of zero and is described by Eq. (7).

$$[P] = \nu_s t + \frac{\nu_{i-}\nu_s}{k_{obs}}[1 - \exp(-k_{obs}t)] \qquad (12)$$

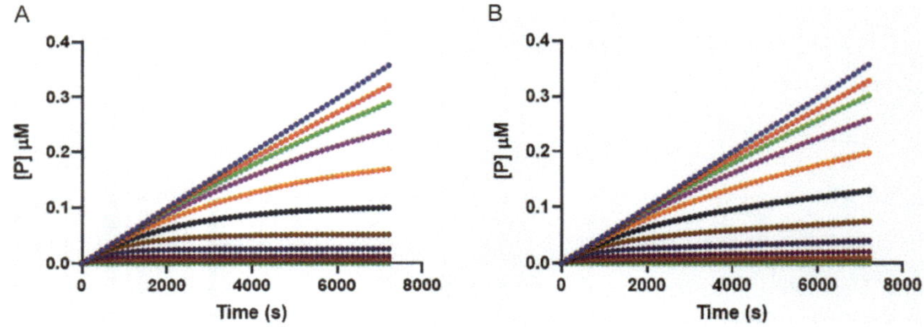

Fig. 7 Simulated time-course data showing the differences in final steady-state velocity for irreversible and reversible covalent inhibitors following mechanism 1b (Fig. 1). (A) An irreversible covalent inhibitor with a K_I of 50 nM and k_{inact} 0.01 s^{-1}. (B) A reversible covalent inhibitor with a K_I of 50 nM k_{inact} 0.01 s^{-1} and k^{-2} of 0.0001 s^{-1}. Simulated assay conditions [S] = 10 µM (K_M), [E] = 100 pM, k_{cat} = 1 µM s^{-1}. 16 × 2-fold dilutions of inhibitor (10 µM top concentration) and a no inhibitor control were simulated.

Jump dilution, or dialysis in conjunction with crystallography or mass spectrometry, may be used to confirm the formation and irreversibility of the covalent complex.[11,51]

3.5.5 Determination of mechanism-based inactivation

Lack of substrate protection, inactivation kinetics that are not saturable, and a stoichiometry of inactivation that is >1:1 with enzyme, can be used to rule out mechanism-based inactivation.[16] However, for enzymes that utilize the formation of a covalent bond during substrate catalysis it may be difficult to determine if a covalent inhibitor is a two-step or mechanism-based inhibitor. This is because both mechanisms give a hyperbolic plot of k_{obs} as a function of inhibitor concentration and additional experiments such as mass-spectrometry and crystallography are required to discriminate between the two.[7,11,16,45,53] As the reactive intermediate is only unmasked as part of the catalytic cycle, mechanism-based inactivators are not inherently reactive toward other nucleophiles, which may give an indication that inhibition is occurring via mechanism-based inhibition.[7,11,16]

4. Experimental techniques

There are numerous experimental techniques that may be employed to characterize covalent inhibitors, either to measure the kinetic parameters of covalent inhibition or to aid in identifying the mechanism and/or site of covalent bond formation. The choice of assay methodologies will be influenced by factors including the target protein of interest (if it is an enzyme or protein-protein interaction), whether a continuous assay may be developed, the availability of automation, the speed of reaction, required throughput, etc.

With recent advances in plate-reader speed/capacity and increased use of automation in many laboratories, it is often possible to run kinetic assays at high enough capacity that the k_{inact}/K_I value determined from a kinetic assay should be the default metric as early as possible and certainly in the lead optimization phase of a covalent inhibitor program.

4.1 Measurements of residual enzyme activity

In the original 1962 paper on covalent inhibition, Kitz and Wilson outlined a method to determine k_{obs} whereby enzyme and inhibitor are preincubated at various time points and then the residual enzyme activity determined.[19] Assessment of activity is determined either by large dilution of enzyme into substrate to prevent additional covalent modification of the protein while the enzyme activity is determined, or by assaying over a short time-course (typically 10 min or less).[16] k_{obs} values for each inhibitor concentration are then obtained from a linear fit of the natural logarithm of percentage activity remaining vs preincubation time.

This method of analysis for covalent drug discovery has mostly been superseded by time-course data generated from continuous enzyme assays but the method is still in use, particularly for mechanism-based inactivation or assay systems where it is not possible to develop a continuous assay.[45] In addition, measurement of residual enzyme activity is still the most commonly utilized experimental technique for safety assessment of time-dependent CYP inactivation.[54–58]

4.2 Continuous enzymatic assays

Advances in plate-reader speed and capacity over the past two decades mean that continuous enzyme activity assays are currently one of the most commonly used techniques for kinetic characterization of irreversible inhibitors. To determine the kinetic parameters using a continuous assay the enzyme, inhibitor, and substrate are mixed with no prior preincubation step and the loss of enzyme activity with time is measured. While these assays are generally carried out in a standard plate reader, for particularly fast kinetic systems a stopped (or quench) flow system may be required to fully characterize the kinetics.[33] In cases where a continuous assay format is not available, it is often possible to measure the time-course using a series of stopped assays. The increased use of laboratory automation in drug discovery makes both this and the stopped measurements of residual enzyme activity feasible to run as a reasonably high throughput process.[59]

4.2.1 Substrate concentration

In the early stages of a drug discovery project screening assays are often run at a substrate concentration equivalent to K_M to ensure that the assay is balanced and can find inhibitors with competitive, non-competitive and un-competitive mechanisms of action.[60,61] Running assays at a substrate concentration equivalent to K_M, has the advantage that for a rapidly equilibrating inhibitor, regardless of the mechanism, the apparent K_I will be within twofold of the true K_I.[38,62] The same principle applies to covalent inhibition as although the protein is irreversibly inhibited once the covalent *E*-I complex has formed, the initial reversible binding event to form the EI complex will still be affected by the substrate concentration and the inhibitor's mechanism of action. If the mechanism of action for the reversible step of covalent inhibition is known, the apparent K_I may be corrected for substrate concentration.[32,62,63]

Due to limitations with solubility of the substrate or sensitivity of the detection system (which may require the use of comparatively low or high concentrations of the substrate relative to its K_M) it is not always possible to run an assay at a substrate concentration equivalent to K_M.[32,33] For high affinity competitive inhibitors, which are commonly encountered late in the development stage of drug discovery projects, it may be advantageous to run the assay system at high concentrations of the competing substrate to avoid tight binding.[32,41]

4.3 Competition experiments

Probe competition experiments are routinely used to characterize the binding of reversible compounds.[14,64–68] While they are less common than catalytic assays for the characterization of covalent binding, it is possible to use a similar approach to characterize covalent inhibition using either reversible or covalent probes.[69–74]

When a reversible probe is used, the probe is preincubated with protein. Addition of the covalent inhibitor causes loss of signal over time as the probe

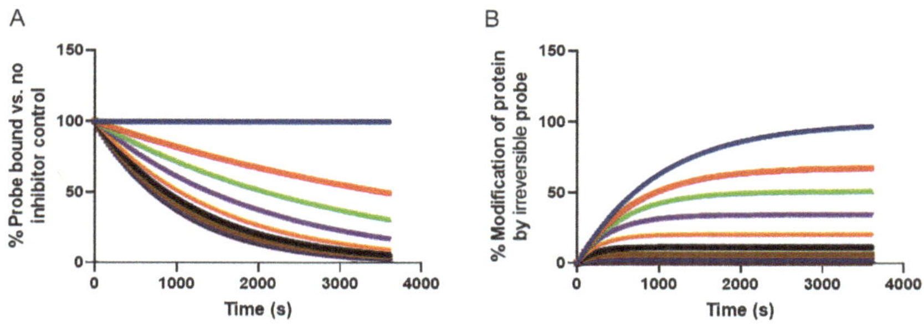

Fig. 8 Probe competition experiments. (A) Simulated time-course data showing the decrease in signal over time for a reversible fluorescent probe ([Probe] = K_I) that has previously been preincubated with enzyme upon the addition of varying concentrations of irreversible inhibitor with a k_{inact} of 0.01 s^{-1} and a K_I of 50 nM. 12 × 2-fold dilutions of inhibitor (10 μM top concentration) and a no inhibitor control were simulated. (B) Simulated time-course data showing the increase in signal over time for an irreversible covalent probe mixed at time = 0 with an irreversible inhibitor and enzyme. Simulated assay conditions [Probe] = 150 nM (K_I), [E] = 1 nM, irreversible covalent inhibitor K_I = 50 nM and k_{inact} = 0.01 s^{-1}, irreversible covalent probe K_I = 150 nM and k_{inact} = 0.002 s^{-1}. 12 × 2-fold dilutions of inhibitor (10 μM top concentration) and a no inhibitor control were simulated.

dissociates and cannot re-bind to the covalently modified protein (Fig. 8A). The time-course data may then be analysed in a similar manner to a continuous enzymatic assay but with the time-course plots fitted to a one-phase decay to determine the k_{obs} value for each inhibitor concentration.[73]

If both the probe and inhibitor are covalent then protein, enzyme, and probe are mixed with no preincubation step (Fig. 8B). The fraction of probe bound may either be monitored over time or the k_{inact}/K_I value determined as an end point IC_{50} assay.[69,72,74] These methods may be particularly useful for covalent inhibitors of protein-protein interactions for which it is not possible to build an enzymatic assay.[75]

4.4 Jump dilution

Jump dilution can be used as a qualitative technique to confirm whether the covalent inhibition is irreversible over the duration of the assay. Compounds are preincubated with protein at a concentration of at least 10 × the measured K_I in the assay and for a sufficiently long time for an irreversible covalent inhibitor to have fully covalently modified the protein. A large dilution (usually 100-fold) into substrate is performed to initiate the reaction and recovery of enzyme activity is monitored.[76] For irreversible inhibitors no recovery of activity will be observed, whereas for slowly reversible non-covalent inhibitors there will be recovery of activity with time (Fig. 9). As with slowly dissociating reversible non-covalent inhibitors, the rate of recovery in a jump

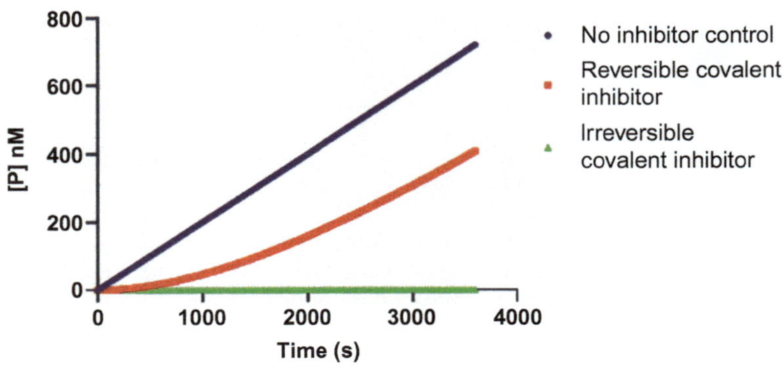

Fig. 9 Jump dilution experiments. Simulated time-course data showing no recovery of activity for a covalent irreversible inhibitor when jump diluted. A slowly reversible covalent inhibitor (residence time = 30 mins, $\nu_o = 0$, $\nu_s = 0.2$ nM s^{-1}) recovers activity over the duration of the assay. It is assumed that the dilution factor is sufficiently large enough that the enzyme activity would recover back to the rate of the control reaction (0.2 nM s^{-1}).

dilution assay can be used to determine the residence time of the reversible covalent inhibitor by fitting to Eq. (12). The residence time is the reciprocal of k_{obs}.[51,76]

4.5 Mass spectrometry

Mass spectrometry is widely utilized to characterize covalent inhibitors as it directly detects the formation of a chemical bond between the inhibitor and target protein. Comparison of the native protein and covalent complex can also be used to determine the mass of the covalent adduct.

The most common use of mass spectrometry in the characterization of inhibitors is intact mass spectrometry, which can determine if the compound forms a single or multiple adducts with the protein of interest.[71,77] The ability to directly detect covalent adduct formation makes it a useful technique for screening covalent fragment libraries to identify start points for covalent drug discovery programs.[78]

Non-specific covalent inhibitors are generally highly reactive and label multiple residues within a protein and so mass spectrometry can be a useful technique to identify non-specific inhibitors during the hit validation stage of a drug discovery program.[71,77]

In addition to intact-mass measurements, tryptic-digest mass spectrometry is often used in covalent drug discovery programs as it can aid in identifying the residue(s) within a protein that is covalently modified.

While it can be used to directly measure the formation of a covalent complex over time, mass spectrometry tends not to be used for kinetic characterization as the technique requires relatively high concentrations of protein (typically in the μM range) whereas clinically relevant inhibitors tend to have affinities in the low nM range and so the inhibitors will be tight binding.

4.6 Protein mutants

Mutation of the putative target residue in the protein of interest may be used to aid the identification of the site that is chemically modified by the covalent inhibitor.[72,79,80] Loss of the nucleophilic residue will often result in an increase in IC_{50} in a primary screening assay due to the lack of additional potency from the formation of the irreversible covalent bond, as well as a loss of curvature in the inhibition time-course.[81] Mass spectrometry can be used to confirm that the mutated protein no longer forms a covalent adduct on incubation with the inhibitor.

Mutation of the specific nucleophilic residue that is being targeted is often also used to identify non-specific covalent inhibitors identified during screening campaigns.[82] As the mutated protein lacks the ability to form a covalent bond, it may also be used to estimate the initial K_i for the reversible binding event.[6] However, this should only be used as an approximation for the affinity of the initial reversible binding event as mutation of the residue may affect the binding affinity of the inhibitor as well as preventing the formation of the covalent complex.[83]

In addition a drug discovery strategy known as covalent tethering has been developed where a cystine residue is engineered into or near a protein binding site and screened against thiol containing covalent fragment libraries under partially reducing conditions which promote rapid thiol exchange.[84,85] Molecules with affinity for the target protein will stabilize the disulfide bond whereas reactive fragments with no intrinsic affinity will be reduced. Modification of the target protein is then detected by mass spectrometry. This technique has been shown to identify molecules with weak affinity that would have been missed by a reversible fragment screen. Once the affinity of inhibitors has been optimized, the reactive functional group is removed from the molecule.

5. Summary

There are multiple mechanisms by which a covalent inhibitor can react with and inhibit a target protein. These mechanisms may often be distinguished by kinetic characterization of time-course data in conjunction with additional experimental data, for example, mass spectrometry to confirm the site of modification and mass of the covalent adduct, or jump dilution to confirm the irreversibility of covalent bond formation.

Multiple experimental and analytical techniques may be used to characterize covalent inhibitors. The selection of which techniques to use for a particular protein target will depend on a range of factors including: if the target is an enzyme or protein-protein interaction, if the control reaction in an enzyme assay system is non-linear, the speed of data collection required to fully define the inhibition time-course, and the required throughput of the assay system.

As covalent inhibition depends on both the affinity and reactivity of the inhibitor, biochemical IC_{50} values should not be used for ranking covalent inhibitors as the rank order may change over time depending on the specific combination of affinity and reactivity of each inhibitor. Instead, the second-order rate constant k_{inact}/K_I, which accounts for both the affinity of the initial reversible encounter complex and the maximal rate of covalent bond formation, should be used.

Acknowledgment

I thank Petr Kuzmič for numerous helpful discussions on the mechanisms and characterization of covalent inhibition.

References

1. Singh, J.; Petter, R. C.; Baillie, T. A.; Whitty, A. The Resurgence of Covalent Drugs. *Nat. Rev. Drug Discov.* **2011**, *10* (4), 307–317. https://doi.org/10.1038/nrd3410.
2. Johnson, D. S.; Weerapana, E.; Cravatt, B. F. Strategies for Discovering and Derisking Covalent, Irreversible Enzyme Inhibitors. *Future Med. Chem.* **2010**, *2* (6), 949–964. https://doi.org/10.4155/fmc.10.21.
3. Copeland, R. A. The Drug–Target Residence Time. *Nat. Rev. Drug Discov.* **2016**, *15* (February), 87–95. Ref 4.
4. Tonge, P. J. Drug-Target Kinetics in Drug Discovery. *ACS Chem. Nerosci.* **2018**, *9* (1), 29–39. https://doi.org/10.1021/acschemneuro.7b00185.
5. Bandyopadhyay, A.; Gao, J. Targeting Biomolecules with Reversible Covalent Chemistry. *Curr. Opin. Chem. Biol.* **2016**, *34*, 110–116. https://doi.org/10.1016/j.cbpa.2016.08.011.
6. Strelow, J. M. A Perspective on the Kinetics of Covalent and Irreversible Inhibition. *SLAS Discov. Adv. Sci. Drug Discov.* **2016**, *22* (1), 3–20. https://doi.org/10.1177/1087057116671509.
7. Copeland, R. A. *Evaluation of Enzyme Inhibitors in Drug Discovery: A Guide for Medicinal Chemists and Pharmacologists*; Wiley: Hoboken, N.J., 2013.
8. Ehmann, D. E.; Jahić, H.; Ross, P. L.; Gu, R. F.; Hu, J.; Kern, G.; Walkup, G. K.; Fisher, S. L. Avibactam Is a Covalent, Reversible, Non-β-Lactam β-Lactamase Inhibitor. *Proc. Natl. Acad. Sci. U. S. A.* **2012**, *109* (29), 11663–11668. https://doi.org/10.1073/pnas.1205073109.
9. Serafimova, I. M.; Pufall, M. A.; Krishnan, S.; Duda, K.; Cohen, M. S.; Maglathlin, R. L.; McFarland, J. M.; Miller, R. M.; Frödin, M.; Taunton, J. Reversible Targeting of Noncatalytic Cysteines with Chemically Tuned Electrophiles. *Nat. Chem. Biol.* **2012**, *8* (5), 471–476. https://doi.org/10.1038/nchembio.925.
10. Forster, M.; Chaikuad, A.; Bauer, S. M.; Holstein, J.; Robers, M. B.; Corona, C. R.; Gehringer, M.; Pfaffenrot, E.; Ghoreschi, K.; Knapp, S.; Laufer, S. A. Selective JAK3 Inhibitors with a Covalent Reversible Binding Mode Targeting a New Induced Fit Binding Pocket. *Cell Chem. Biol.* **2016**, *23* (11), 1335–1340. https://doi.org/10.1016/j.chembiol.2016.10.008.
11. Tuley, A.; Fast, W. The Taxonomy of Covalent Inhibitors. *Biochemistry* **2018**, *57* (24), 3326–3337. https://doi.org/10.1021/acs.biochem.8b00315.
12. Copeland, R. A.; Pompliano, D. L.; Meek, T. D. Drug–Target Residence Time and its Implications for Lead Optimization. *Nat. Rev. Drug Discov.* **2006**, *5* (9), 730–739. https://doi.org/10.1038/nrd2082.
13. Swinney, D. Biochemical Mechanisms of New Molecular Entities (NMEs) Approved by United States FDA During 2001-2004: Mechanisms Leading to Optimal Efficacy and Safety. *Curr. Top. Med. Chem.* **2006**, *6* (5), 461–478. https://doi.org/10.2174/156802606776743093.

14. Georgi, V.; Schiele, F.; Berger, B.-T.; Steffen, A.; Marin Zapata, P. A.; Briem, H.; Menz, S.; Preusse, C.; Vasta, J. D.; Robers, M. B.; Brands, M.; Knapp, S.; Fernández-Montalván, A. Binding Kinetics Survey of the Drugged Kinome. *J. Am. Chem. Soc.* **2018**, *140* (46), 15774–15782. https://doi.org/10.1021/jacs.8b08048.
15. De Bruin, G.; Barf, T. CHAPTER 4: Covalent Inhibition of Kinases. In *RSC Drug Discovery Series*; Ward, R. A., Goldberg, F. W., Eds.; Vol. 2019-January; Royal Society of Chemistry, 2019; pp. 61–96. https://doi.org/10.1039/9781788013093-00061.
16. Silverman, R. B. [10] Mechanism-Based Enzyme Inactivators. *Methods Enzymol.* **1995**, *249*, 240–283. https://doi.org/10.1016/0076-6879(95)49038-8.
17. Cornish-Bowden, A. *Fundamentals of Enzyme Kinetics*; Wiley-Blackwell: Weinheim, Germany, 2014.
18. Malcolm, A. D. B.; Radda, G. K. The Reaction of Glutamate Dehydrogenase with 4-Iodoacetamido Salicylic Acid. *Eur. J. Biochem.* **1970**, *15* (3), 555–561. https://doi.org/10.1111/j.1432-1033.1970.tb01040.x.
19. Kitz, R.; Wilson, I. B. Esters of Methanesulfonic Acid as Irreversible Inhibitors of Acetylcholinesterase. *J. Biol. Chem.* **1962**, *237* (10), 3245–3249.
20. Segel, I. H. *Enzyme Kinetics: Behavior and Analysis of Rapid Equilibrium and Steady-State Enzyme Systems*; Wiley: New York; Chichester, 1993.
21. Ulusu, N. N. Evolution of Enzyme Kinetic Mechanisms. *J. Mol. Evol.* **2015**, *80*, 251–257, Springer New York LLC June 20. https://doi.org/10.1007/s00239-015-9681-0.
22. Sauvage, E.; Kerff, F.; Terrak, M.; Ayala, J. A.; Charlier, P. The Penicillin-Binding Proteins: Structure and Role in Peptidoglycan Biosynthesis. *FEMS Microbiol. Rev.* **2008**, *32* (2), 234–258. https://doi.org/10.1111/j.1574-6976.2008.00105.x.
23. Tooke, C. L.; Hinchliffe, P.; Bragginton, E. C.; Colenso, C. K.; Hirvonen, V. H. A.; Takebayashi, Y.; Spencer, J. β-Lactamases and β-Lactamase Inhibitors in the 21st Century. *J. Mol. Biol.* **2019**, *431* (18), 3472–3500. https://doi.org/10.1016/j.jmb.2019.04.002.
24. Binda, C.; Hubálek, F.; Li, M.; Herzig, Y.; Sterling, J.; Edmondson, D. E.; Mattevi, A. Binding of Rasagiline-Related Inhibitors to Human Monoamine Oxidases: A Kinetic and Crystallographic Analysis. *J. Med. Chem.* **2005**, *48* (26), 8148–8154. https://doi.org/10.1021/jm0506266.
25. Powers, J. C.; Asgian, J. L.; Ekici, Ö. D.; James, K. E. Irreversible Inhibitors of Serine, Cysteine, and Threonine Proteases. *Chem. Rev.* **2002**, *102* (12), 4639–4750. https://doi.org/10.1021/cr010182v.
26. Li, D.; Ambrogio, L.; Shimamura, T.; Kubo, S.; Takahashi, M.; Chirieac, L. R.; Padera, R. F.; Shapiro, G. I.; Baum, A.; Himmelsbach, F.; Rettig, W. J.; Meyerson, M.; Solca, F.; Greulich, H.; Wong, K.-K. BIBW2992, an Irreversible EGFR/HER2 Inhibitor Highly Effective in Preclinical Lung Cancer Models. *Oncogene* **2008**, *27* (34), 4702–4711. https://doi.org/10.1038/onc.2008.109.
27. Pan, Z.; Scheerens, H.; Li, S.-J.; Schultz, B. E.; Sprengeler, P. A.; Burrill, L. C.; Mendonca, R. V.; Sweeney, M. D.; Scott, K. C. K.; Grothaus, P. G.; Jeffery, D. A.; Spoerke, J. M.; Honigberg, L. A.; Young, P. R.; Dalrymple, S. A.; Palmer, J. T. Discovery of Selective Irreversible Inhibitors for Bruton's Tyrosine Kinase. *ChemMedChem* **2007**, *2* (1), 58–61. https://doi.org/10.1002/cmdc.200600221.
28. Ostrem, J. M.; Peters, U.; Sos, M. L.; Wells, J. A.; Shokat, K. M. K-Ras(G12C) Inhibitors Allosterically Control GTP Affinity and Effector Interactions. *Nature* **2013**, *503* (7477), 548–551. https://doi.org/10.1038/nature12796.
29. Ghosh, A. K.; Samanta, I.; Mondal, A.; Liu, W. R. Covalent Inhibition in Drug Discovery. *ChemMedChem* **2019**, *14* (9), 889–906. https://doi.org/10.1002/cmdc.201900107.

30. Liu, Q.; Sabnis, Y.; Zhao, Z.; Zhang, T.; Buhrlage, S. J.; Jones, L. H.; Gray, N. S. Developing Irreversible Inhibitors of the Protein Kinase Cysteinome. *Chem. Biol.* **2013**, *20*, 146–159, Elsevier Ltd February 21. https://doi.org/10.1016/j.chembiol.2012.12.006.
31. Leproult, E.; Barluenga, S.; Moras, D.; Wurtz, J. M.; Winssinger, N. Cysteine Mapping in Conformationally Distinct Kinase Nucleotide Binding Sites: Application to the Design of Selective Covalent Inhibitors. *J. Med. Chem.* **2011**, *54* (5), 1347–1355. https://doi.org/10.1021/jm101396q.
32. Schwartz, P. A.; Kuzmic, P.; Solowiej, J.; Bergqvist, S.; Bolanos, B.; Almaden, C.; Nagata, A.; Ryan, K.; Feng, J.; Dalvie, D.; Kath, J. C.; Xu, M.; Wani, R.; Murray, B. W. Covalent EGFR Inhibitor Analysis Reveals Importance of Reversible Interactions to Potency and Mechanisms of Drug Resistance. *Proc. Natl. Acad. Sci. U. S. A.* **2013**, *111* (1), 173–178. https://doi.org/10.1073/pnas.1313733111.
33. Zhai, X.; Ward, R. A.; Doig, P.; Argyrou, A. Insight into the Therapeutic Selectivity of the Irreversible EGFR Tyrosine Kinase Inhibitor Osimertinib through Enzyme Kinetic Studies. *Biochemistry* **2020**, *59* (14), 1428–1441. https://doi.org/10.1021/acs.biochem.0c00104.
34. Liclican, A.; Serafini, L.; Xing, W.; Czerwieniec, G.; Steiner, B.; Wang, T.; Brendza, K. M.; Lutz, J. D.; Keegan, K. S.; Ray, A. S.; Schultz, B. E.; Sakowicz, R.; Feng, J. Y. Biochemical Characterization of Tirabrutinib and Other Irreversible Inhibitors of Bruton's Tyrosine Kinase Reveals Differences in on—And off—Target Inhibition. *Biochim. Biophys. Acta Gen. Subj.* **2020**, *1864* (4), 129531. https://doi.org/10.1016/j.bbagen.2020.129531.
35. Canon, J.; Rex, K.; Saiki, A. Y.; Mohr, C.; Cooke, K.; Bagal, D.; Gaida, K.; Holt, T.; Knutson, C. G.; Koppada, N.; Lanman, B. A.; Werner, J.; Rapaport, A. S.; San Miguel, T.; Ortiz, R.; Osgood, T.; Sun, J. R.; Zhu, X.; McCarter, J. D.; Volak, L. P.; Houk, B. E.; Fakih, M. G.; O'Neil, B. H.; Price, T. J.; Falchook, G. S.; Desai, J.; Kuo, J.; Govindan, R.; Hong, D. S.; Ouyang, W.; Henary, H.; Arvedson, T.; Cee, V. J.; Lipford, J. R. The Clinical KRAS(G12C) Inhibitor AMG 510 Drives Anti-Tumour Immunity. *Nature* **2019**, *575* (7781), 217–223. https://doi.org/10.1038/s41586-019-1694-1.
36. Fell, J. B.; Fischer, J. P.; Baer, B. R.; Blake, J. F.; Bouhana, K.; Briere, D. M.; Brown, K. D.; Burgess, L. E.; Burns, A. C.; Burkard, M. R.; Chiang, H.; Chicarelli, M. J.; Cook, A. W.; Gaudino, J. J.; Hallin, J.; Hanson, L.; Hartley, D. P.; Hicken, E. J.; Hingorani, G. P.; Hinklin, R. J.; Mejia, M. J.; Olson, P.; Otten, J. N.; Rhodes, S. P.; Rodriguez, M. E.; Savechenkov, P.; Smith, D. J.; Sudhakar, N.; Sullivan, F. X.; Tang, T. P.; Vigers, G. P.; Wollenberg, L.; Christensen, J. G.; Marx, M. A. Identification of the Clinical Development Candidate MRTX849, a Covalent KRASG12CInhibitor for the Treatment of Cancer. *J. Med. Chem.* **2020**, *63* (13), 6679–6693. https://doi.org/10.1021/acs.jmedchem.9b02052.
37. Dougall, I. G. CHAPTER 11. In Vitro Biology: Measuring Pharmacological Activity. In *The Handbook of Medicinal Chemistry*; Davis, A., Ward, S. E., Eds.; Royal Society of Chemistry, 2015; pp. 292–309. https://doi.org/10.1039/9781782621836-00292.
38. Holdgate, G. A.; Meek, T. D.; Grimley, R. L. Mechanistic Enzymology in Drug Discovery: A Fresh Perspective. *Nat. Rev. Drug Discov.* **2017**, *17* (2), 115–132. https://doi.org/10.1038/nrd.2017.219.
39. Resnick, E.; Bradley, A.; Gan, J.; Douangamath, A.; Krojer, T.; Sethi, R.; Geurink, P. P.; Aimon, A.; Amitai, G.; Bellini, D.; Bennett, J.; Fairhead, M.; Fedorov, O.; Gabizon, R.; Gan, J.; Guo, J.; Plotnikov, A.; Reznik, N.; Ruda, G. F.; Díaz-Sáez, L.; Straub, V. M.; Szommer, T.; Velupillai, S.; Zaidman, D.; Zhang, Y.; Coker, A. R.; Dowson, C. G.; Barr, H. M.; Wang, C.; Huber, K. V. M.; Brennan, P. E.; Ovaa, H.; von Delft, F.; London, N. Rapid Covalent-Probe

Discovery by Electrophile-Fragment Screening. *J. Am. Chem. Soc.* **2019**, *141* (22), 8951–8968. https://doi.org/10.1021/jacs.9b02822.
40. Krippendorff, B.-F.; Neuhaus, R.; Lienau, P.; Reichel, A.; Huisinga, W. Mechanism-Based Inhibition: Deriving KIand Kinact Directly from Time-Dependent IC50Values. *J. Biomol. Screen.* **2009**, *14* (8), 913–923. https://doi.org/10.1177/1087057109336751.
41. Morrison, J. F. Kinetics of the Reversible Inhibition of Enzyme-Catalysed Reactions by Tight-Binding Inhibitors. *Biochim. Biophys. Acta, Enzymol.* **1969**, *185* (2), 269–286. https://doi.org/10.1016/0005-2744(69)90420-3.
42. Janssen, A. P. A.; van Hengst, J. M. A.; Béquignon, O. J. M.; Deng, H.; van Westen, G. J. P.; van der Stelt, M. Structure Kinetics Relationships and Molecular Dynamics Show Crucial Role for Heterocycle Leaving Group in Irreversible Diacylglycerol Lipase Inhibitors. *J. Med. Chem.* **2019**, *62* (17), 7910–7922. https://doi.org/10.1021/acs.jmedchem.9b00686.
43. Kuzmič, P. A Steady-State Algebraic Model for the Time Course of Covalent Enzyme Inhibition. *bioRxiv* **2020**. https://doi.org/10.1101/2020.06.10.144220.
44. Kuzmič, P.; Solowiej, J.; Murray, B. W. An Algebraic Model for the Kinetics of Covalent Enzyme Inhibition at Low Substrate Concentrations. *Anal. Biochem.* **2015**, *484*, 82–90. https://doi.org/10.1016/j.ab.2014.11.014.
45. Pham, T. V.; Murkin, A. S.; Moynihan, M. M.; Harris, L.; Tyler, P. C.; Shetty, N.; Sacchettini, J. C.; Huang, H.l.; Meek, T. D. Mechanism-Based Inactivator of Isocitrate Lyases 1 and 2 from Mycobacterium Tuberculosis. *Proc. Natl. Acad. Sci. U. S. A.* **2017**, *114* (29), 7617–7622. https://doi.org/10.1073/pnas.1706134114.
46. Beechem, J. M. [2] Global Analysis of Biochemical and Biophysical Data. *Methods Enzymol.* **1992**, *210*, 37–54. https://doi.org/10.1016/0076-6879(92)10004-w.
47. Bevc, S.; Konc, J.; Stojan, J.; Hodoščeк, M.; Penca, M.; Praprotnik, M.; Janežič, D. ENZO: A Web Tool for Derivation and Evaluation of Kinetic Models of Enzyme Catalyzed Reactions. *PLoS One* **2011**, *6* (7), e22265. https://doi.org/10.1371/journal.pone.0022265.
48. Johnson, K. A.; Simpson, Z. B.; Blom, T. Global Kinetic Explorer: A new Computer Program for Dynamic Simulation and Fitting of Kinetic Data. *Anal. Biochem.* **2009**, *387* (1), 20–29. https://doi.org/10.1016/j.ab.2008.12.024.
49. Kuzmič, P. Program DYNAFIT for the Analysis of Enzyme Kinetic Data: Application to HIV Proteinase. *Anal. Biochem.* **1996**, *237* (2), 260–273. https://doi.org/10.1006/abio.1996.0238.
50. Buckwrrz, D.; Holzhotter, H.-G. A New Method to Discriminate Between Enzyme-Kinetic Models. *Comput. Math. Appl.* **1990**, *20*, 117–126.
51. Bradshaw, J. M.; McFarland, J. M.; Paavilainen, V. O.; Bisconte, A.; Tam, D.; Phan, V. T.; Romanov, S.; Finkle, D.; Shu, J.; Patel, V.; Ton, T.; Li, X.; Loughhead, D. G.; Nunn, P. A.; Karr, D. E.; Gerritsen, M. E.; Funk, J. O.; Owens, T. D.; Verner, E.; Brameld, K. A.; Hill, R. J.; Goldstein, D. M.; Taunton, J. Prolonged and Tunable Residence Time Using Reversible Covalent Kinase Inhibitors. *Nat. Chem. Biol.* **2015**, *11* (7), 525–531. https://doi.org/10.1038/nchembio.1817.
52. Morrison, J. F.; Walsh, C. T. The Behavior and Significance of Slow-Binding Enzyme Inhibitors. *Adv. Enzymol. Relat. Areas Mol. Biol.* **1988**, *61*, 201–301, John Wiley & Sons, Ltd. https://doi.org/10.1002/9780470123072.ch5.
53. Marangoni, A. G. *Enzyme Kinetics*; John Wiley & Sons, Inc.: Hoboken, NJ, USA, 2002. https://doi.org/10.1002/0471267295.
54. Riley, R. J.; Grime, K.; Weaver, R. Time-Dependent CYP Inhibition. *Expert Opin. Drug Metab. Toxicol.* **2007**, *3*, 51–66, February. https://doi.org/10.1517/17425255.3.1.51.
55. Ghanbari, F.; Rowland-Yeo, K.; Bloomer, J.; Clarke, S.; Lennard, M.; Tucker, G.; Rostami-Hodjegan, A. A Critical Evaluation of the Experimental Design of Studies

of Mechanism Based Enzyme Inhibition, with Implications for in Vitro-in Vivo Extrapolation. *Curr. Drug Metab.* **2006**, 7 (3), 315–334. https://doi.org/10.2174/138920006776359293.
56. Stresser, D. M.; Mao, J.; Kenny, J. R.; Jones, B. C.; Grime, K. Exploring Concepts of in Vitro Time-Dependent CYP Inhibition Assays. *Expert Opin. Drug Metab. Toxicol.* **2014**, *10*, 157–174, Taylor & Francis February. https://doi.org/10.1517/17425255.2014.856882.
57. Kosaka, M.; Kosugi, Y.; Hirabayashi, H. Pharmacokinetics, Pharmacodynamics and Drug Transport and Metabolism Risk Assessment Using Cytochrome P450 Time-Dependent Inhibition Assays at Single Time and Concentration in the Early Stage of Drug Discovery. *J. Pharm. Sci.* **2017**, *106*, 2839–2846. https://doi.org/10.1016/j.xphs.2017.04.077.
58. Filppula, A. M.; Parvizi, R.; Mateus, A.; Baranczewski, P.; Artursson, P. Improved Predictions of Time-Dependent Drug-Drug Interactions by Determination of Cytosolic Drug Concentrations. *Sci. Rep.* **2019**, *9* (1), 1–14. https://doi.org/10.1038/s41598-019-42051-x.
59. Schneider, G. Automating Drug Discovery. *Nat. Rev. Drug Discov.* **2018**, *17*, 97–113, Nature Publishing Group February 1. https://doi.org/10.1038/nrd.2017.232.
60. Copeland, R. A. Mechanistic Considerations in High-Throughput Screening. *Anal. Biochem.* **2003**, *320*, 1–12, Academic Press Inc. September 1. https://doi.org/10.1016/S0003-2697(03)00346-4.
61. Yang, J.; Copeland, R. A.; Lai, Z. Defining Balanced Conditions for Inhibitor Screening Assays That Target Bisubstrate Enzymes. *J. Biomol. Screen.* **2009**, *14* (2), 111–120. https://doi.org/10.1177/1087057108328763.
62. Brandt, R. B.; Laux, J. E.; Yates, S. W. Calculation of Inhibitor Ki and Inhibitor Type from the Concentration of Inhibitor for 50% Inhibition for Michaelis-Menten Enzymes. *Biochem. Med. Metab. Biol.* **1987**, *37* (3), 344–349. https://doi.org/10.1016/0885-4505(87)90046-6.
63. Yung-Chi, C.; Prusoff, W. H. Relationship between the Inhibition Constant (KI) and the Concentration of Inhibitor which Causes 50 per Cent Inhibition (I50) of an Enzymatic Reaction. *Biochem. Pharmacol.* **1973**, *22* (23), 3099–3108. https://doi.org/10.1016/0006-2952(73)90196-2.
64. Motulsky, H. J.; Mahan, L. C. The Kinetics of Competitive Radioligand Binding Predicted by the Law of Mass Action. *Mol. Pharmacol.* **1984**, *25* (1), 1–9.
65. Schiele, F.; Ayaz, P.; Fernández-Montalván, A. A Universal Homogeneous Assay for High-Throughput Determination of Binding Kinetics. *Anal. Biochem.* **2015**, *468*, 42–49. https://doi.org/10.1016/j.ab.2014.09.007.
66. Georgi, V.; Dubrovskiy, A.; Steigele, S.; Fernández-Montalván, A. E. Considerations for Improved Performance of Competition Association Assays Analysed with the Motulsky–Mahan's "Kinetics of Competitive Binding" Model. *Br. J. Pharmacol.* **2019**, *176* (24), 4731–4744. https://doi.org/10.1111/bph.14841.
67. Sykes, D. A.; Moore, H.; Stott, L.; Holliday, N.; Javitch, J. A.; Robert Lane, J.; Charlton, S. J. Extrapyramidal Side Effects of Antipsychotics Are Linked to their Association Kinetics at Dopamine D2 Receptors. *Nat. Commun.* **2017**, *8* (1), 1–11. https://doi.org/10.1038/s41467-017-00716-z.
68. Sykes, D. A.; Jain, P.; Charlton, S. J. Investigating the Influence of Tracer Kinetics on Competition-Kinetic Association Binding Assays: Identifying the Optimal Conditions for Assessing the Kinetics of Low-Affinity Compounds. *Mol. Pharmacol.* **2019**, *96* (3), 378–392. https://doi.org/10.1124/mol.119.116764.
69. Miyahisa, I.; Sameshima, T.; Hixon, M. S. Rapid Determination of the Specificity Constant of Irreversible Inhibitors (Kinact/KI) by Means of an Endpoint Competition Assay. *Angew. Chem. Int. Ed.* **2015**, *54* (47), 14099–14102. https://doi.org/10.1002/anie.201505800.

70. Craven, G. B.; Affron, D. P.; Allen, C. E.; Matthies, S.; Greener, J. G.; Morgan, R. M. L.; Tate, E. W.; Armstrong, A.; Mann, D. J. High-Throughput Kinetic Analysis for Target-Directed Covalent Ligand Discovery. *Angew. Chem. Int. Ed.* **2018**, *57* (19), 5257–5261. https://doi.org/10.1002/anie.201711825.
71. Pettinger, J.; Le Bihan, Y.-V.; Widya, M.; van Montfort, R. L. M.; Jones, K.; Cheeseman, M. D. An Irreversible Inhibitor of HSP72 that Unexpectedly Targets Lysine-56. *Angew. Chem. Int. Ed.* **2017**, *56* (13), 3536–3540. https://doi.org/10.1002/anie.201611907.
72. Hunter, J. C.; Gurbani, D.; Ficarro, S. B.; Carrasco, M. A.; Lim, S. M.; Choi, H. G.; Xie, T.; Marto, J. A.; Chen, Z.; Gray, N. S.; Westover, K. D. In Situ Selectivity Profiling and Crystal Structure of SML-8-73-1, an Active Site Inhibitor of Oncogenic K-Ras G12C. *Proc. Natl. Acad. Sci. U. S. A.* **2014**, *111* (24), 8895–8900. https://doi.org/10.1073/pnas.1404639111.
73. Goedken, E. R.; Argiriadi, M. A.; Banach, D. L.; Fiamengo, B. A.; Foley, S. E.; Frank, K. E.; George, J. S.; Harris, C. M.; Hobson, A. D.; Ihle, D. C.; Marcotte, D.; Merta, P. J.; Michalak, M. E.; Murdock, S. E.; Tomlinson, M. J.; Voss, J. W. Tricyclic Covalent Inhibitors Selectively Target Jak3 through an Active Site Thiol. *J. Biol. Chem.* **2015**, *290* (8), 4573–4589. https://doi.org/10.1074/jbc.M114.595181.
74. Xiong, Y.; Lu, J.; Hunter, J.; Li, L.; Scott, D.; Choi, H. G.; Lim, S. M.; Manandhar, A.; Gondi, S.; Sim, T.; Westover, K. D.; Gray, N. S. Covalent Guanosine Mimetic Inhibitors of G12C KRAS. *ACS Med. Chem. Lett.* **2017**, *8* (1), 61–66. https://doi.org/10.1021/acsmedchemlett.6b00373.
75. Cheng, S. S.; Yang, G. J.; Wang, W.; Leung, C. H.; Ma, D. L. The Design and Development of Covalent Protein-Protein Interaction Inhibitors for Cancer Treatment. *J. Hematol. Oncol.* **2020**, *13*, 26, BioMed Central Ltd. March 30. https://doi.org/10.1186/s13045-020-00850-0.
76. Copeland, R. A.; Basavapathruni, A.; Moyer, M.; Scott, M. P. Impact of Enzyme Concentration and Residence Time on Apparent Activity Recovery in Jump Dilution Analysis. *Anal. Biochem.* **2011**, *416* (2), 206–210. https://doi.org/10.1016/j.ab.2011.05.029.
77. West, J. M.; Zvonok, N.; Whitten, K. M.; Vadivel, S. K.; Bowman, A. L.; Makriyannis, A. Biochemical and Mass Spectrometric Characterization of Human N-Acylethanolamine-Hydrolyzing Acid Amidase Inhibition. *PLoS One* **2012**, 7 (8), e43877. https://doi.org/10.1371/journal.pone.0043877.
78. Keeley, A.; Petri, L.; Ábrányi-Balogh, P.; Keserű, G. M. Covalent Fragment Libraries in Drug Discovery. *Drug Discov. Today* **2020**, *25* (6), 983–996. https://doi.org/10.1016/j.drudis.2020.03.016.
79. Wymann, M. P.; Bulgarelli-Leva, G.; Zvelebil, M. J.; Pirola, L.; Vanhaesebroeck, B.; Waterfield, M. D.; Panayotou, G. Wortmannin Inactivates Phosphoinositide 3-Kinase by Covalent Modification of Lys-802, a Residue Involved in the Phosphate Transfer Reaction. *Mol. Cell. Biol.* **1996**, *16* (4), 1722–1733. https://doi.org/10.1128/mcb.16.4.1722.
80. Lee, C. U.; Grossmann, T. N. Reversible Covalent Inhibition of a Protein Target. *Angew. Chem. Int. Ed.* **2012**, *51* (35), 8699–8700. https://doi.org/10.1002/anie.201203341.
81. Gurbani, D.; Du, G.; Henning, N. J.; Rao, S.; Bera, A. K.; Zhang, T.; Gray, N. S.; Westover, K. D. Structure and Characterization of a Covalent Inhibitor of Src Kinase. *Front. Mol. Biosci.* **2020**, 7, 81. https://doi.org/10.3389/fmolb.2020.00081.
82. Bueno, A. B.; Showalter, A. D.; Wainscott, D. B.; Stutsman, C.; Marín, A.; Ficorilli, J.; Cabrera, O.; Willard, F. S.; Sloop, K. W. Positive Allosteric Modulation of the Glucagon-Like Peptide-1 Receptor by Diverse Electrophiles. *J. Biol. Chem.* **2016**, *291* (20), 10700–10715. https://doi.org/10.1074/jbc.m115.696039.

83. Harris, C. M.; Foley, S. E.; Goedken, E. R.; Michalak, M.; Murdock, S.; Wilson, N. S. Merits and Pitfalls in the Characterization of Covalent Inhibitors of Bruton's Tyrosine Kinase. *SLAS Discov. Adv. Sci. Drug Discov.* **2018**, *23* (10), 1040–1050. https://doi.org/10.1177/2472555218787445.
84. Erlanson, D. A.; Braisted, A. C.; Raphael, D. R.; Randal, M.; Stroud, R. M.; Gordon, E. M.; Wells, J. A. Site-Directed Ligand Discovery. *Proc. Natl. Acad. Sci. U. S. A.* **2000**, *97* (17), 9367–9372. https://doi.org/10.1073/pnas.97.17.9367.
85. Kathman, S. G.; Statsyuk, A. V. Covalent Tethering of Fragments for Covalent Probe Discovery. *MedChemComm* **2016**, 7, 576–585, Royal Society of Chemistry April 1. https://doi.org/10.1039/c5md00518c.

CHAPTER TWO

Covalent drugs in development for immune-mediated diseases

Leonard Sung*
Principia Biopharma, A Sanofi Company, South San Francisco, CA, United States
*Corresponding author: e-mail address: leonard.sung@principiabio.com

Contents

1. Introduction

The discovery and development of novel small molecule immune-mediated drugs has been a formidable challenge for the drug discovery community. Immune-mediated diseases are challenging to treat because numerous symptoms could result from multiple signaling pathways. Mechanistically, it is difficult to deconvolute these diseases down to one biological target or pathway. Significant progress for the treatment of immune-mediated diseases, such as rheumatoid arthritis and multiple sclerosis, have been achieved with new biologic agents, such as TNF-α and anti-CD20 antibodies. Within the past decade a handful of orally-administered small molecule immune-mediated drugs were approved.[1] Fingolimod (Gilenya®), a sphingosine-1-phosphate receptor (S1P) modulator, was approved for the

Annual Reports in Medicinal Chemistry, Volume 56
ISSN 0065-7743
https://doi.org/10.1016/bs.armc.2021.03.001

treatment of the relapsing form of multiple sclerosis (MS) in 2010. Tofacitinib (Xeljanz®), a pan-JAK inhibitor, was approved in 2012 for the treatment of rheumatoid arthritis (RA). Apremilast (Otezla®), a PDE4 inhibitor, was approved for the treatment of adults with psoriatic arthritis and severe plaque psoriasis in 2014. Finally, fostamatinib (Tavalisse®), a SYK inhibitor, was approved in 2018 for the treatment of chronic immune thrombocytopenia for patients who have failed previous treatments.

While the approvals of these small molecule immune-mediated drugs are encouraging achievements, these marketed drugs have not been well-received by providers and patients compared to the complementary biological agents for those immune-mediated diseases. One of the reasons may be off-target selectivity for these small-molecules. For example, fingolimod modulates all five isoforms of the S1P receptors, which could lead to bradycardia or a decrease in heart rate.[2] Tofacitinib also inhibits JAK2, in addition to JAK1 and JAK3, which could lead to an increased risk of blood clots at high doses of 10 mg.[3] Encouragingly, several selective S1P modulators and selective JAK1 or JAK3 inhibitors are currently advancing in the clinic or have been recently approved. After tofacitinib, baricitinib (Olumiant®), a JAK1 & JAK2 inhibitor and upadacitinib (Rinvoq®), a selective JAK1 inhibitor, were approved for the treatment of patients with rheumatoid arthritis in 2018 and 2019, respectively. Recently in 2020, selective S1P1 & S1P5 modulator ozanimod (Zeposia®) was approved for the treatment of adults with relapsing forms of MS.

Would a designed covalent approach toward the development of an immune-mediated drug be beneficial to overcome the issues of selectivity described earlier? Targeting amino acid residues (primarily cysteine) that are shared by a small group of kinases could lead to the development of drugs with higher selectivity.[4] In addition, inactivation of a target protein through covalency would minimize systemic exposure of the drug until the target protein could be resynthesized.[5] In the treatment of chronic immune-mediated diseases, a successful covalent inhibitor would have the potential to minimize drug exposure while maintaining high target coverage for efficacy. If applied correctly, the use of covalent inhibitors could increase the therapeutic window between toxicity and efficacy of orally administered small molecule drugs for the treatment of chronic immune-mediated diseases.

To date, no designed covalent inhibitor has been approved for the treatment of a chronic immune-mediated disease, with the exception of ibrutinib for chronic graft versus host disease (cGvHD).[6] On the other hand, several covalent inhibitors, targeting for example, EGFR/HER2, BTK, and FGFR, have been approved for the treatment of cancer. There are limited examples of drugs that have been approved for an oncological indication and then later for an immune-mediated disease. The safety requirements, such as high on-target selectivity and minimal side effects, are extremely high for disease-modifying agents attempting to treat chronic immune-mediated diseases. For example, acalabrutinib and ibrutinib have been approved for treating B-cell malignancies for oncology, but they both cause platelet dysfunction, which increase the risk of bleeding for patients.[7] Ibrutinib also inhibits wild-type EGFR, resulting in a skin rash in 30% of patients.

In this book chapter, the features of an immune-mediated drug target for a druggable covalent inhibition strategy will be discussed. Is there a unique nucleophilic residue on the immune-mediated drug target? Can it be targeted selectively? What is the target turnover rate?

In the next section of this book chapter, an in-depth analysis of immune-mediated covalent drug programs in the clinic will be presented. What adjustments are needed to convert the designed covalent inhibitors for oncology into acceptable candidates for the treatment of immune-mediated diseases? What key safety experiments would need to be performed before submitting the covalent inhibitor candidate into the clinic for immune-mediated diseases?

2. Developing covalent kinase inhibitors for immune-mediated diseases

As discussed earlier, several small-molecule drugs that inhibit immune-mediated targets (JAK, S1P, PDE4 & SYK) have been approved within the last 10 years.[8] Two of these targets (JAK & SYK) are kinases.[9] Kinases act like switches that relay signals from the extracellular environment to cellular processes inside the cell. These ATP binding pockets have been blocked successfully with inhibitors developed by the medicinal chemistry community, but it is extremely challenging to develop inhibitors with exquisite selectivity. Thus, aberrant signaling pathways in immune-mediated diseases can be blocked through the inhibition of specific kinases. There are over 500 different human protein kinases that share a common ATP binding pocket.[10] Traditionally to achieve selectivity, inhibitors would interact alternative conformations of the kinase, such as the inactive non-ATP bound conformation of the kinase versus the commonly ATP-bound active form.

Designing covalent inhibitors to interact with a less-common nucleophilic group (most often cysteine) in the ATP binding pocket could impart the necessary level of selectivity.[11] For example, a study identified approximately 200 kinases with an accessible cysteine inside or near the ATP binding pocket.[12] The reactive thiol moiety (–SH) would need to be in an environment where deprotonation could readily occur—revealing the reactive anionic thiolate for nucleophilic addition. Also, under high intracellular ATP concentrations, engagement of a non-conserved cysteine residue with a covalent inhibitor provides an advantage over a reversible noncovalent inhibitor, which would have to compete with the natural ATP ligand. In addition, the activity of the target protein is restored only upon its resynthesis, and thus prolonged target inhibition could be maintained with a covalent inhibitor without the requirement for constant systemic exposure.

To measure potency, a biochemical assay would typically use an IC_{50} value to calculate the percent of inhibited target protein at equilibrium. For measuring the potency of covalent inhibitors, the rate of covalent modification, defined by k_{inact}/K_I, accurately assesses the reactivity and potency of the inhibitor.[13] The k_{inact}/K_I value is a rate constant that incorporates the rate of covalent bond formation along with drug concentration. Conceptually, the rate of irreversible covalent bond formation is described by k_{inact} and the contribution of reversible binding non-covalent interactions is roughly defined by K_I. Unfortunately, calculating k_{inact}/K_I can be time consuming.

The rate of covalent bond formation is influenced somewhat by the reactivity of the covalent electrophile. There are a number of recent reviews and papers that discuss the reactivity of various electrophiles and rank order them based on intrinsic reactivity.[14] In addition, positioning of the electrophile in relation to the nucleophilic group on the protein is critical. While finding the optimal position of the electrophile may be difficult to predict prospectively, retrospective analysis of the binding pocket with X-ray crystallography along with molecular dynamic simulations of the protein have been used to provide an analysis for the high rate of covalent bond formation. A recent example of this analysis was performed to describe the reactivity of AZD9291 (osimertinib) for EGFR T790M over EGFR WT.[15]

These highly reactive electrophiles could also be a detriment because of safety concerns. The development of covalent drugs to irreversibly modify proteins raises the possibility of idiosyncratic adverse drug reactions (IADRs).[16] Bound covalent drugs or metabolites of covalent drugs increase the risk of haptenization or the generation of auto-antibodies against the foreign entities. Drugs dosed below 10 mg/day are less likely to cause IADRs and another report calculated the in vivo daily covalent binding burden (DCBB) to be similar, suggesting drugs dosed less than 10 mg/day are more likely to be clinically safe.[17] Since irreversible binding to other endogenous thiols is a concern, depletion of free drug after incubation in whole blood, without metabolic activation, can serve as an indicator of indiscriminate binding of a covalent inhibitor to proteins. Customary safety protocols for developing non-covalent drugs (genotoxicity, drug-drug interactions, CYP & hERG inhibition, time-dependent inhibition, etc.) can also be applied for the safety evaluation of covalent inhibitors.

In the next section, two kinase targets (JAK3 & BTK) will be highlighted, both of which contain a targetable cysteine residue and play key roles in critical immune-mediated pathways. Then, specific clinical candidates for each target are discussed with their attributes as covalent inhibitors to treat immune-mediated diseases.

2.1 JAK3

The Janus kinase family, composing of JAK1, JAK2, JAK3, and TYK2, are involved in cytokine signaling in response to stressors.[18] When a cytokine binds to a cell-surface receptor, the JAK receptors become activated and form pairs, resulting in phosphorylation of the cell-surface receptor and downstream transcription factors such as signal transducers and activators of transcription (STAT) molecules. The JAK-STAT pathway relays extracellular chemical signals from cytokines to the nucleus (via STAT) to initiate gene transcription.

Among the approved JAK inhibitors for immune-mediated diseases, some are pan-JAK inhibitors like tofacitinib and peficitinib, while others are JAK1 and JAK2 selective inhibitors, such as ruxolitinib and baricitinib (see Fig. 1). However, these approved JAK inhibitors have a narrow therapeutic index.[19] The next generation of non-covalent JAK inhibitors such as filgotinib, upadacitinib, and abrocitinib are JAK1 selective non-covalent inhibitors.

Unlike JAK1, JAK2, and TYK2 which are expressed ubiquitously, JAK3 is expressed primarily in lymphocytes. JAK3 is associated with the common gamma (γc) receptor chain shared by interleukins IL-2, IL-4, IL-7, IL-9, IL-15 and IL-21 receptors. Genetic defects in JAK3 can result in X-linked human severe combined immunodeficiency (X-SCID).

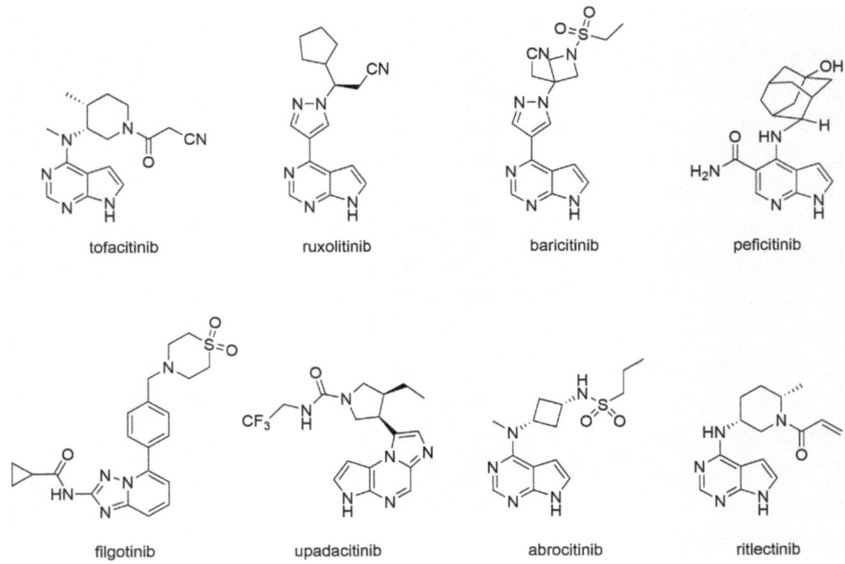

Fig. 1 JAK inhibitors in clinical trials.

Targeting JAK3 selectively would spare the clinical effects of JAK1 and JAK2 inhibition, which could have an improved overall safety profile for the treatment of autoimmune diseases.

Despite the high amino-acid sequence homology (>80%) in the active sites of the JAKs and TYK2, JAK3 contains a cysteine residue at position 909 (Cys909) while the others contain a serine at this position. Another difference between the JAKs is their affinity for ATP. JAK3 has a ~10-fold higher affinity for ATP with a K_m of ~4 μM for JAK3 and ~40 μM for JAK1.[20] Thus, designing a non-covalent selective JAK3 inhibitor would be quite challenging under typical cellular concentrations of ATP.[21] Several groups have published work toward the development of a covalent JAK3 inhibitor[22]; however only one, PF-06651600 (ritlecitinib), has advanced into the clinic.

2.2 PF-06651600 (Ritlecitinib) in clinical trials

With crystal structures of tofacitinib in the kinase domain of JAK3, researchers from Pfizer designed acrylamides on the tofactinib core, which later resulted in the discovery of PF-06651600. The covalent interaction between the Cys909 in JAK3 with PF-06651600 was confirmed by X-ray crystal structure analysis (see Fig. 2, PDB: 5TOZ).[23] The 2-methyl moiety of PF-06651600 imparted more stability than its des-methyl

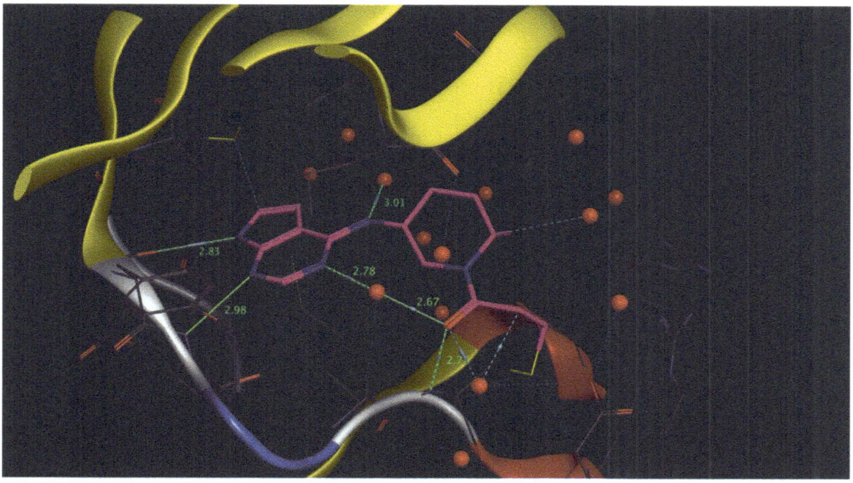

Fig. 2 Crystal structure of PF-06651600 (magenta ball and stick) bound to JAK3 kinase domain (green; PDB: 5TOZ).

Table 1 In vitro and in vivo pharmacokinetic properties of PF-06651600 in various species.

	Mouse	Rat	Monkey	Dog	Human
Hepatocytes $CL_{int,u}$ ((μM/min)/10^6 cells)	53	14	7.6	2.1	2.8
Blood $t_{1/2}$ (min)	>360	161	37	>360	>360
In vivo blood CL (mL min^{-1} kg^{-1})	48	53	42	8.6	
V_{ss} (L/kg)	0.8	1.4	2.6	1.1	
In vivo plasma $t_{1/2}$ (h)	1.3	0.3	0.7	1.1	
F_{po} bioavailability (%)	62	85	56	109	

derivative when evaluated in human liver microsomes, hepatocytes and blood stability assays.[24] The high oral bioavailability of PF-00651600 in rat and dog corresponded to its high aqueous solubility (>2 mg/mL), low hepatic clearance, and high in vitro passive permeability (RRCK, 15×10^{-6} cm/s) (see Table 1). Oxidative metabolism of the acrylamide, glutathione conjugation on the acrylamide and ring oxidation through CYP3A4 were determined as pathways for the metabolic clearance of PF-06651600.

Excellent PK properties with low clearance of PF-06651600 is desired because of the short protein half-life of JAK3. In a previous report, the resynthesis rates for JAK1, JAK2, and TYK2 were determined to be around 2–4 h from pulse-chase experiments.[25] The half-life for JAK3 protein turnover was in the 3–4 h range in human primary CD4 + T cells from a pulse-chase experiment run by Pfizer scientists (see Fig. 3).[23]

Since inhibition of JAK1 and JAK3 block signaling through the γc cytokine receptors, the question remained whether the blockade of either one (JAK1 or JAK3) would affect all of cytokines associated with γc cytokine signaling. Using selective JAK3 inhibitor PF-06651600 in macrophages and human whole blood, the signaling pathways of IL-10, IL-27 and IL-35 were preserved. Inhibition of IL-10, IL-27 and IL-35 are associated with the blockade of JAK1. Thus, blocking the γc cytokine receptors with a selective JAK3 covalent inhibitor spares immunoregulatory cytokine signaling through IL-10, IL-27 and IL-35, resulting in a differentiated cytokine profile from JAK1 or pan-JAK inhibition.

The selectivity of PF-06651600 was measured in a panel against 305 kinases from Invitrogen (see Table 2). PF-066515600 appeared to be quite selective for JAK3 and showed activity against eight targets within the TEC family. In addition to measuring the K_i and K_{inact} values of PF-06651600 against these eight targets, the binding activity of PF-06651600 was measured at high ATP (1 mM) concentrations. PF-06651600 was found to weakly bind to SLK, FGR and FLT3, but none of those kinases were inhibited in a covalent manner even though there are cysteines present in their respective ATP binding sites.

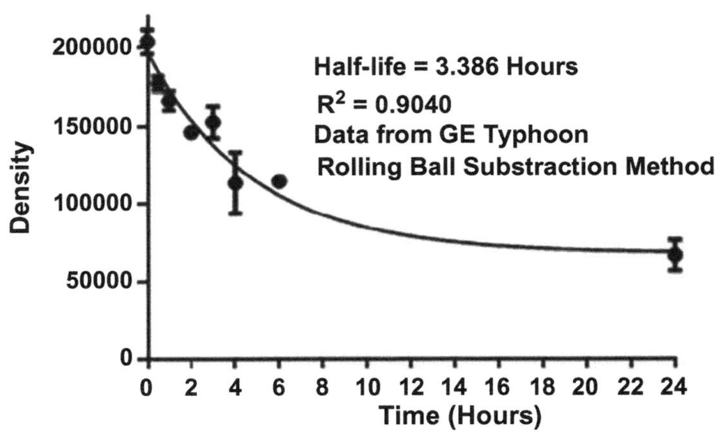

Fig. 3 Resynthesis rate of JAK3 determined by pulse chase experiment.

Table 2 Kinase inhibition of TEC family kinases measured as IC_{50} with 1 mM ATP (nM) and inactivation rates of PF-06651600 (K_{inact} (s^{-1}), K_i (μM), and K_{inact}/K_i ($M^{-1}\ s^{-1}$)).

Kinase	Tofacitinib	JTE-051	Fenebrutinib	PF-06651600	K_{inact} (s^{-1})	K_i (μM)	K_{inact}/K_i ($M^{-1}\ s^{-1}$)
BLK	NA	5120	>10,000	>10,000	0.0278	32.4	8.58×10^2
BMX	>10,000	174	8181	666	0.00564	0.545	1.03×10^4
BTK	>10,000	469	74	4–4	0.124	62.3	1.99×10^3
EGFR	NA	>10,000	>10,000	>10,000	ND	ND	ND
HER2	NA	>10,000	>10,000	>10,000	ND	ND	ND
HER4	NA	>10,000	>10,000	>10,000	ND	ND	ND
ITK	>10,000	4.4	>10,000	395	0.000144	0.0269	5.35×10^3
JAK3	55	>10,000	>10,000	33.1	2.32	6.31	3.68×10^5
TEC	>10,000	486	1676	403	0.00156	0.679	2.30×10^3
RLK/TXK	>10,000	323	>10,000	155	0.000487	0.131	3.79×10^3
MAP2K7	NA	>10,000	>10,000	>10,000	ND	ND	ND

The effects of PF-066516600 on immune cell functions by blocking TEC kinase family members were measured using two selective inhibitors, JTE-051 and fenebrutinib, respectively, for ITK and BTK, as benchmark compounds (see Table 3).[26] PF-066516600 appears to have some BTK inhibition activity by blocking CD69 inhibition on B cells, following BCR activation with antibodies. In addition, PF-066516600 inhibits degranulation and IFN-γ production in CD8+ cells and NK cells. Blocking degranulation and IFN-γ production in CD8+ cells suggest inhibition of ITK or possibly TEC & RLK. In contrast, inhibiting degranulation and IFN-γ production in NK cells hint at the possibility of selective TEC or RLK blockage because the selective ITK inhibitor JTE-051 does not affect NK cells. Thus, PF-066516600 is a covalent inhibitor that blocks immune functions related to JAK3 and several TEC kinase family members.

PF-06651600 was evaluated in the rat adjuvant-induced arthritis (AIA) model and it was demonstrated to reduce paw swelling with an unbound EC of 169 nM. In addition, PF-06651600 was investigated in the experimental autoimmune encephalomyelitis (EAE) mouse model. PF-06651600 was dosed either prophylactically at 20 and 60 mg/kg or therapeutically at 30 or 100 mg/kg and it was shown to reduce disease severity in that mouse model.[23]

A small, double-blind, placebo-controlled Phase II study of PF-06651600 was conducted in 70 patients with RA.[27] Patients were randomized 3:2 to receive oral PF-06651600 (200 mg once daily) or a placebo for 8 weeks. PF-06651600 was generally well-tolerated with the most common treatment-emergent adverse effects (TEAEs) being infections and infestations and skin and subcutaneous tissue disorders, where 11.9% of the treatment group and 3.6% of the placebo group were affected. Clinical effects of JAK1 inhibition include elevation of cholesterol and liver enzyme levels, rapid reduction of CRP levels, anemia or neutropenia—no clinically significant cases were observed in the patients who received PF-06651600. However, in some patients, reduction in platelet levels was observed. This reduction, also found in the use of clinically approved BTK inhibitors, may result from inhibition of certain TEC kinase family members that play a role in platelet aggregation. PF-06651600 is currently being studied for the treatment of alopecia areata (NCT02974868 and NCT03732807), vitiligo (NCT03715829), ulcerative colitis (NCT02958865), and Crohn's disease (NCT03395184).

Table 3 Cellular effects of PF-06651600, tofacitinib, fenebrutinib, and JTE-051 on immune cell functions (IC_{50} values (nM)).

	IL-15 induced p-STAT5 in PBMC (JAK1/3)	B lymphocyte anti IgD-induced CD69	T lymphocyte anti CD2/3/28-induced CD69	CD8+ T lymphocyte CD3/28-induction		NK cell K562 stimulation	
Readout	**p-STAT5**	**CD69**	**CD69**	**CD107a**	**IFNγ**	**CD107a**	**IFNγ**
PF-06651600	52	344	380	213	185	509	188
Tofactinib (JAK 1/3)	17	>20,000	326	22,800	21,200	>30,000	15,500
Fenebrutinib (BTK)	4900	15	8200	9100	12,100	>30,000	12,100
JTE-051 (ITK)	3600	3300	190	120	160	2200	790

2.3 BTK examples for immune-mediated diseases

Bruton's tyrosine kinase (BTK) is a non-receptor tyrosine kinase. It is expressed in all hematopoietic cells, with the exception of T cells and mature plasma cells. Genetic defects of BTK results in X-linked agammaglobulinemia (XLA), a disease in which sufferers cannot generate mature B cells to fight infections. BTK is an essential signaling node downstream of the B cell receptor (BCR) and it is involved in activation of other hematopoietic cells, resulting in cell proliferation, antibody and cytokine production, and co-stimulatory molecule expression. BTK is required for the maturation, function and differentiation of B-cells. BTK is also involved in the signaling pathway of Fcγ receptors (FcγR) for immunoglobulin G (IgG) complex in monocytes and macrophages, as well as the signaling pathway of Fcε receptors (FcεR), which activates mast cells and basophils, resulting in the expression of cytokines, chemokines and cell adhesion molecules.

As a member of the Tec family of non-receptor tyrosine kinases, BTK contains a cysteine (Cys481) located near the ATP binding pocket.[28] Other members of the TEC family (BMX, TEC, ITK and TXK) share a conserved cysteine at this position, along with other kinases such as EGFR, ERBB2 (HER2), ERBB4 (HER4), JAK3, BLK and MKK7—for a total number 10 kinases and BTK.

From one report, the half-life of BTK was measured to be somewhere between 48 and 72 h based on purified human B cells.[29] Another report suggests that resynthesis half-life of BTK to be longer, between 115 and 154 h, from human whole blood.[30] This long protein half-life makes BTK an ideal target for covalent inhibition with designed inhibitors.

The clinical development of BTK inhibitors were initially approved for the treatment of patients in the oncology setting (see Fig. 4). Covalent irreversible BTK inhibitors ibrutinib,[31] acalabrutinib,[32] and zanubrutinib have been approved by the FDA for the treatment of certain hematological malignancies, such as chronic lymphocytic leukemia (CLL), mantle cell lymphoma (MCL), Waldenstrom's macroglobulinemia, and small lymphocytic lymphoma (SLL), while tirabrutinib was recently approved in Japan. In contrast, the covalent BTK inhibitors for the treatment of immune-mediated diseases are currently in clinical trials. In the next sections, four of these BTK inhibitors will be profiled in their clinical development journey for the treatment of immune-mediated diseases.

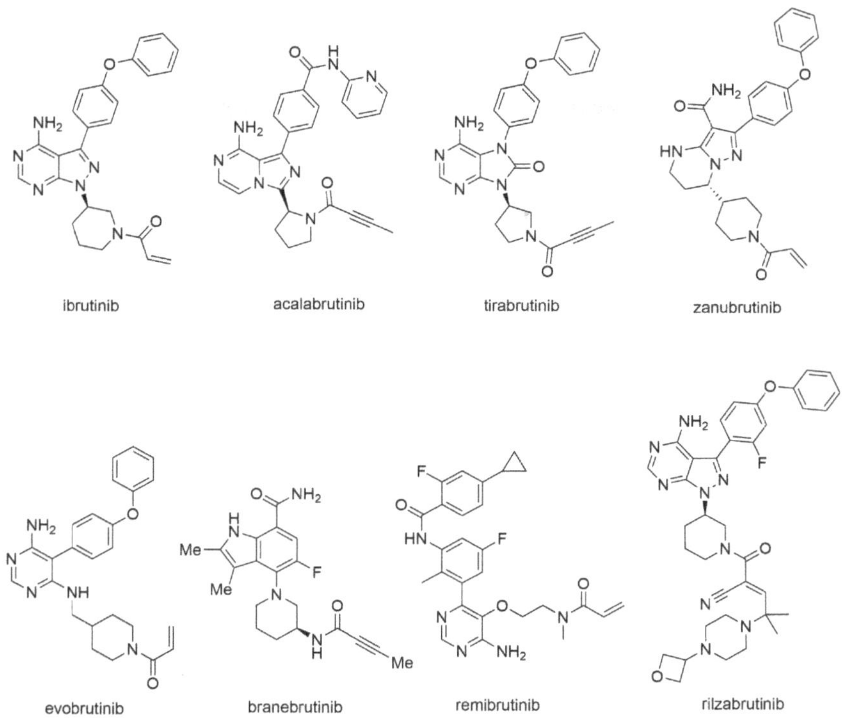

Fig. 4 BTK inhibitors in clinical trials.

3. Evobrutinib—Multiple sclerosis

The goal for the EMD Serono Research group was to identify a highly selective covalent BTK inhibitor, particularly against EGFR and for it to possess moderate clearance.[33] Starting with an X-ray crystal structure of a moderately potent BTK inhibitor (see Fig. 5), the group replaced the pyrrolopyrimidine core and found the corresponding pyrimidine core to provide a vector amenable for covalent BTK inhibition, which in turn reduced inhibition of EGFR. Optimization of the linker and warhead resulted in the selection of evobrutinib as their clinical candidate.

Evobrutinib's kinase selectivity was determined by the EMD Millipore KinaseProfiler screen.[34] From a panel of 267 kinases at a concentration of 1 μM, evobrutinib inhibited BTK, BMX, and TEC by 90%, 93%, and 82%, respectively.[35] Other kinases such as TXK, BLK, RSK1, PAK3,

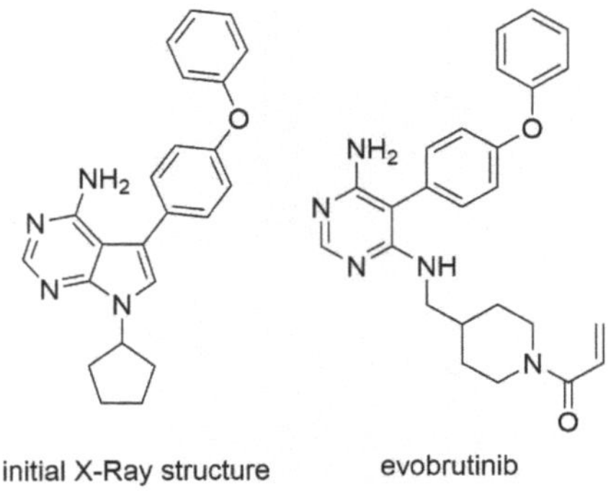

Fig. 5 EMD Serono group BTK inhibitors.

Her4/ErbB4 and ITK were weakly inhibited by 36%, 36%, 31%, 30%, 29% and 13% respectively.[33] Evobrutinib requires the acrylamide warhead to inhibit BTK. The analog of evobrutinib containing an acetamide instead of the acrylamide was inactive. In contrast, the ibrutinib scaffold, without the acrylamide warhead, displayed similar potency toward BTK and comparable inhibition profile against its off-target kinases.

The in vivo PK properties of evobrutinib were evaluated in three species and it was determined to have moderate to high clearance, a short plasma half-life (~1 h), and it was rapidly absorbed after oral administration with C_{max} reached between 0.25 and 1 h (see Table 4).[33]

Evobrutinib was evaluated in a rat collagen-induced arthritis (CIA) model where rats with type II collagen arthritis were treated once daily for 11 days (days 6–16) with evobrutinib, vehicle or methotrexate. Disease-induced ankle-swelling was significantly reduced for rates treated with evobrutinib at doses higher than 3 mg/kg (10 and 30 mg/kg), but not doses less than 3 mg/kg (0.3 and 1 mg/kg).[33] A similar result was found when a mouse CIA model was performed with evobrutinib—the difference was that lower doses of evobrutinib (1 mg/kg) showed a positive effect in mice.[35] Evobrutinib also showed efficacy at low doses (3 mg/kg) in a NZB/W lupus-prone mouse model, where mice develop a disease like systemic lupus erythematosus (SLE) and lupus nephritis (LN) in humans.

Evobrutinib was evaluated in a double-blind, placebo-controlled, single- and multiple-ascending dose (SAD; MAD) Phase I study (see Fig. 6).[36] In the SAD group, 48 healthy participants were randomized 3:1 to receive a single dose of evobrutinib (25, 50, 100, 200, 350, or 500 mg; 6 participants in each group) or placebo (12 participants). In the MAD group, 36 participants were randomized 3:1 to receive evobrutinib (25, 75 or 200 mg; 9

Table 4 Species comparison of plasma-derived PK parameters of evobrutinib after single administration.

	Mouse, 1 mg/kg iv; 3 mg/kg, po (susp.)	Rat, 2 mg/kg iv; 10 mg/kg, po (susp.)	Dog 1 mg/kg, iv; 3 mg/kg, po (susp.)
C_{max} (ng/mL) (iv/po)	(754–806)/(145–179)	(559–717)/(22–75)	787/185
$AUC_{0\text{-}inf}$/dose (ng·h/mL/mg/kg) (iv/po)	(360–434)/(64–67)	(114–143)/(2–6)	658/80
$T_{1/2}$ (h), iv/po	0.52/2.9 ± 0.7	128 ± 15/1.46	1.03/0.88
CL (mL min^{-1} kg^{-1}), iv	38.5–46.5	119–146	25.5
V_{ss} (L/kg), iv	1.10–1.16	2.26–2.82	1.47
F_{po} (%)	12.7–1.37	1.28–4.81	11.4

participants in each group) QD for 14 days or placebo (9 participants). Evobrutinib was well tolerated with mild treatment-emergent adverse events (TEAEs) occurred in 25% of subjects after single dosing and 48% after multiple dosing. There most common TEAEs were headache, skin irritation, upper respiratory tract infection and fatigue. The effect of evobrutinib on QT interval was also investigated, but no prolongation of QT/QTc interval was observed in the healthy subjects.

In the pharmacokinetics (PK) assessment of evobrutinib, it was rapidly absorbed after oral administration and eliminated quickly with a geometric mean terminal half-life ($t_{1/2}$) of 1.8–2.6 h at low doses (25–200 mg), but around 6.6–6.8 h at higher doses (350 and 500 mg) (see Fig. 7). At those high single doses, evobrutinib concentrations above the lower limit of quantitation (LLOQ) were measurable up to 48 h, but it was undetectable when measured 8 h after dosages less than 200 mg. To measure the degree of BTK occupancy by evobrutinib, the total BTK and free BTK were measured in PBMC protein lysates using anti-BTK antibodies and a biotinylated probe that irreversibly binds to the active site of BTK.

A dose-dependent increase in BTK occupancy was observed in the single dose administration groups for evobrutinib. In the 200-mg dose group and above (350 and 500 mg), greater than 99% BTK occupancy, described by maximum effect (E_{max}), was observed after a single dose.

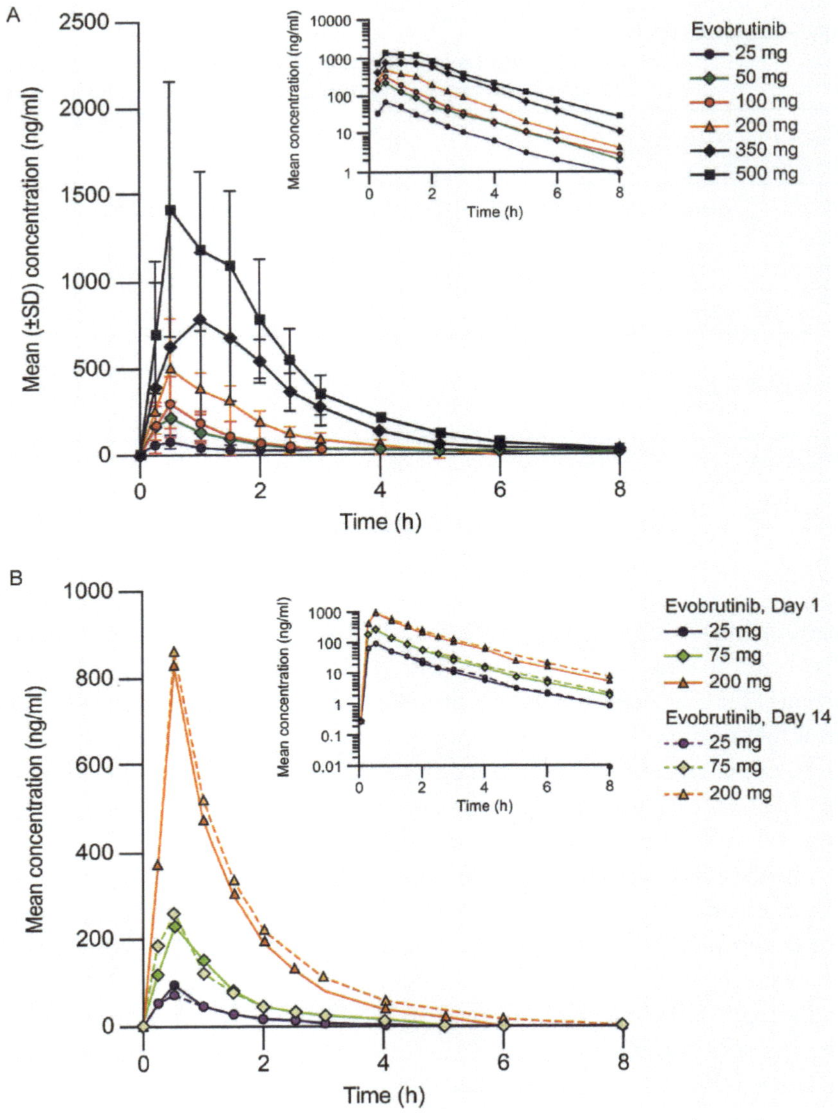

Fig. 6 Pharmacokinetics of evobrutinib. (A) Arithmetic mean (SD) evobrutinib serum exposure vs time after single dosing on day 1 of part 1 (single ascending dose study) and (B) mean evobrutinib serum exposure vs time on day 1 and day 14 of part 2 (multiple ascending dose study). *From Becker, A.; et al. Safety, Tolerability, Pharmacokinetics, Target Occupancy, and Concentration-QT Analysis of the Novel BTK Inhibitor Evobrutinib in Healthy Volunteers,* Clin. Transl. Sci. ***2020,*** 13, *325–336. doi:10.1111/cts.12713.*

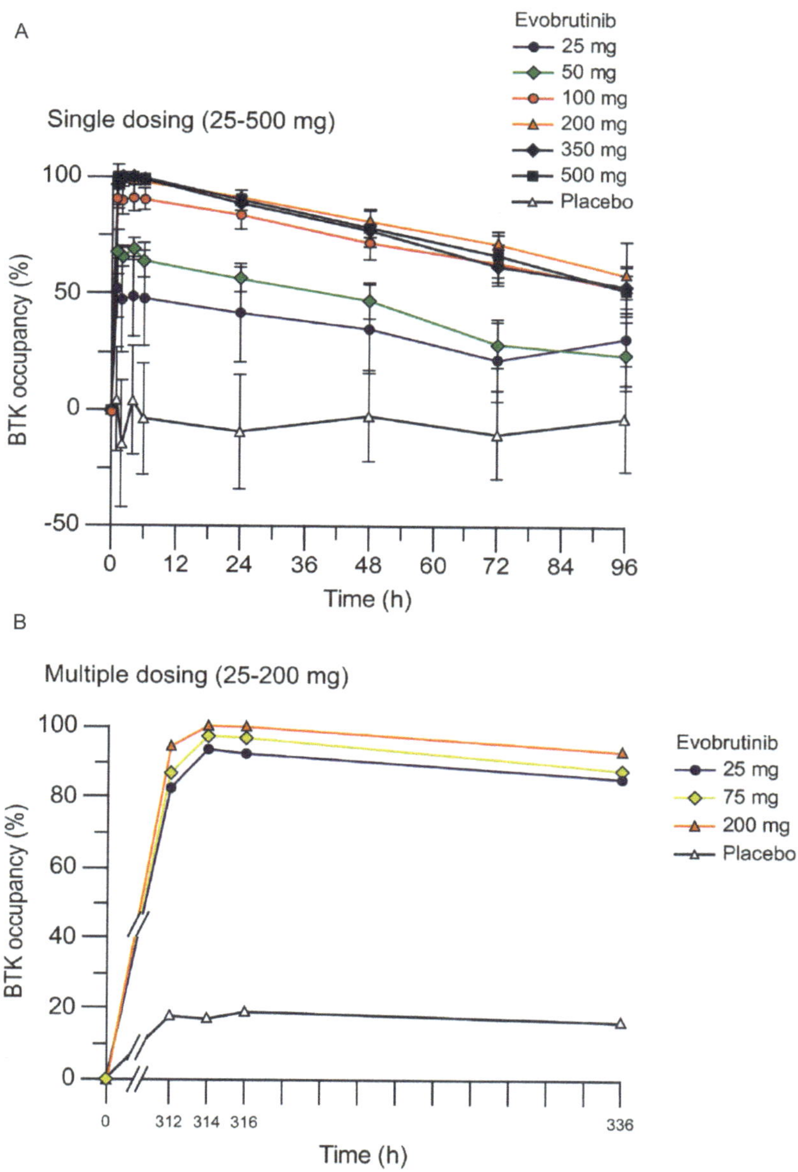

Fig. 7 Arithmetic mean (SD) of % Bruton's tyrosine kinase (BTK) receptor occupancy* vs time after administration of evobrutinib for (A) single dosing and (B) multiple dosing. *% BTK occupancy = (1 − (BTKF/BTKT)/(BTKF pre-dose/BTKT pre-dose))*100. *BTKF*, free Bruton's tyrosine kinase; *BTKT*, total Bruton's tyrosine kinase. *From Becker, A.; et al. Safety, Tolerability, Pharmacokinetics, Target Occupancy, and Concentration-QT Analysis of the Novel BTK Inhibitor Evobrutinib in Healthy Volunteers,* Clin. Transl. Sci. ***2020,*** 13, *325–336. doi:10.1111/cts.12713.*

The median E_{max} for the other dose groups were 92% at 100 mg, 73% at 50 mg, and 58% at 25 mg. The E_{max} was met rapidly within 1–4 h for all dose groups and declined slowly post dose, although BTK occupancy was still detectable 96 h post dose. For the multiple dosing groups, median E_{max} for the 25 mg dose increased from 70% on day 1 to 95% on day 14. The day 14 E_{max} values for the 75 mg and 200 mg groups were 98% and 100%, respectively. However, low median levels of BTK occupancy (~20%) were observed in the placebo group, which may be attributed to assay variability.

A double-blind, randomized, phase 2 trial with three groups of patients with relapsing multiple sclerosis on evobrutinib (at a dose of 25 mg once daily, 75 mg once daily, or 75 mg twice daily) was reported in 2019 (NCT02975349).[37] A Phase III clinical trial to study evobrutinib in patients with relapsing multiple sclerosis is ongoing (NCT04338022).

4. BMS-986195 (Branebrutinib): Preclinical safety and efficacy determination to Ph I clinical trials

The motivation for BMS to develop a covalent BTK inhibitor came from a hypothesis that coverage as high as IC_{90} might be needed for clinical efficacy in the treatment of rheumatoid arthritis.[38] Previously, BMS-986142 was developed as a potent, non-covalent BTK inhibitor and it was advanced into clinical trials for the treatment of rheumatoid arthritis (see Fig. 9).[39] In addition, another objective was to develop a covalent BTK inhibitor that required a dosage of less than 10 mg in humans. This low dose was desired to minimize the potential for idiosyncratic adverse drug reactions that may result from non-selective covalent inhibition.[17]

With guidance from their non-covalent BTK inhibitor, BMS-986142, Watterson et al. identified BMS-986195 (branebrutinib) as their clinical candidate.[38] Key design features include the selection of an electrophilic but-2-ynamide acceptor, which is also featured in the BTK inhibitors tirabrutinib and acalabrutinib. The but-2-ynamide is positioned in the BTK pocket to have the ideal geometry for covalent bond formation with Cys 481 (see Fig. 8). Another key discovery was the addition of the fluoro substituent at the C5 position of the indole core. Compared with the des-fluoro analog, BMS-986195 had better biochemical and cellular potency, higher plasma concentrations in a mouse PK study (C_{max} and AUC) and more potent inactivation of BTK in the human whole blood assay.

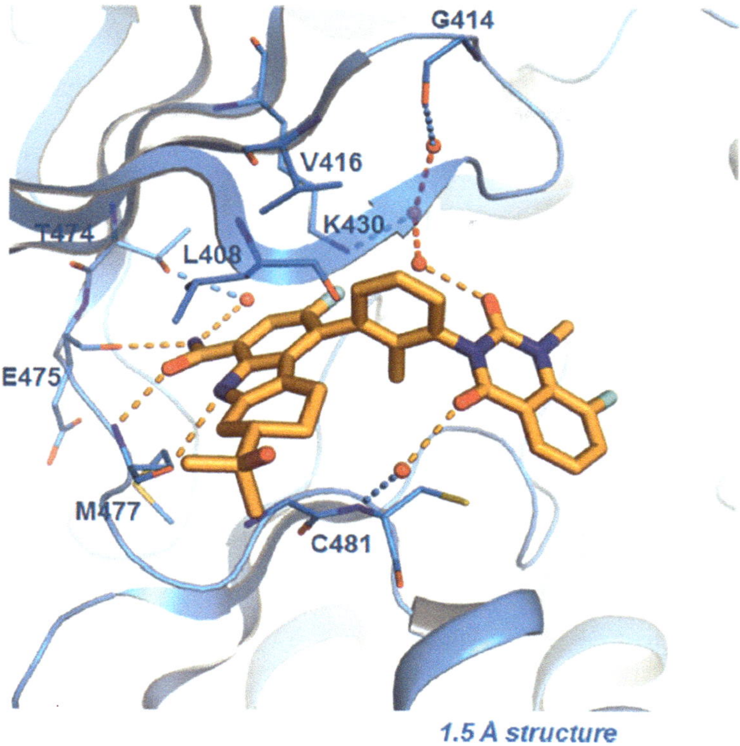

Fig. 8 X-ray crystal structure of BMS-986195 bound to the kinase domain of BTK (1.5 Å structure) (PDB code 5T18).

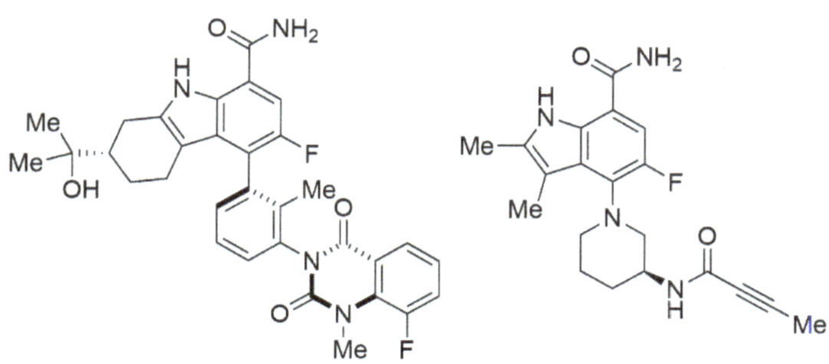

Fig. 9 Structures of BMS-986142 and BMS-986195 (Branebrutinib).

Table 5 In vitro selectivity data for BMS-986142 and BMS-986195.

	BMS-986142 (non-covalent)		BMS-986195 (covalent)	
Kinase	**Biochemical IC_{50} (nM)**	**Kinase/BTK fold selectivity**	**Biochemical IC_{50} (nM)**	**Kinase/BTK fold selectivity**
BTK (Tec family)	0.50		0.10	
TEC (Tec family)	10	20×	0.90	9×
BMX (Tec family)	32	64×	1.5	15×
TXK (Tec family)	28	56×	5.0	50×
ITK (Tec family)	15	30×	100	1000×

To understand the preclinical safety profile for BMS-986195, Watterson et al. evaluated its kinase selectivity and assessed its off-target reactivity (see Table 5). BMS-986195 was screened against a panel of 245 kinases (Cell Signaling Technology, Inc.) and it was determined to be highly selective for BTK over all the kinases screened (at least >1000×-fold selectivity) with the exception of Tec family kinases TEC (9-fold selectivity), BMX (15-fold selectivity) and TXK (50-fold selectivity). Comparing the covalent BTK inhibitor BMS-986195 to the non-covalent BTK inhibitor, BMS-986142, the covalent BTK inhibitor had an improved selectivity profile. For example, the selectivity of BMS-986195 for BTK over Tec family member ITK was 1000-fold compared with a 30-fold selectivity of BMS-986142 for BTK over ITK.

Another potential concern for BTK covalent inhibitors is their selectivity for BTK over members in the EGFR family. In the biochemical assay, BMS-986195 was selective for BTK over EGFR family members by >10,000-fold. BMS-986195 was also evaluated against an EGFR-dependent AU565 breast cancer cell line to determine functional EGFR kinase activity and no inhibition for this assay was observed with inhibitor concentrations up to 25 μM. Furthermore, in the single- and multiple-ascending dose (SAD and MAD) Phase I Study of BMS-986195 in healthy volunteers,[30] very little adverse events associated with EGFR inhibition (skin rash) were observed compared to placebo (1 instance out of 8 in MAD) and active dosing groups (2 instances out of 30 in SAD group).

To assess the off-target reactivity of BMS-986195 to proteins and thiols, a radiolabeled analog of BMS-986195 was evaluated in human hepatocytes (concentration of 10 micromolar) and very low levels of protein were measured to be covalently bound (1% of 131 pmol/mg of protein). Low levels of covalent binding were also observed in human hemoglobin, protein mixtures containing BTK, human serum albumin and GSH in enzyme free phosphate buffer. Finally, a rat quantitative whole-body autoradiography study was performed with the radiolabeled analog of BMS-986195. The study revealed that after 10 mg/kg p.o. dosing, less than 0.7% radioactivity remained in the carcass and low covalent binding was observed in multiple organs (liver, spleen and pancreas) with a single dose of the radiolabeled analog of BMS-986195. Most of the radiolabeled material was recovered in the urine and feces.

With BMS-986195's clean off-target safety profile, the on-target in vitro activity for BMS-986195 was determined (see Table 6). BMS-986195 inhibited calcium flux signaling at $IC_{50}=7$ nM, after stimulating the Ramos B cells through the B cell receptor (BCR). In addition, stimulation of the human peripheral B cells through the BCR saw inhibition for the surface expression of costimulatory molecule CD86 and proliferation of those B cells in the presence of BMS-986195 (IC_{50} values of <1 nM for each assay). As a negative control for functional selectivity, expression of CD69

Table 6 In vitro cell activity data and whole blood data for BMS-986142 and BMS-986195.

Cellular assays	**Receptor/ Stimulation**	**BMS-986142 IC_{50} (nM)**	**BMS-986195 IC_{50} (nM)**
Calcium flux in ramos B cells	BCR/Anti-IgM	9 ± 5	7.2 ± 2.3
Proliferation of human peripheral B cells	BCR/Anti-IgM/ IgG	3 ± 1	0.04 ± 0.01
CD86 surface expression in peripheral B cells	BCR/Anti-IgM/ IgG	4 ± 3	0.3
CD69 surface expression in peripheral B cells	CD40/CD40L	>300	>300
TNFα from human PBMC cells	FCγR/Immune complex	3 ± 3	0.3
HWB CD69 expression in peripheral B cells	BCR/Anti-IgM	90 ± 40	11 ± 8

on peripheral B cells through CD40 stimulation is not affected by BTK kinase activity. Thus, BMS-986195 does not impact CD69 expression in peripheral B cells when stimulated through CD40 ($IC_{50} > 300$ nM). In human whole blood assays, BMS-986195 inhibited CD69 expression on BCR-stimulated B cells with an IC_{50} of 11 nM. The in vitro activity profile for BMS-986195 is comparable to the BMS non-covalent BTK inhibitor BMS-986142.

The pharmacokinetic profiles and physical properties between the non-covalent and covalent BMS BTK inhibitors are quite similar (see Table 7). Both BTK inhibitors have high plasma protein binding across five species (98.3% or higher), moderate PAMPA permeability, no hERG liability and no mutagenicity was observed in the Ames test. The pharmacokinetic parameters between the two BTK inhibitors were also similar, but the covalent inhibitor BMS-986195 had a shorter plasma $T_{1/2}$ and T_{max} compared with the non-covalent inhibitor BMS-986142. This shorter plasma $T_{1/2}$

Table 7 Pharmacokinetic parameters for BMS-986142 and BMS-986195.

	BMS-986142		**BMS-986195**	
Protein binding (bound)	99.1% hu, 98.5% mouse, 98.3% rat		98.8% hu, 99.6% mouse, 99.8% rat	
PAMPA permeability	673/725 nm/s (pH 5.5/7.4)		859/978 nm/s (pH 5.5/7.4)	
	Rat, 2 mg/kg iv; 5 mg/kg, po	Monkey, 2 mg/kg iv; 4 mg/kg, po	Rat 5 mg/kg, iv po	Monkey, 1 mg/kg iv; 2 mg/kg, po
C_{max} (μM), po	2.3 ± 0.3	0.85 ± 0.68	9.6 ± 1.0	2.3 ± 1.5
T_{max} (h), po	3.0 ± 1.2	3.7 ± 2.3	0.58 ± 0.4	1.0 ± 0.3
AUC (μM·h), po	23 ± 4.9	5.0 ± 2.8	18 ± 0.8	4.9 ± 3.1
$T_{1/2}$ (h), iv	3.6 ± 0.7	5.1 ± 1.4	4.3 ± 0.5	3.2 ± 2.5
CL (mL min^{-1} kg^{-1}), iv	4.2 ± 0.9	10 ± 7.9	8.7 ± 0.4	9.4 ± 3.7
V_{ss} (L/kg), iv	1.3 ± 0.3	3.0 ± 2.2	0.40 ± 0.10	0.45 ± 0.1
F_{po} (%)	67 ± 15	33 ± 21	74 ± 3	46 ± 18

and T_{max} of BMS-986195 is a benefit of covalent inhibition where any unbound inhibitor is cleared rapidly to limit off-target exposure.

Despite the rapid metabolic clearance of BMS-986195, the advantages of covalent inhibition were observed during the oral administration of BMS-986195 to BALB/c mice and measuring the amount of inactivated BTK in peripheral blood (see Fig. 10). A single dose of BMS-986195 (0.5 mg/kg) to BALB/c mice resulted in 90% inactivation of BTK, 4 h after dosing and 86% inactivation at 24 h post-dose. In cynomolgus monkeys, a single dose of BMS-986195 (0.5 mg/kg) resulted in 100% inactivation of BTK at the 4 h time point and 74% inactivation remaining at 72 h. Inactivation of BTK remained after the compound was cleared within 4–6 h in both species.

BMS-986195 was evaluated in two animal disease models to determine the efficacious dose in vivo. The first disease model, mouse collagen-induced arthritis (CIA), is a B-cell and FcγR-dependent model of human RA (see Fig. 11). Robust protection against clinically evident disease (83% and 93% inhibition) in this disease model was observed after daily oral dosing of BMS-986195 at dose levels of 0.5 and 2.5 mg/kg p.o. q.d. for 38 days, while lower dose levels (0.1 mg/kg) resulted in less protection (21% inhibition). This dose-dependent inhibition was confirmed in measurements of BTK inactivation of ex vivo blood samples collected 24 h after the last dose (see figure). In the second disease model, called NZB/W lupus-prone mouse model, mice develop a disease like systemic lupus erythematosus (SLE) and lupus nephritis (LN) in humans. In this disease model, complete protection from disease-related death occurred at all dose levels (0.2, 0.5, 1.5 mg/kg) of BMS-986195, which was dosed by once daily oral gavage for 14 weeks. For comparison, a 5 mg/kg dose of reversible BMS-986142 in the CIA model resulted in 72% reduction of clinically evident disease, and a 30 mg/kg dose of BMS-986142 in the NZB/W lupus-prone mouse model resulted in complete protection from disease-related death over the course of treatment (16 weeks). A lower dose of BMS-986195 compared to BMS-986142 was enough to suppress disease in both animal disease models. Measurements of BTK inactivation in blood samples collected 24 h after the last dose also confirmed high levels of inactivation with low doses of BMS-986195 compared to BMS-986142.

BMS-986195 was advanced to a double-blind, placebo-controlled, single- and multiple-ascending dose (SAD; MAD) Phase I study. In the

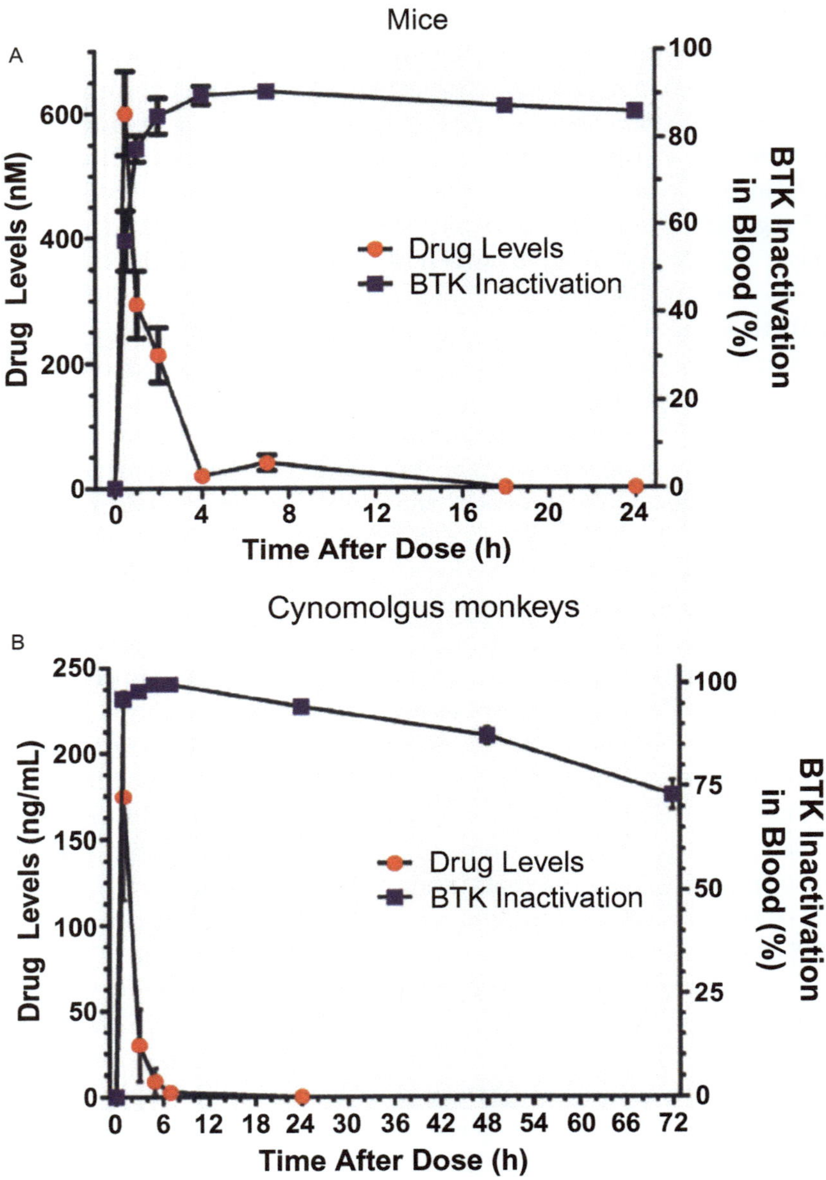

Fig. 10 A single low dose of BMS-986195 (0.5 mg/kg orally, once daily) resulted in >90% BTK inactivation in mice (A) and 100% inactivation in cynomolgus monkey (B).

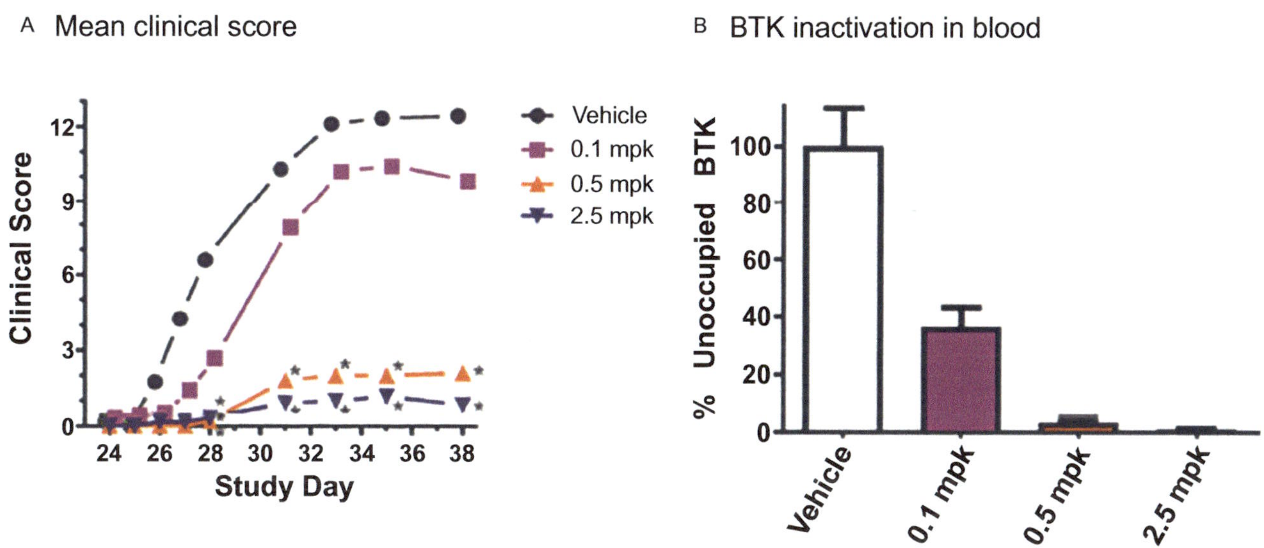

Fig. 11 Potent and robust activity against in vivo CIA model with BMS-986195 (A), and blood samples collected 24 h after the last dose (B).

SAD group, 40 healthy participants were randomized 3:1 to receive a single dose of BMS-986195 (0.3, 1, 3, 10 or 30 mg; 6 participants in each group) or placebo (10 participants). In the MAD group, 32 participants were randomized 3:1 to receive BMS-986195 (0.3, 1, 3 or 10 mg; 6 participants in each group) QD for 14 days or placebo (8 participants). BMS-986195 was well tolerated with only mild to moderate adverse events (AEs). There was one serious AE that led to a discontinuation, but it was later determined to be unrelated to BMS-986195 exposure.

In the pharmacokinetics (PK) assessment, BMS-986195 was rapidly absorbed after oral administration and eliminated quickly with a $t_{1/2}$ of 1.2–1.7 h (see Fig. 12). The drug was undetectable when measured 24 h after single or multiple doses. The key pharmacodynamic (PD) endpoint is the proportion of the BTK protein occupied by BMS-986195. The more reactive propiolamide derivative of BMS-986195 was used as a quencher to occupy any free, unbound BTK protein remaining.[40] A novel immunocapture LC-MS/MS-based assay was developed to measure real-time occupancy of the BTK inhibitor.

Occupancy of BMS-986195 on the BTK receptor reached steady state at different times depending on dose amounts (see Fig. 13). At steady state, differences between the peak and trough BTK occupancy are minimal. In the 10-mg dose group, 100% BTK occupancy was observed from Day 1 and steady state was achieved. In the 3 mg dose group, 90% BTK occupancy was observed after 24 h. After 1 day of dosing, the 3 mg dose group reached a steady state of 100% BTK occupancy. For the 1 mg dose group, 50% BTK occupancy was measured after 24 h. After 4 days of dosing, the 1 mg group reached steady state of >95% BTK occupancy. However, in the 0.3 mg dose group, ~20% BTK occupancy was seen after 24 h and it took 14 days before this dose group reached steady state at around 70% BTK occupancy. Thus, doses greater than or equal to 1 mg of BMS-986195 are projected to reach IC_{90} coverage of BTK.

An open Ph II clinical trial enrolling patients with active systemic lupus erythematosus (SLE), primary Sjorgren's syndrome (pSS), or active rheumatoid arthritis (RA) (NCT04186871) is ongoing. With the advancement of BMS-986195 into Phase II clinical trials, the BMS investigators reached their goal of developing a covalent BTK inhibitor that required a dosage of less than 10 mg in humans and maintained BTK target coverage as high as IC_{90} at steady state. Their previous BTK candidate, BMS-986142 had a mean $t_{1/2}$ of 7–11 h and steady-state inhibition of BTK was achieved at doses at or above 350 mg QD for 50% inhibition of ex vivo CD69 expression

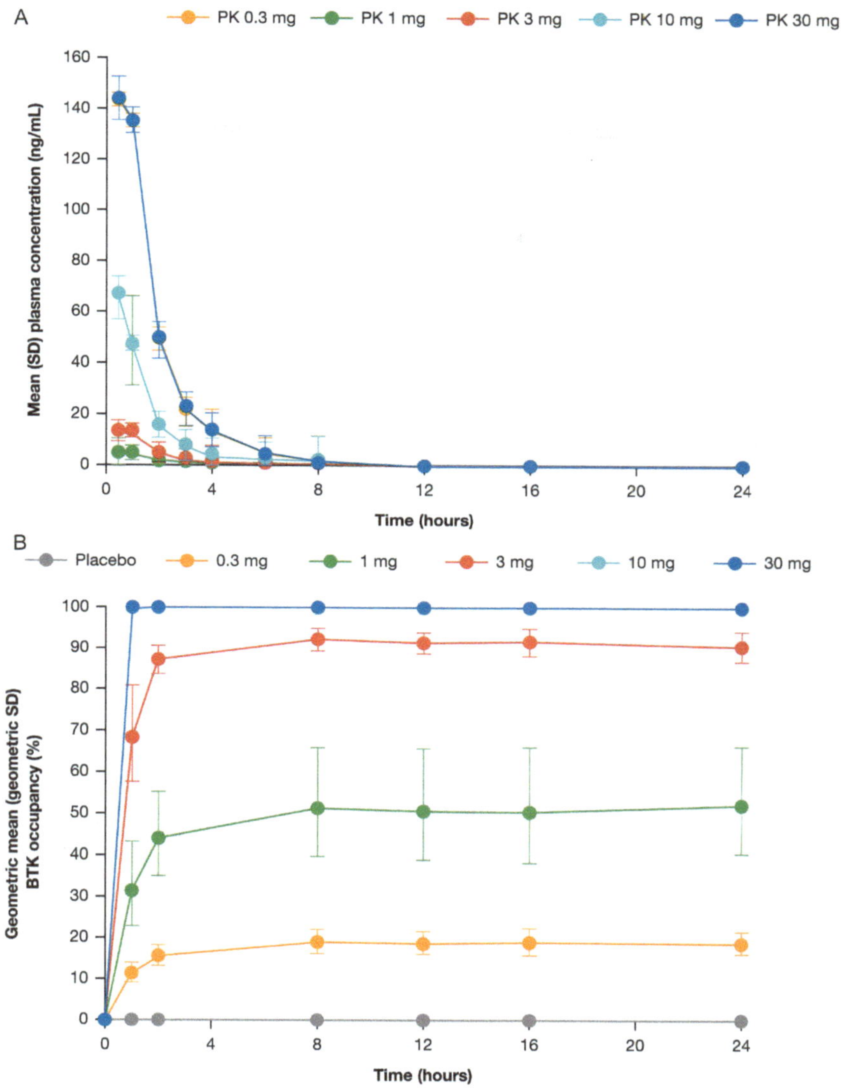

Fig. 12 Plasma concentration (A) and BTK occupancy (B) following a single dose of BMS-986195.

(0.145 μM or 83 ng/mL). To reach IC_{90} coverage of BTK, a dosage of 350 mg or more of BMS-986142 would have been required.[41] With no ongoing studies of BMS-986142 in the clinic, the properties of covalent inhibitor BMS-986195 (low dose, high IC_{90} coverage) appear to be more promising for BMS to continue pursuing as a treatment of immune-mediated diseases.

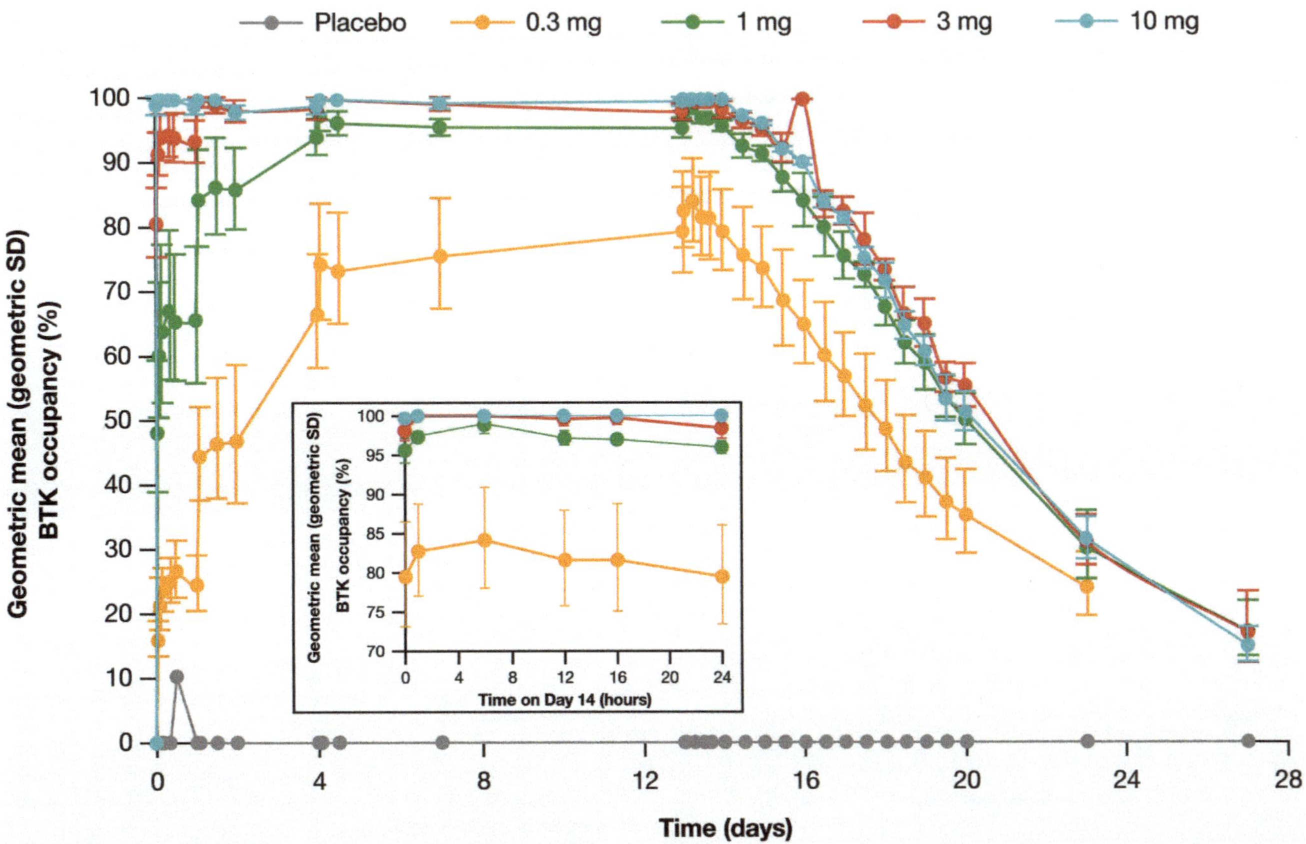

Fig. 13 BTK occupancy following multiple doses of BMS-986195.

5. LOU064 (Remibrutinib): Preclinical safety and efficacy determination to Ph I clinical trials

In contrast to ibrutinib, evobrutinib and branebrutinib, which binds to the active site of BTK, researchers at Novartis wanted to design a covalent BTK inhibitor that bound to the inactive BTK conformation. All activated kinases share a common homology around the ATP binding pocket, but the inactive conformation are often unique to individual kinases.[42] CGI-1746 was the first reported reversible BTK inhibitor that bound to the inactive conformation of BTK (see Fig. 14). In the inactive conformation, phosphorylation of Tyr551 in the activation loop of BTK is prevented. CGI-1746 was later optimized by researchers at Genentech and they developed GDC-0853 (fenebrutinib), a selective reversible BTK inhibitor that is bound to BTK in the inactive conformation.[43] Compared to other BTK inhibitors, GDC-0853 is an extremely selective BTK inhibitor. GDC-0853 inhibited only 3 (BMX, FGR, SRC) of 286-off target kinases and was >100 fold against each. To minimize off-target kinase inhibition and maximize a durable pharmacodynamic effect, Novartis pursued a strategy to develop a covalent BTK inhibitor targeting the inactive conformation of BTK.

Novartis found an internal pyrrolopyrimidine screening hit and combined it with the tail portion of CGI-1746, which led to the discovery of the selective reversible BTK inhibitor (see Fig. 16).[44] An X-ray co-crystal structure of the inhibitor with human BTK kinase domain confirmed the binding of the inactive conformation of BTK. Truncating the pyrrolopyrimidine core into the 6-aminopyrimidine hinge binder motif resulted in molecules with better PK profiles. Addition of a fluoro group adjacent to the amide on the cyclopropylphenyl ring improved the permeability of the molecule. This improvement in permeability may be due to the formation of an internal hydrogen bond between the fluoro substituent and the amide NH proton. Optimization of the linker and electrophilic warhead off the 5-position of the pyrimidine moiety resulted in the discovery of LOU064. Positioning of the warhead near Cys481 and sequestration of Tyr551 away from phosphorylation by upstream kinases in the inactive BTK conformation were confirmed through an X-ray crystal structure of LOU064 (see Fig. 15).

The selectivity of LOU064 was evaluated through three different methods. In a competition binding assay with a proprietary promiscuous probe (KINOMEscan, DiscoverX) against 395 human kinases, 1 μM concentration of LOU064 showed full inhibition to BTK (0.25% remaining

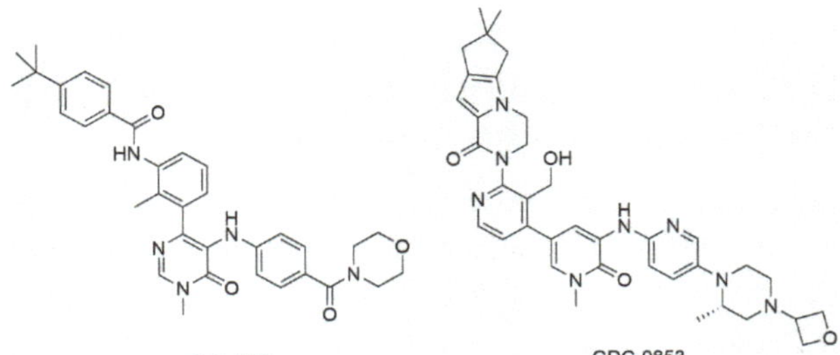

CGI-1746 GDC-0853
Fenebrutinib

Fig. 14 Structures of CGI-1746 and GDC-0853 (fenebrutinib).

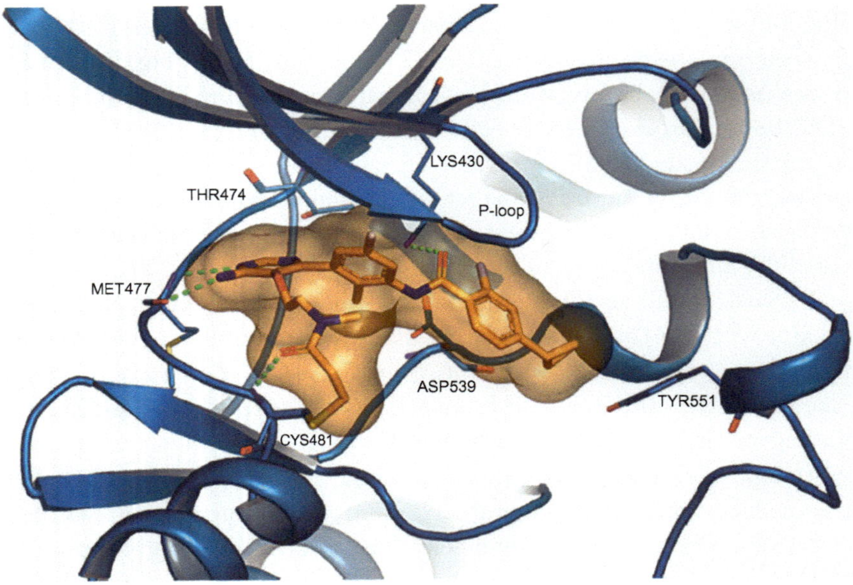

Fig. 15 X-ray complex structure of LOU064 bound to the BTK kinase domain (PDB code 6TFP).

probe binding) with partial inhibition only to TEC (24% remaining probe binding) and BMX (53% remaining probe binding). In a second assay (see Table 8), LOU064 was compared with other clinical covalent BTK inhibitors by measuring K_d values in a competition binding assay with other TEC family kinase members (BMX, TEC, ITK, EGFR, ERBB2, ERBB4, JAK3). LOU064 inhibited BTK with a K_d of 0.6 nM and with a selectivity of 175-fold against TEC (K_d of 110 nM) and 857-fold against BMX

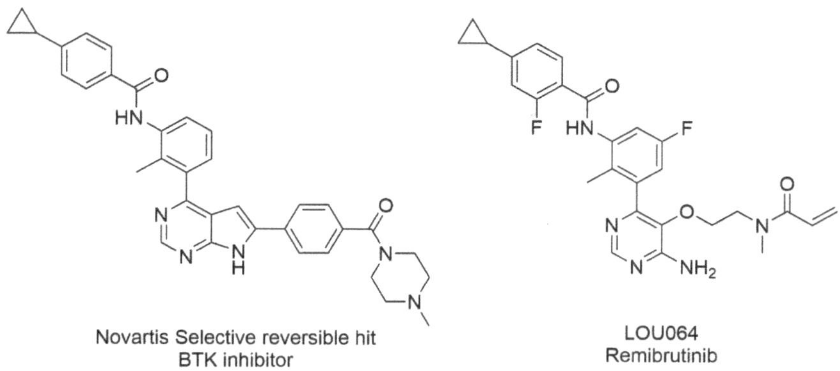

Fig. 16 Structures of Novartis internal screening BTK hit and LOU064 (remibrutinib).

(K_d of 540 nM). Acalabrutinib, tirabrutinib, and evobrutinib are less potent than LOU064 on BTK and less selective. Ibrutinib showed similar potency with LOU064, but its selectivity was poorer. Branebrutinib showed high inhibition of BTK (K_d of 0.56 nM), comparable to LOU064, but only 1.8-fold selectivity versus TEC (K_d of 0.99 nM) and 39-fold against BMX (K_d of 22 nM). Finally, the cellular selectivity of LOU064 was evaluated against a panel of primary human cells from various tissues (BioMAP Diversity plus panel, DiscoverX). LOU064 showed no toxicity across the cellular systems. LOU064 displayed inhibition of cellular BTK function in B cells in the corresponding panels with the appropriate immune cells. More importantly, no off-target effects on EGFR was observed in the dermal fibroblast system with LOU064, which is not the case with other non-selective BTK inhibitors such as ibrutinib.

Finally, the safety profile of LOU064 was investigated against commonly used toxicity assays in preclinical development. LOU064 showed no issues in the in vitro Ames and in vivo micronucleus assays to measure genotoxicity and mutagenicity, respectively. LOU064 also showed no activity in the 3T3 NRU phototoxicity assay. No significant inhibition was observed in the evaluation of LOU064 against an in vitro panel of 85 G protein-coupled receptors, transporters, nuclear receptors, enzymes, and ion channels. LOU064 was found to have an IC_{50} of 1.4 μM in the in vitro hERG patch clamp assay, but long-term dosing in dog telemetry studies revealed no effects on any cardiovascular parameters. The non-covalent version of LOU064 also showed activity on hERG (IC_{50} = 2.9 μM in Qpatch assay). Thus, the interaction of LOU064 with hERG appears to be noncovalent.

Table 8 Binding constants (K_d in nM) of covalent BTK inhibitors and selectivity vs off-target kinases ($K_{d,\ off\text{-}target}/K_{d,\ BTK}$).

		BTK	BMX	TEC	ITK	EGFR	ERBB2	ERBB4	JAK3
Remibrutinib	K_d	0.63	540	110	>10,000	>10,000	>10,000	>10,000	>10,000
	fold		857	175	>15,873	>15,873	>15,873	>15,873	>15,873
Ibrutunib	K_d	1.9	1.6	1.7	57	6.9	1.2	2.4	37
	fold	1	0.6	0.9	30	3.6	0.6	1.3	19
Acalabrutinib	K_d	21	610	14	>10,000	5600	7.4	140	>10,000
	fold	1	29	0.7	>476	267	0.4	140	>10,000
Tirabrutinib	K_d	14	47	6.2	>10,000	>10,000	610	>10,000	>10,000
	fold	1	3.4	0.4	>714	>714	44	>714	>714
Evobrutinib	K_d	16	31	4.5	3700	7100	>10,000	5600	>10,000
	fold	1	1.9	0.3	231	443	>625	350	>625
Branebrutinib	K_d	0.56	22	0.99	320	>10,000	>10,000	2500	1300
	fold	1	39	1.8	571	>17857	>17853	4464	2321

The in vivo PK properties of LOU064 were evaluated in three species and it was determined to have high clearance, eliminated quickly, and possessed high permeability—all desired properties for a rapidly absorbed covalent inhibitor where minimal exposure is required for sustained inhibition (see Table 9).

LOU064 was measured in human whole blood at various concentrations from 3 to 300 nM for cellular BTK occupancy (see Fig. 17). Covalent binding of LOU064 to BTK was determined with immunoassays to calculate binding of free BTK, which was measured with a biotinylated covalent BTK probe, to total BTK protein, which was measured through binding with an anti-BTK antibody. The rate of BTK inhibition for LOU064 in human whole blood was compared with other covalent BTK inhibitors at a concentration of 300 nM. Only branebrutinib (BMS-986165) displayed faster BTK binding kinetics than LOU064 even though it shares the same electrophile as tirabrutinib and acalabrutinib.

The on-target BTK activity of LOU064 was evaluated and compared with other covalent BTK inhibitors across several enzymatic and cellular assays (see Table 10). The PK/PD relationship of LOU064 was evaluated in rats with BTK occupancy in spleen homogenates as the PD readout. At 1 mg/kg dose of LOU064, 78% BTK occupancy was observed from the spleen of rats, 5 h after oral dosing. Systemic exposure of LOU064 was 5.6 nM after 5 h post-dose and 49.1 nM at half hour post-dose.

LOU064 was examined in two animal disease models to determine the efficacious dose in vivo. The first disease model, rat collagen-induced

Table 9 In vitro and in vivo PK parameters of LOU064 in rat, mouse, and dog.

	Rat, 1 mg/kg iv; 3 mg/kg, po	Mouse, 1 mg/kg iv; 3 mg/kg, po	Dog, 1 mg/kg, iv; 3 mg/kg po
C_{max} d.n. (μM), po	0.148 ± 0.033	0.053 ± 0.014	0.019
T_{max} (h), po	0.3	0.3 ± 0.1	0.8
AUC (μM·h), po	0.225 ± 0.027	0.054 ± 0.015	55
$T_{1/2}$ (h), iv	0.5 ± 0.03	1.0 ± 0.1	0.9 ± 0.1
CL (mL min^{-1} kg^{-1}), iv	44 ± 12	97 ± 11	29 ± 4
V_{ss} (L/kg), iv	1.4 ± 0.3	1.0 ± 0.1	1.0 ± 0.1
F_{po} (%)	29 ± 3	16 ± 4	5

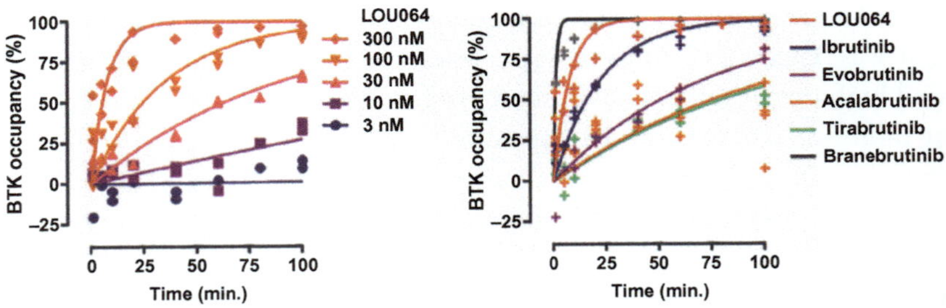

Fig. 17 Time & concentration-dependent in vitro engagement of cellular BTK with LOU064 and competitors in human blood from different donors.

Table 10 Comparative potency of LOU064 and covalent BTK inhibitors in cellular assays.

BTK enzyme & cellular assays	Stimulation	Evobrutinib IC_{50} (nM)	BMS-986195 IC_{50} (nM)	LOU064 IC_{50} (nM)
BTK enzyme inhibition Caliper assay	None	62 ± 4	0.1 ± 0.05	1.3 ± 0.9
IL8	FCγR	61 ± 37	6.0	2.5 ± 0.7
CD69 expression on hu blood B cells	BCR/Anti-IgM/IL-4	320 ± 74	18 ± 4	18 ± 5
CD63 expres. on hu blood basophils	FcεR/Anti-IgE	1182 ± 819	40 ± 10	67 ± 54

arthritis (CIA), is a B-cell and FcγR-dependent model of human RA. Histopathological analysis of bone and cartilage erosion was inhibited by LOU064 for the 3 and 10 mg/kg groups. Paw swelling were also reduced for these groups (see Fig. 18). Levels of BTK occupancy were measured on day 26, the last day of dosing, 24 h after the last dose. At 0.3 mg/kg dose, the trough BTK occupancy was around 40%, increasing to around 80% in the 3.0 mg/kg dose group. For the higher dose groups, (10 and 30 mg/kg), the trough BTK occupancy levels were greater than 90%. Thus, the 3 mg/kg dose group could sustain peak to trough BTK occupancy levels around 90%. In the second disease model, female rats were immunized with sheep red blood cells (SRBC) to generate an IgM antibody response through the B cell receptor pathway. After immunization with SRBC, the rats were treated for 4 days with one daily oral dose of LOU064. On day 5, SRBC antibody formation and BTK occupancy were assessed ex vivo on splenocytes. In this disease model, the IgM antibody response was inhibited

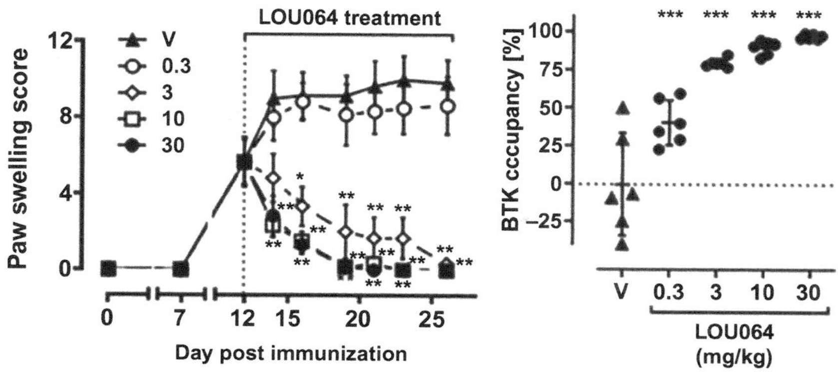

Fig. 18 Efficacy of LOU064 in the rat CIA model. Left: Paw swelling scores. Right: BTK occupancy in spleen 24 h after the last dose of LOU064.

by 96% at a dose of 3 mg/kg q.d., which corresponded to a trough spleen BTK occupancy after 24 h at 71.5%.

In human subjects, LOU064 was evaluated in healthy volunteers where single doses between 0.5 and 600 mg (SAD) and multiple doses (MAD) between 10 and 600 mg/day over 12 days.[45] No SAEs or AEs were reported during this study. Skin pharmacodynamics was determined by a skin prick test in response to a known allergen in select atopic healthy volunteers. LOU064 showed a dose-dependent inhibition of the SPT reaching ~50% reduction in allergen wheal size with maximal efficacy at doses beginning at 100 mg.

LOU064 is being evaluated in several Ph II clinical trials for patients with chronic spontaneous urticaria (NCT04109313, NCT03926611) and patients with moderate to severe Sjogren's syndrome (NCT04035668).

6. PRN1008 in clinical trials

Pioneering work by Taunton demonstrated that a non-catalytic cysteine in RSK2 could be targeted with a reversible-covalent cyanoacrylamide inhibitor.[46] Reversible covalent inhibitors tend to be more reactive to nucleophilic attack because of the electron-withdrawing nature of the cyano-group. On the other hand, the presence of the electron-withdrawing cyano-group increases the acidity of the alpha-C–H bond, favoring rapid beta elimination of the nucleophilic adduct. To identify reversible covalent inhibitors with longer prolonged residence times,[47] the non-covalent interactions between the inhibitor and the desired target would need to be optimized.[48] This process involves washout experiments where weak binders rapidly dissociate from the target while those optimal reversible covalent inhibitors maintain sustained engagement with the target protein. In contrast to irreversible inhibitors, off rates of reversible covalent inhibitors can be optimized for specific kinases that share a homologous cysteine. Thus, for a highly-tuned reversible covalent inhibitor, the on-target versus off-target selectivity window could be significantly increased (Fig. 19).

From this optimization process, PRN1008 (rilzabrutinib) was discovered as a reversible covalent BTK inhibitor. PRN1008 was advanced to a double-blind, placebo-controlled, randomized, single- and multiple-ascending dose (SAD; MAD) Phase I study.[49] In the SAD group, 40 healthy volunteers were randomized 3:1 to receive a single dose of PRN1008 (50, 150, 300, 600 or 1200 mg; 6 participants in each group) or placebo (10 participants). In the MAD group, 40 healthy volunteers were randomized 4:1 to receive PRN1008 (300 mg QD, 300 mg BID, 600 mg QD or 450 mg BID; 8 participants in each group) for 10 days or placebo (8 participants). All doses of PRN1008 were administered as a liquid formulation under fasting conditions with the exception of one MAD group. PRN1008 was safe and well tolerated in single doses up to 1200 mg and over 10 days of multiple dosing. The most common adverse events (AEs) were gastrointestinal (GI) in nature, which included diarrhea, nausea, throat irritation, vomiting and oropharyngeal discomfort. These GI-related AEs were observed most frequently in the

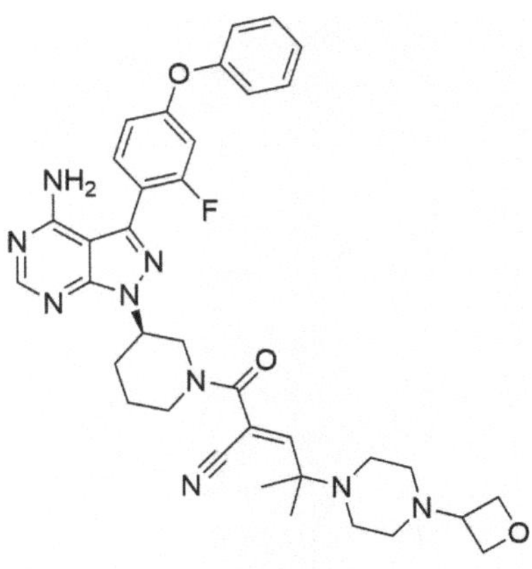

PRN1008
Rilzabrutinib

Fig. 19 Structure of PRN1008 (rilzabrutinib).

highest dose levels (1200 mg) compared to lower doses administered in either the SAD or MAD group. It is anticipated that administering PRN1008 via a solid dosage form will improve the GI tolerability of PRN1008.

In the pharmacokinetics (PK) assessment, PRN1008 was rapidly absorbed after oral administration under fasted conditions and eliminated quickly with a mean apparent terminal elimination half-life ($t_{1/2}$) of 1.4 to 3.9 h for all cohorts (see Fig. 20). The T_{max} ranged from a median of 0.5 h at 50 mg to 2.5 h at 600 and 1200 mg. Plasma concentrations generally increased with dose over the range of 50–600 mg. The PK of multiple dose PRN1008 were consistent with the single dose results. The key pharmacodynamic (PD) endpoint is the proportion of the BTK protein occupied by PRN1008. A BTK selective boron dipyrromethene-labeled probe was used to measure BTK occupancy from PBMCs extracted from whole blood samples at 0, 4, 12 and 24 h post-dosing in the SAD groups. The cell lysates from the PBMCs were measured by gel electrophoresis for BTK occupancy along with a primary mouse anti-human BTK antibody.

Occupancy of BTK by PRN1008 in SAD occurred rapidly at doses of 150 mg and above, in a dose dependent manner, with the maximum occupancy observed withing the first 4 h post-dose. Greater than 90% occupancy of BTK was observed after 4 h in the 600 mg and 1200 mg dose groups with very low interpatient variability. For the other SAD groups, ~50% BTK occupancy was observed in the 150 mg dose group and ~75% BTK occupancy was found in the 300 mg dose group after 4 h. The rate of occupancy decline after single doses of PRN1008 was very consistent across cohorts, with all doses exhibiting a ~30–35% reduction in occupancy between 4 and 24 h. With multiple dose administrations of PRN1008, "steady-state" occupancy levels were observed on Day 3 where both the 300 mg twice daily and the 600 mg once daily regimens achieved >90% occupancy 4 h after dosing (see Fig. 21). The pre-dose occupancy on Day 10 was ~75% for the twice daily regimen (300 or 450 mg) and 58% for the once daily regimen (600 mg QD), respectively. Maximal occupancy was achieved at C_{max} concentrations of greater than 100 ng/mL, suggesting that higher doses or higher C_{max} concentrations would not be beneficial.

PRN1008 is under evaluation in Ph II clinical trials for patients IgG4-related disease (NCT04520451) and adult patients with immune thrombocytopenia (ITP) on a 400 mg twice daily dose regimen. PRN1008 is being studied in two Ph III trials with patients with persistent or chronic ITP (NCT04562766) and patients with pemphigus (NCT03762265).

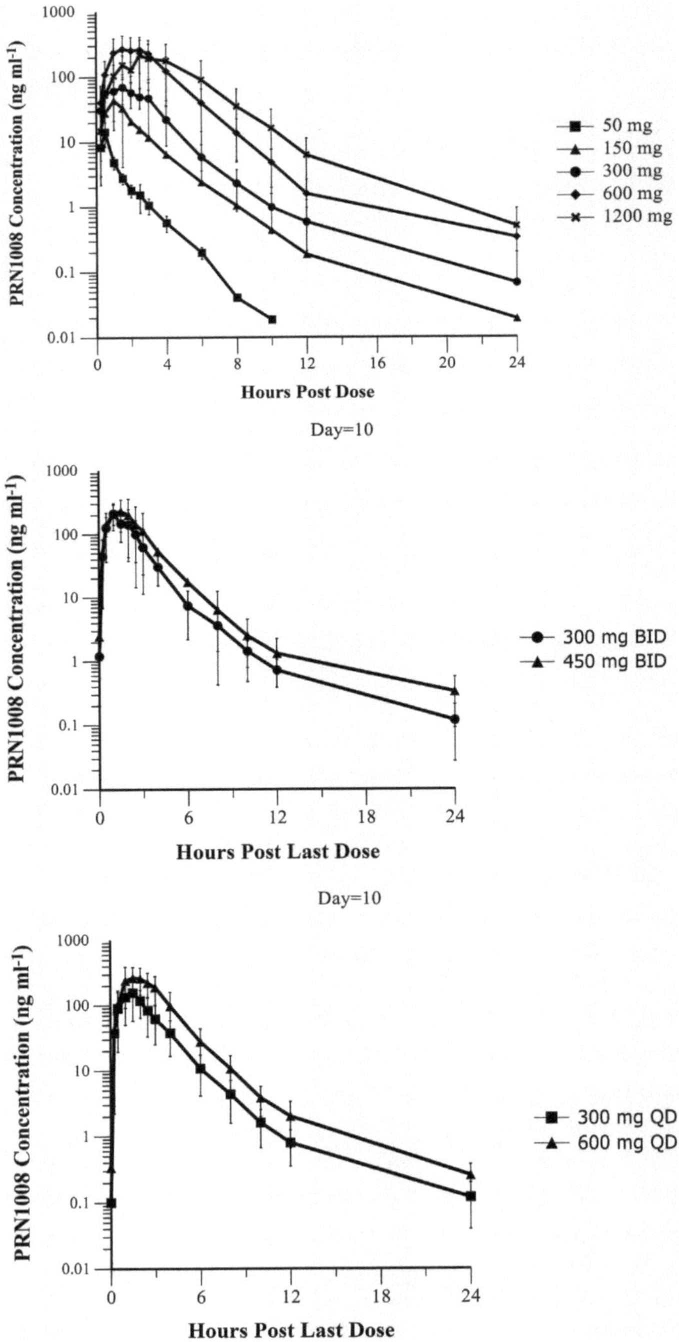

Fig. 20 Mean concentration time profiles of PRN1008 following single (Part A) and multiple (Part B, Day 10) oral doses in healthy volunteers.

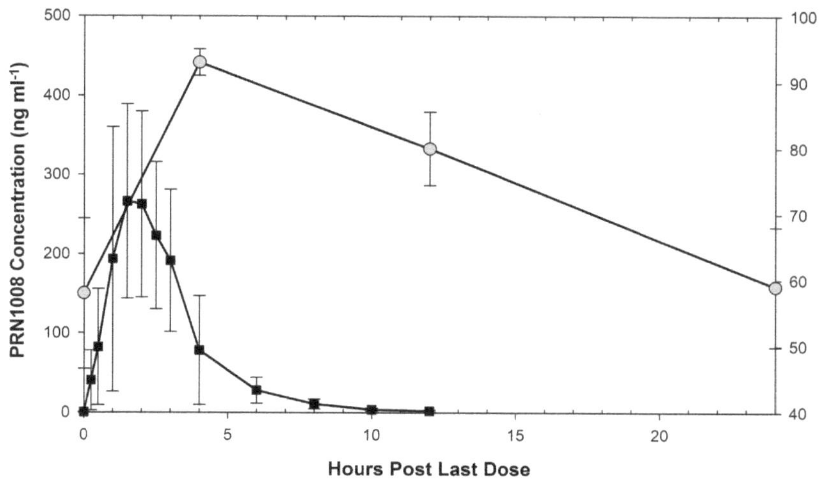

Fig. 21 Duration of BTK occupancy (squares) in relation to the plasma concentration profile of PRN1008 (circles), following final dose on day 10 of a 600 mg once daily dose regimen in MAD study.

7. Conclusion

In this book chapter, the development stories of five covalent inhibitors in clinical trials for immune-mediated diseases are highlighted. A slowly metabolized JAK3-TEC inhibitor ritlecitinib (PF-06651600) from Pfizer was developed which maintains coverage despite the rapid turnover of the JAK3 protein. Numerous efforts have led to covalent BTK inhibitors that are currently in advanced clinical studies for immune-mediated diseases. Evobrutinib, a covalent BTK inhibitor developed by EMD Serono, and tolebrutinib (SAR442168) are both in phase 3 trials for relapsing multiple sclerosis. The comparison between two BTK inhibitors from BMS, branebrutinib (BMS-986195) and BMS-986142, highlights the advantages of covalent inhibitor compared with a non-covalent, reversible inhibitor, respectively. Branebrutinib is being studies in SLE, RA and pSS. Remibrutinib (LOU-064) is a potent BTK covalent inhibitor that binds to the inactive conformation of BTK and is being evaluated in urticaria, pSS and asthma. Finally, rilzabrutinib (PRN1008) is a reversible-covalent BTK inhibitor currently in phase 3 clinical trials for the treatment of pemphigus and ITP.

Other covalent inhibitors are currently being evaluated in the clinic for immune-mediated diseases, such as KZR-616 for Lupus (NCT03393013), and more will undoubtedly continue to emerge. This chapter highlights the challenges and medicinal chemistry efforts toward developing covalent inhibitors of validated targets BTK & JAK3. Publications highlighting the unique discovery paths and clinical development into Phase I safety and dose-dependent trials in humans are particularly instructive as the community awaits results of ongoing clinical trials in patients. Even though these five inhibitors are highly selective for their intended targets, they all have some varying activity against a couple of the TEC family members such as ITK, BMX, TXK and TEC. It will be interesting to see how their distinct off-target profiles begin to appear in the clinic from these five inhibitors, such as the reduction of platelet levels, due to their effect on other TEC family kinase members.

References

1. Hanke, T.; Merk, D.; Steinhilber, D.; Geisslinger, G.; Schubert-Zsilavecz, M. Small Molecules With Anti-Inflammatory Properties in Clinical Development. *Pharmacol. Ther.* **2016**, *157*, 163–187.
2. Park, S.-J.; Im, D.-S. Sphingosine 1-Phosphate Receptor Modulators and Drug Discovery. *Biomol. Ther.* **2017**, *25*, 80–90.
3. Verden, A.; Dimbil, M.; Kyle, R.; Overstreet, B.; Hoffman, K. B. Analysis of Spontaneous Postmarket Case Reports Submitted to the FDA Regarding Thromboembolic Adverse Events and JAK Inhibitors. *Drug Saf.* **2018**, *41*, 357–361.
4. Bauer, R. A. Covalent Inhibitors in Drug Discovery: From Accidental Discoveries to Avoided Liabilities and Designed Therapies. *Drug Discov. Today* **2015**, *20*, 1061–1073.
5. Lonsdale, R.; Ward, R. A. Structure-Based Design of Targeted Covalent Inhibitors. *Chem. Soc. Rev.* **2018**, *47*, 3816–3830.
6. Jaglowski, S. M.; Blazar, B. R. How Ibrutinib, a B-Cell Malignancy Drug, Became an FDA-Approved Second-Line Therapy for Steroid-Resistant Chronic GVHD. *Blood Adv.* **2018**, *2*, 2012–2019.
7. Series, J.; et al. Differences and Similarities in Ibrutinib and Acalabrutinib Effects on Platelet Functions. *Haematologica* **2019**, *104*, 2292–2299.
8. Zarrin, A. A.; Bao, K.; Lupardus, P.; Vucic, D. Kinase Inhibition in Autoimmunity and Inflammation. *Nat. Rev. Drug Discov.* **2020**, *20*, 1–25.
9. Bryan, M. C.; Rajapaksa, N. S. Kinase Inhibitors for the Treatment of Immunological Disorders: Recent Advances. *J. Med. Chem.* **2018**, *61*, 9030–9058.
10. Go, Y.-M.; Chandler, J. D.; Jones, D. P. The Cysteine Proteome. *Free Radic. Biol. Med.* **2015**, *84*, 227–245.
11. Liu, Q.; et al. Developing Irreversible Inhibitors of the Protein Kinase Cysteinome. *Chem. Biol.* **2013**, *20*, 146–159.
12. Zhang, J.; Yang, P. L.; Gray, N. S. Targeting Cancer with Small Molecule Kinase Inhibitors. *Nat. Rev. Cancer* **2009**, *9*, 28–39.
13. Strelow, J. M. A Perspective on the Kinetics of Covalent and Irreversible Inhibition. *J. Biomol. Screen.* **2016**, *22*, 3–20.
14. Gehringer, M.; Laufer, S. A. Emerging and Re-Emerging Warheads for Targeted Covalent Inhibitors: Applications in Medicinal Chemistry and Chemical Biology. *J. Med. Chem.* **2018**, *62*, 5673–5724.
15. Yan, X.-E.; et al. Structural Basis of AZD9291 Selectivity for EGFR T790M. *J. Med. Chem.* **2020**, *63*, 8502–8511.

16. Johnson, D. S.; Weerapana, E.; Cravatt, B. F. Strategies for Discovering and Derisking Covalent, Irreversible Enzyme Inhibitors. *Future Med. Chem.* **2010**, *2*, 949–964.
17. Dahal, U. P.; Obach, R. S.; Gilbert, A. M. Benchmarking In Vitro Covalent Binding Burden as a Tool to Assess Potential Toxicity Caused by Nonspecific Covalent Binding of Covalent Drugs. *Chem. Res. Toxicol.* **2013**, *26*, 1739–1745.
18. Laurence, A.; Pesu, M.; Silvennoinen, O.; O'Shea, J. JAK Kinases in Health and Disease: An Update. *Open Rheumatol. J.* **2012**, *6*, 232–244.
19. Wrobleski, S. T.; et al. Highly Selective Inhibition of Tyrosine Kinase 2 (TYK2) for the Treatment of Autoimmune Diseases: Discovery of the Allosteric Inhibitor BMS-986165. *J. Med. Chem.* **2019**, *62*, 8973–8995.
20. Thorarensen, A.; et al. ATP-Mediated Kinome Selectivity: The Missing Link in Understanding the Contribution of Individual JAK Kinase Isoforms to Cellular Signaling. *ACS Chem. Biol.* **2014**, *9*, 1552–1558.
21. Dai, J.; Yang, L.; Addison, G. Current Status in the Discovery of Covalent Janus Kinase 3 (JAK3) Inhibitors. *Mini-Rev. Med. Chem.* **2019**, *19*, 1531–1543.
22. Goedken, E. R.; et al. Tricyclic Covalent Inhibitors Selectively Target Jak3 through an Active Site Thiol. *J. Biol. Chem.* **2015**, *290*, 4573–4589.
23. Telliez, J.-B.; et al. Discovery of a JAK3-Selective Inhibitor: Functional Differentiation of JAK3-Selective Inhibition over pan-JAK or JAK1-Selective Inhibition. *ACS Chem. Biol.* **2016**, *11*, 3442–3451.
24. Thorarensen, A.; et al. Design of a Janus Kinase 3 (JAK3) Specific Inhibitor 1-((2 S,5 R)-5-((7 H -Pyrrolo[2,3- d]pyrimidin-4-yl)amino)-2-methylpiperidin-1-yl)prop-2-en-1-one (PF-06651600) Allowing for the Interrogation of JAK3 Signaling in Humans. *J. Med. Chem.* **2017**, *60*, 1971–1993.
25. Siewert, E.; Müller-Esterl, W.; Starr, R.; Heinrich, P. C.; Schaper, F. Different Protein Turnover of Interleukin-6-Type Cytokine Signalling Components. *Eur. J. Biochem.* **2017**, *265*, 251–257.
26. Xu, H.; et al. PF-06651600, a Dual JAK3/TEC Family Kinase Inhibitor. *ACS Chem. Biol.* **2019**, *14*, 1235–1242.
27. Robinson, M. F.; et al. Efficacy and Safety of PF-06651600 (Ritlecitinib), a Novel JAK3/TEC Inhibitor, in Patients With Moderate-to-Severe Rheumatoid Arthritis and an Inadequate Response to Methotrexate. *Arthritis Rheumatol.* **2020**, *72*, 1621–1631.
28. Pan, Z.; et al. Discovery of Selective Irreversible Inhibitors for Bruton's Tyrosine Kinase. *ChemMedChem* **2007**, *2*, 58–61.
29. Evans, E. K.; et al. Inhibition of Btk with CC-292 Provides Early Pharmacodynamic Assessment of Activity in Mice and Humans. *J. Pharmacol. Exp. Ther.* **2013**, *346*, 219–228.
30. Catlett, I. M.; et al. Safety, Pharmacokinetics and Pharmacodynamics of Branebrutinib (BMS-986195), a Covalent, Irreversible Inhibitor of Bruton's Tyrosine Kinase: Randomised Phase I, Placebo-Controlled Trial in Healthy Participants. *Br. J. Clin. Pharmacol.* **2020**, *86*, 1849–1859.
31. Byrd, J. C.; et al. Ibrutinib versus Ofatumumab in Previously Treated Chronic Lymphoid Leukemia. *N. Engl. J. Med.* **2014**, *371*, 213–223.
32. Byrd, J. C.; et al. Acalabrutinib (ACP-196) in Relapsed Chronic Lymphocytic Leukemia. *N. Engl. J. Med.* **2016**, *374*, 323–332.
33. Caldwell, R. D.; et al. Discovery of Evobrutinib: An Oral, Potent, and Highly Selective, Covalent Bruton's Tyrosine Kinase (BTK) Inhibitor for the Treatment of Immunological Diseases. *J. Med. Chem.* **2019**, *62*, 7643–7655.
34. Anastassiadis, T.; Deacon, S. W.; Devarajan, K.; Ma, H.; Peterson, J. R. Comprehensive Assay of Kinase Catalytic Activity Reveals Features of Kinase Inhibitor Selectivity. *Nat. Biotechnol.* **2011**, *29*, 1039–1045.
35. Haselmayer, P.; et al. Efficacy and Pharmacodynamic Modeling of the BTK Inhibitor Evobrutinib in Autoimmune Disease Models. *J. Immunol.* **2019**, *202*, ji1800583.

36. Becker, A.; et al. Safety, Tolerability, Pharmacokinetics, Target Occupancy, and Concentration-QT Analysis of the Novel BTK Inhibitor Evobrutinib in Healthy Volunteers. *Clin. Transl. Sci.* **2020**, *13*, 325–336.
37. Montalban, X.; et al. Placebo-Controlled Trial of an Oral BTK Inhibitor in Multiple Sclerosis. *N. Engl. J. Med.* **2019**, *380*, 2406–2417.
38. Watterson, S. H.; et al. Discovery of Branebrutinib (BMS-986195): A Strategy for Identifying a Highly Potent and Selective Covalent Inhibitor Providing Rapid in Vivo Inactivation of Bruton's Tyrosine Kinase (BTK). *J. Med. Chem.* **2019**, *62*, 3228–3250.
39. Watterson, S. H.; et al. Discovery of 6-Fluoro-5-(R)-(3-(S)-(8-fluoro-1-methyl-2, 4-dioxo-1,2-dihydroquinazolin-3(4 H)-yl)-2-methylphenyl)-2-(S)-(2-hydroxypropan-2-yl)-2,3,4,9-tetrahydro-1 H -carbazole-8-carboxamide (BMS-986142): A Reversible Inhibitor of Bruton's Tyrosine Kinase (BTK) Conformationally Constrained by Two Locked Atropisomers. *J. Med. Chem.* **2016**, *59*, 9173–9200.
40. Zheng, N.; et al. Determination of Real Time In Vivo Drug Receptor Occupancy for a Covalent Binding Drug as a Clinical Pharmacodynamic Biomarker by Immunocapture-LC-MS/MS. *Anal. Chem.* **2019**, *91*, 8443–8452.
41. Lee, S. K.; et al. Safety, Pharmacokinetics, and Pharmacodynamics of BMS-986142, a Novel Reversible BTK Inhibitor, in Healthy Participants. *Eur. J. Clin. Pharmacol.* **2017**, *73*, 689–698.
42. Noble, M. E. M.; Endicott, J. A.; Johnson, L. N. Protein Kinase Inhibitors: Insights Into Drug Design From Structure. *Science* **2004**, *303*, 1800–1805.
43. Crawford, J. J.; et al. Discovery of GDC-0853: A Potent, Selective, and Noncovalent Bruton's Tyrosine Kinase Inhibitor in Early Clinical Development. *J. Med. Chem.* **2018**, *61*, 2227–2245.
44. Angst, D.; et al. Discovery of LOU064 (Remibrutinib), a Potent and Highly Selective Covalent Inhibitor of Bruton's Tyrosine Kinase. *J. Med. Chem.* **2020**, *63*, 5102–5118.
45. Kaul, M.; Storim, J. LOU064: A Highly Selective and Potent Covalent Oral BTK Inhibitor With Promising Pharmacodynamic Effects in Skin. *Allergy* **2019**, *74*, 333–376.
46. Serafimova, I. M.; et al. Reversible Targeting of Noncatalytic Cysteines with Chemically Tuned Electrophiles. *Nat. Chem. Biol.* **2012**, *8*, 471–476.
47. Copeland, R. A.; Pompliano, D. L.; Meek, T. D. Drug–Target Residence Time and Its Implications for Lead Optimization. *Nat. Rev. Drug Discov.* **2006**, *5*, 730–739.
48. Bradshaw, J. M.; et al. Prolonged and Tunable Residence Time Using Reversible Covalent Kinase Inhibitors. *Nat. Chem. Biol.* **2015**, *11*, 525–531.
49. Smith, P. F.; et al. A Phase I Trial of PRN1008, a Novel Reversible Covalent Inhibitor of Bruton's Tyrosine Kinase, in Healthy Volunteers. *Br. J. Clin. Pharmacol.* **2017**, *83*, 2367–2376.

CHAPTER THREE

Chemistry perspectives of reversible covalent drugs

Jianmin Gao* and Vincent Nobile
Department of Chemistry, Boston College, Chestnut Hill, MA, United States
*Corresponding author: e-mail address: jianmin.gao@bc.edu

Contents

1. Introduction

Covalent drugs refer to therapeutic agents that engage their corresponding target via covalent bond formation.[1] Covalent drugs are appealing over their noncovalent counterparts as the covalent mechanism can potentially afford much enhanced potency, selectivity and prolonged time of action. Covalent drugs are not new to medicinal chemistry—there are over 40 covalent drugs approved by the Food and Drug Administration (FDA) to date. Examples of covalent drugs include the well-known names such as aspirin and penicillin, as well as several anticancer therapeutics such as afatinib and ibrutinib. Despite their significant presence in our drug collection, most covalent drugs were discovered through serendipity with their covalent mechanism of action confirmed only after clinical use. Medicinal chemists had been wary of the potential problems of covalent drugs that manifest as idiosyncratic toxicity, for which the underlying causes include, but are likely not limited to, drug reaction with unintended proteins (off-target effect) and immunogenicity of modified proteins. Within the past decade, there has been a sharp increase of interest in the strategic development

Annual Reports in Medicinal Chemistry, Volume 56
ISSN 0065-7743
https://doi.org/10.1016/bs.armc.2020.10.004

of targeted covalent drugs. Efforts on this front are fueled by the rapidly growing collection of chemoselective and site-selective reactions for modifying native proteins (as needed for covalent drugs), as well as by the much improved technologies for high-throughput screening and target identification. A number of new concepts for covalent drug design have emerged and shown promise in overcoming the roadblocks thwarting covalent drug discovery. An active area of investigation focuses on developing reversible covalent drugs (RCDs) that target therapeutically important proteins.[2] Similar to irreversible covalent drugs, RCDs also engage their intended target through formation of a covalent bond. However, different from irreversible covalent drugs, RCDs form a weak bond to the target protein, which readily falls apart under permissive conditions (e.g., upon denaturation or degradation of the protein) to regenerate the intact protein (Fig. 1A), thereby avoiding the undesirable activation of the immune system caused by modified proteins. Importantly, having a reversible covalent linkage is also expected to minimize the off-target effect of covalent drugs as a RCD can fall off unintended proteins and keep searching for the desired protein target (Fig. 1B).

This chapter does not mean to be a comprehensive review of RCD development, for which there are excellent review articles in recent literature.[3] Instead, we intend to present an analysis of RCDs from a physical organic chemist's perspective, seeking to articulate the general principles that would serve as useful guidelines for developing novel reversible covalent drugs. Note that an absolute majority of RCDs function by inhibiting relevant proteins, and hence the abbreviation RCI (reversible covalent inhibitor) is used at appropriate contexts of this chapter.

2. Prominent examples of RCDs

Several prominent examples of RCDs are shown in Fig. 2. These drug molecules inhibit or, in one case, activate their protein targets via reversible conjugation to a nucleophilic side chain, which could be a serine, threonine, cysteine or lysine. These nucleophilic residues carry a hydroxyl, a thiol or an amine nucleophile. Vildagliptin (VIL) and saxagliptin (SAX) are covalent drugs used to treat type 2 diabetes mellitus. They function by inhibiting the dipeptidyl peptidase-4 (DPP-4), a serine protease that rapidly degrades incretin hormones in plasma. The nitrile group serves as the warhead for these two covalent drugs, which react with the catalytic serine to give an imidate linkage.[4] The covalent conjugation between SAX and DPP-4 is known to fully reverse to regenerate the drug and the protein in their

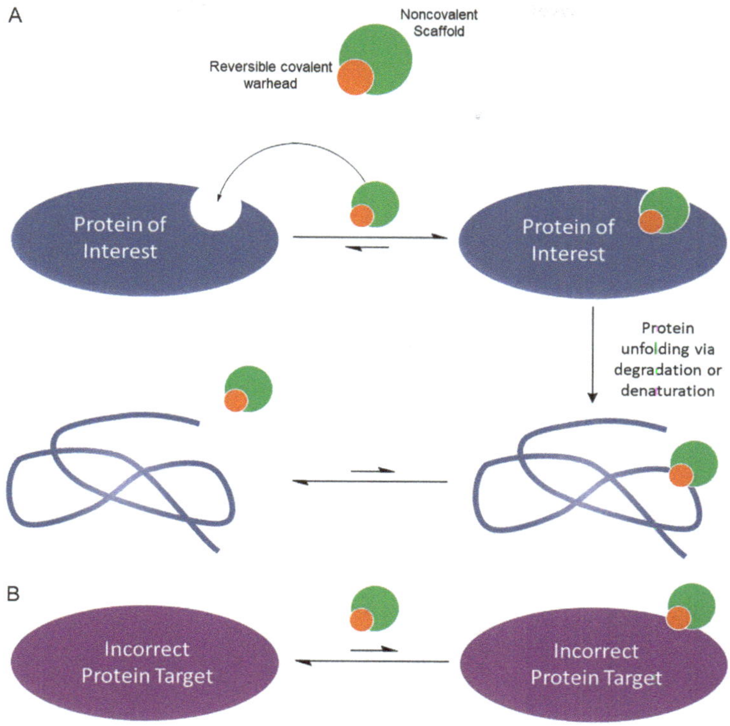

Fig. 1 Cartoon illustration of the key features of a RCD. (A) A noncovalent scaffold allows the RCD to fit into the binding pocket of a protein of interest. This places the reversible covalent warhead in close proximity with the target amino acid, allowing for covalent bond formation. Bond formation leads to prolonged inhibition of the protein. Upon eventual degradation of the protein, which abolishes the noncovalent interactions, the covalent linkage undergoes the reverse reaction to regenerate the intact protein. (B) RCDs are expected to show minimal off-target binding and inhibition as a mistakenly conjugated drug can readily dissociate from the protein.

original forms. However, the dissociation of VIL primarily follows a hydrolysis pathway to give an inactive analogue that has the nitrile group converted to a carboxylate. Similar to VIL and SAX, two FDA approved antiviral drugs telaprevir and boceprevir are also RCIs of serine proteases.[5] However, these two drug molecules utilize a α-ketoamide warhead to react with the catalytic serine residue of their target enzyme, the hepatitis C NS3 serine protease. The covalent conjugation gives a hemiketal linkage, which is well known to undergo the reverse reaction to give back the ketone and serine respectively. In contrast to the RCIs of serine proteases, the FDA approved anticancer drug bortezomib targets the N-terminal threonine of

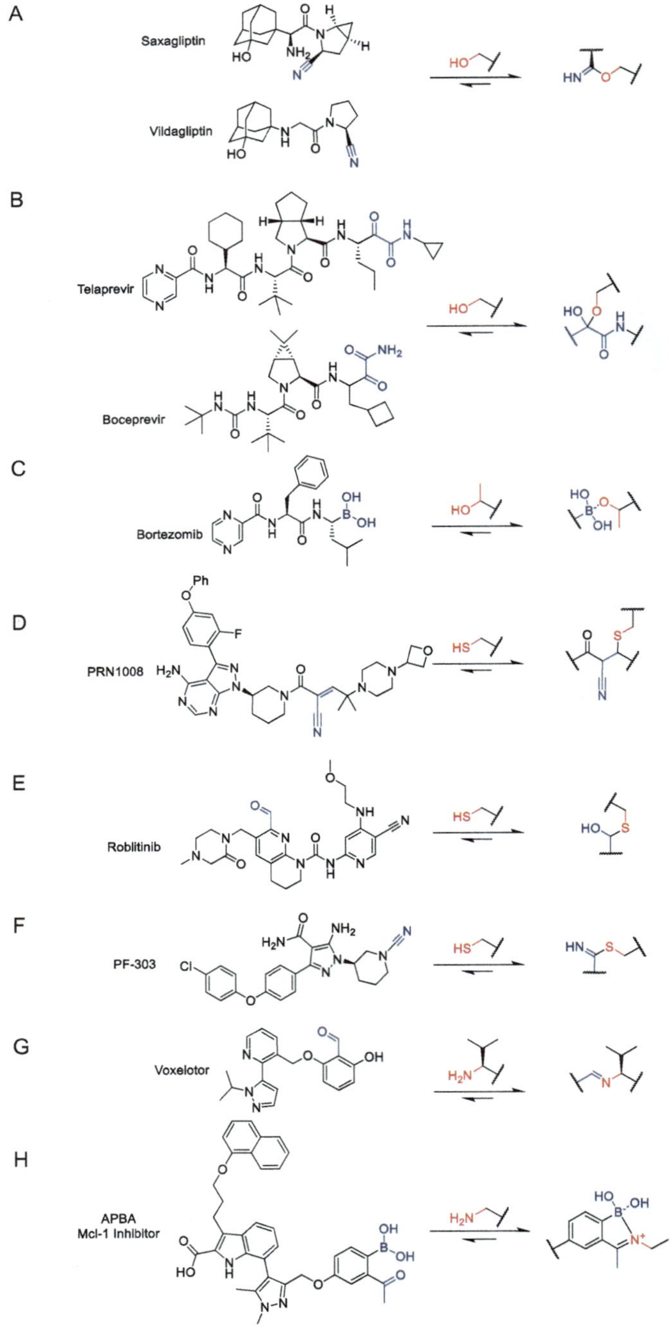

Fig. 2 Prominent reversible covalent drugs. (A) Saxagliptin (Sax) and Vildagliptin (VIL), clinically approved for the treatment of type 2 diabetes mellitus. (B) Teleprivir and boceprivir, clinically approved for treatment of hepatitis C. (C) Bortezomib, clinically approved for treatment of multiple myeloma. (D) PRN1008, developed by Taunton et al., in Phase II clinical trials as a BTK inhibitor. (E) Roblitinib, inhibitor of FGFR4 in Phase I/II clinical trials for solid tumor treatment. (F) PF-303, a published novel BTK inhibitor that shows promise as an oncologic treatment. (G) Voxelotor, clinically approved for the treatment of sickle-cell disease. (H) An exemplary APBA-based Mcl-1 inhibitor designed by an AstraZeneca team to target lysines via iminoboronate chemistry.

the 26s proteasome.[6] Specifically, the boronic acid moiety serves as a warhead to conjugate with the threonine side chain. The resulting boronate ester linkage, which is labile for hydrolysis, can dissociate to regenerate the drug and unmodified protein.

Several reversible reactions have been successfully developed and implemented to target cysteine residues in therapeutically important proteins. PRN1008, a BTK (Bruton's tyrosine kinase) inhibitor currently undergoing phase II/III clinical trials, features a α-cyanoacrylamide warhead that reversibly conjugates with a cysteine side chain. Michael acceptors such as acrylamides have found wide applications in the development of irreversible covalent inhibitors. Specific examples include the marketed drugs such as afatinib and ibrutinib. Installing a cyano group onto the α-carbon of acrylamides increases the α-CH acidity and reduces the thermodynamic stability of the β-thioether adduct, thereby allowing the reverse reaction to proceed rapidly.[7] Also currently going through clinical trials, roblitinib (also called FGF401) is a highly selective inhibitor of fibroblast growth factor receptor 4 (FGFR4). It inhibits the FGFR4 kinase activity via reversible covalent conjugation to a cysteine residue of the kinase, using the aldehyde functionality as a warhead.[8] The resulting thiohemiacetal structure can undergo quick hydrolysis under physiological conditions, which underlies the reversible nature of this RCI. Activated nitriles have been used in the development of cysteine-targeting RCIs. For example, the compound PF-303 was reported by a Pfizer team to be a RCI of BTK. As a structural analogue of ibrutinib, PF-303 inactivates the kinase via reversible covalent conjugation to the conserved cysteine using the dialkylamine-substituted nitrile as a warhead.[9]

In comparison to RCDs targeting serine/threonine and cysteines, amine-targeting RCDs are rare partly due to the limited availability of reversible covalent chemistry for amines. The only chemistry that has found significant success is to form stabilized imines. Imine formation plays important roles in enzyme catalysis as exemplified by PLP (pyridoxal 5′-phosphate) dependent enzymes. The imine structure formed between the PLP aldehyde is stabilized by hydrogen bond formation with the *ortho*-hydroxyl substituent. Mimicking this imine stabilization mechanism, voxelotor (also called GBT440), a recently approved drug to treat sickle cell disease, binds to the N-terminal valine of its target protein hemoglobin via imine formation.[10] The covalent binding event allosterically enhances the hemoglobin's affinity to oxygen, which delays the polymerization of the protein and thereby prevents sickling of red blood cells. Another imine stabilization strategy takes advantage of a proximal boronic acid moiety. Specifically, aryl aldehydes and ketones that harbor an ortho-boronic acid substituent are found to readily conjugate with amines to give iminoboronates, in which the imine structure is stabilized by the N—B coordination.[11] Enlisting an iminoboronate warhead for RCI design, Akcay et al. developed highly potent RCIs for an important oncogenic protein Mcl-1. These RCIs incorporate either 2-formyl or 2-acetyl phenyl boronic acid (abbreviated as FPBA and APBA respectively) as a covalent warhead, which conjugate with Lys234 of the target protein via iminoboronate formation.[12] Careful SAR studies show that removing either the boronic acid moiety or the carbonyl reduces the Mcl-1 binding affinity by 20- to over 100-fold, highlighting the critical role of the iminoboronate formation for Mcl-1 inhibition.

3. Thermodynamic considerations of RCD design

As seen in the RCD examples shown in Fig. 2, and perhaps also expected for all covalent inhibitors, RCDs consist of a covalent warhead installed on a parent scaffold. The parent scaffold, typically adapted from a known noncovalent ligand of the protein of interest, is expected to bind a target protein reversibly, thereby serving as a directing motif for the covalent warhead to react with the protein. A reversible covalent warhead, when installed properly, can significantly boost the affinity and specificity of the drug molecule. Perhaps not surprisingly though, proper choice and installation of the reversible covalent warhead presents a formidable challenge. RCD development has been heretofore primarily carried out on a trial and error basis. With the following sections of discussion, we try to formulate the emerging rules of RCD design that will be useful in guiding the thought process for the development of novel RCDs.

Proper use of a reversible warhead offers exciting opportunities to enlist novel mechanisms to enhance drug–target interactions. The mainstay of small molecule drugs utilize an array of noncovalent interactions, including hydrophobic interactions, hydrogen bonding and electrostatic interactions, to drive its binding to the target protein. The binding process as a thermodynamic equilibrium is dictated by the overall free energy of binding ($\Delta G_{binding}$), which to the first order of approximation equals the sum of the free energy contributed by the individual noncovalent interactions (Fig. 3). In contrast, most covalent bond formation reactions are irreversible and dictated by the kinetics instead of thermodynamics of the reaction. However, if in rapid equilibrium, a covalent bond formation would be

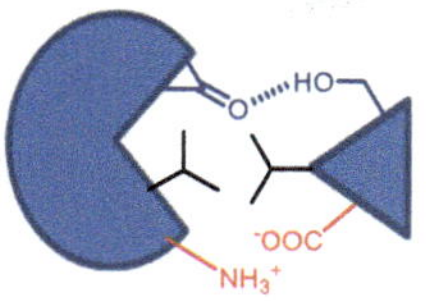

$$\Delta G_{binding} = \Delta G_{H\text{-}bond} + \Delta G_{Hydrophobic\ interactions} + \Delta G_{Electrostatic\ interactions} + \ldots$$

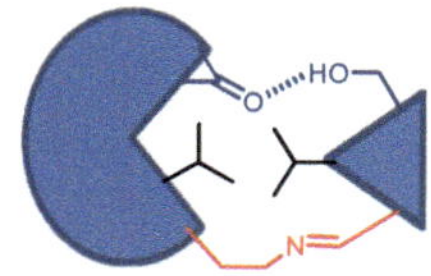

$$\Delta G_{binding} = \Delta G_{H\text{-}bond} + \Delta G_{Hydrophobic\ interactions} + \Delta G_{Electrostatic\ interactions} + \Delta G_{covalent\ binding} + \ldots$$

Fig. 3 Energetic contributions of various molecular interactions in drug–target association. Reversible covalent interactions can be considered analogous to noncovalent interactions in the energetic analysis of drug–target binding interactions.

under thermodynamic control, just like the noncovalent interactions. In other words, *a reversible covalent bond formation can be considered as a "binding" mechanism similar to hydrogen bonds*. The contribution of such a covalent binding mechanism can be understood using the free energy of the reaction (Fig. 3). Note that negative free energy value is generally considered a favorable interaction or a driving force for binding, while a positive free energy is considered an opposing force for binding.

For the design of RCDs, one would ideally like to use covalent warheads that give negative free energies for their conjugation reactions. Reversible conjugations of this sort are still rare. Nevertheless, two recently reported examples nicely illustrate the power of using such reversible conjugations in the development of RCDs. The first example comes from a series of elegant studies of Jack Taunton seeking to develop kinase RCIs targeting noncatalytic cysteines. In neutral aqueous media, α-cyanoacrylates and α-cyanoacrylamides were found to reversibly conjugate with a small molecule thiol at low millimolar concentrations to reach an equilibrium within a minute.[7] A titration with β-mercaptoethanol (BME) revealed a K_d of 9.4 mM for the thiol's addition to a α-cyanoacrylate (Fig. 4A). The K_d value converts to a free energy of about −3 kcal/mol for this reversible conjugation, which presents a strong driving force for binding. A number of highly potent kinase inhibitors, including PRN1008, have been developed that take advantage of this reversible conjugation for cysteine-targeted inhibition. The energetic contribution of this reversible conjugation can be perhaps best estimated by comparing the target enzyme RSK2 with a C436V mutant (RSK2-C436V), which is incapable of covalent bond formation.

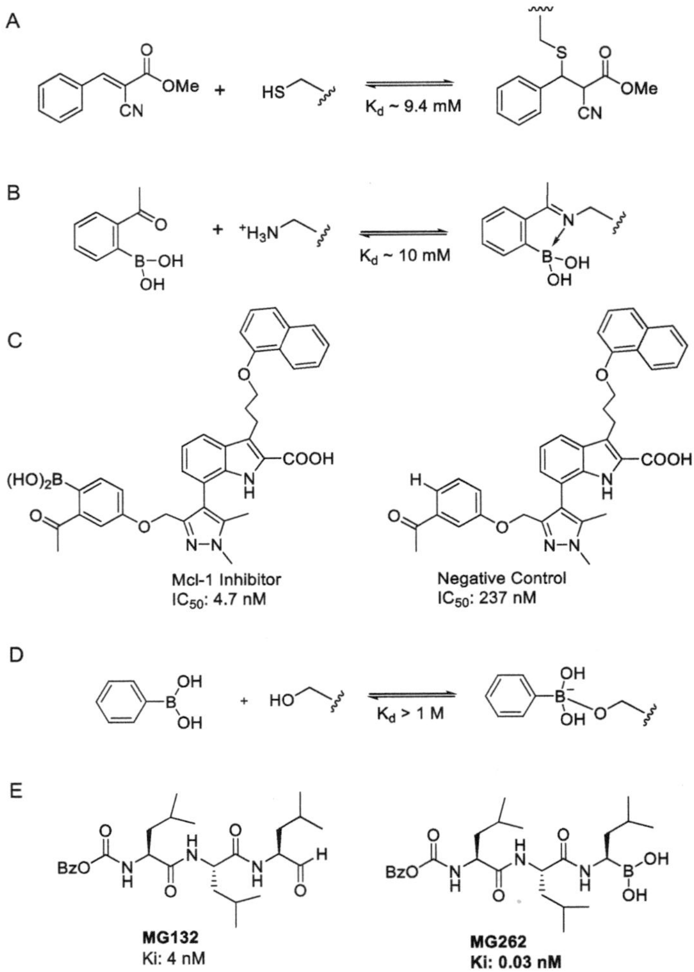

Fig. 4 Energetic contributions of a reversible covalent warhead to binding. Parts (A), (B), and (D) present the thermodynamic equilibrium of three reversible covalent conjugation reactions. Parts (C) and (E) illustrate the activity of a RCD in comparison to a negative control, highlighting the energetic significance of the reversible covalent warhead. The discrepancy between (D) and (E) in terms of the energetic contributions of boronate ester formation can be explained by the elevated reactivity of the threonine residue in proteasome active site, where MG262 binds.

A series of RSK2 inhibitors that carry a α-cyanoacrylamide or its analogues were examined for inhibiting RSK2 and RSK2-C436V respectively.[13] In comparison to the parent protein, the C436V mutant consistently yielded IC_{50} values greater by over 100-fold. The magnitude of binding affinity erosion matches up quite nicely with the missing energetic contribution of the reversible conjugation between the protein and the drug.

Another elegant example of the covalent binding mechanism involves the formation of iminoboronates. While typical imine formation is thermodynamically unfavorable (with positive free energies) in aqueous media,[14] formation of iminoboronates does proceed with −3 to −4 kcal/mol in free energy (Fig. 4B),[15] similar to the thiol addition to α-cyanoacrylates. This reversible conjugation has been successfully implemented to develop RCIs for Mcl-1,[12] an important protein–protein interaction (PPI) target involved in cancer. Specifically, FPBA and APBA were installed onto a previously reported noncovalent inhibitor of Mcl-1 to target Lys234 that sits at the rim of the binding pocket. The FPBA/APBA presenting molecules were found to exhibit single digit nanomolar IC_{50} values against the target Mcl-1. In contrast, a control molecule that removes the boronic acid group afforded IC_{50} values greater by 50-fold respectively in comparison to the one carrying an APBA warhead (Fig. 4C). In addition, the authors examined the RCIs comparatively against Mcl-1 and a K234A mutant, which yielded about 100 times higher IC_{50} in comparison to the wild type, again highlighting the significant energetic contribution that the covalent reaction makes to the protein inhibition. Together with the Taunton report on cysteine-targeting RCIs, this work on the iminoboronate chemistry nicely illustrates the prospect that the reversible covalent binding mechanisms can be understood and rationalized using free energy terms just like how we treat noncovalent interactions.

In contrast to the two examples described above, the reversible covalent chemistry of the anticancer drug bortezomib appears more difficult to understand. Bortezomib uses a boronic acid as warhead to conjugate with the N-terminal threonine of the proteasome. This conjugation reaction gives a boronate ester linkage between the protein and the drug. The use of boronate ester for covalent binding may seem counterintuitive as the reaction per se is thermodynamically unfavorable (Fig. 4D). Specifically, the binding of a phenylboronic acid to glucose gives a dissociation constant around 1 M, which corresponds to a binding free energy of 0 kcal/mol.[16] We note that glucose actually provides a 1,2-diol motif to bind boronic acids and the binding of single –OH group to boronic acids is even less favorable with positive free energies ($\Delta G > 0$ kcal/mol). This is perhaps not surprising as boronate ester formation exchanges a B—O bond with another B—O bond and this enthalpically neutral reaction must be disfavored in bulk water. So how can a thermodynamically unfavorable reaction as such serve as an effective warhead? This paradox can be resolved by noting that the N-terminal threonine of the proteasome is a catalytic residue, which is presumably activated and behaves quite differently from a nonactivated threonine. Specifically, an activated threonine for catalysis is expected to display lowered pK_a and more nucleophilic, which favors the conjugation with the boronic acid warhead. In other words, towards estimating the reaction free energies, the boronate ester formation of the catalytic threonine cannot be modeled by that of nonactivated alcohols. As a matter of fact, the energetic benefit of the boronate ester formation is well documented by the structure-activity studies of bortezomib analogues (Fig. 4E). Specifically, an earlier version of bortezomib MG262 is 100 times more potent in proteasome inhibition than the control molecule MG132,[6] which displays an aldehyde warhead instead of a boronic acid. Even though the aldehyde group could in principle form a hemiacetal linkage with the proteasome, the boronic acid warhead is clearly energetically more favorable to conjugate with the threonine in the proteasome active site.

It is important to further emphasize on the distinction between catalytic (activated) and noncatalytic residues. Although the boronate ester chemistry proves effective in targeting catalytic threonine residues, there have been no examples of RCDs that use a boronic acid warhead to target noncatalytic/nonactivated serines or threonines, for which the boronate ester formation would be thermodynamically unfavorable. In contrast, as discussed above, noncatalytic cysteine and lysine residues have been successfully targeted using reversible covalent binding to enhance the potency of inhibitors. It remains to be seen if suitable chemistries can be developed to target noncatalytic serine/threonine residues.

4. Kinetic considerations of RCD design

The reaction kinetics of a reversible covalent warhead is critically important to RCD design as it dictates two key characteristics of RCDs, namely off-target binding and residence time on the intended target. To minimize off-target binding, one would like to have low on-rate and high off-rate for a RCD. A low on-rate would minimize random conjugation and a high off-rate would allow any accidental conjugation to quickly reverse. On the other hand, a RCD reacting with an intended target is supposed to be directed by a noncovalent binding motif that selectively associates with the target protein and then sets up a pseudo intramolecular reaction for the warhead. To the best of our knowledge, there have been few systematic studies seeking to define the magnitude and range of on-rates suitable for RCD design. However, a series of papers address the significance of the off-rate of a warhead, which is critically important to the residence time of the drug as well as reversing undesired protein modifications.

While the effectiveness of a ligand targeting a specific receptor is often determined by measurements like IC_{50} and K_d, for RCDs, the effective residence time of the ligand is a vital piece of information as well. Simply put, residence time refers to the time a ligand spends bound to its target, which can also be described as the inverse of the off-rate. Designing an inhibitor with a sufficiently long residence time is an added complication not faced when designing irreversible covalent inhibitors. However, since most reversible covalent warheads are installed on a small molecule scaffold, these ligands can be fine-tuned to select for increased residence times. This optimization may prove to be essential as reversible covalent warheads are implemented into drug discovery as residence time has a significant impact on the pharmacodynamics and even safety of a drug.[17] Despite the significance of residence time for this class of inhibitor, many studies of reversible covalent warheads still focus on other factors. A notable exception to this, however, is the design of α-cyanoacrylamide-based BTK inhibitors by Jack Taunton and coworkers. By modulating the kinase-recognizing scaffold attached to the warhead, the previously mentioned, fully reversible molecule PRN1008 was developed. PRN1008 was shown to be kinase specific, dissociate rapidly upon proteolysis of BTK and had a measured residence time of 7 days via kinetic competition assay.[18] This can be compared to dasatinib, a potent and noncovalent reversible inhibitor of BTK, which has a residence time of only 0.46 h. Taunton and coworkers designed an irreversible analogue to their developed RCIs and, as expected, there was no evidence of dissociation from BTK. The percent occupancy based on a competition assay was used to determine the IC_{50} of each synthesized compound. PRN1008 was found to have an IC_{50} of 11 ± 4 nM, which is comparable to the 9 ± 2 nM measured for dasatinib and 5 ± 1 nM measured for the irreversible analogue. This serves to prove that reversible covalent warheads can be installed on a noncovalent scaffold without a significant increase in IC_{50}. PRN1008 was a successful inhibitor that has moved to clinical trials; however, Taunton points out that this molecule most likely would not have been discovered if optimization had been for IC_{50} instead of residence time.[18] This work represents an ideal pathway for the discovery of efficient reversible covalent inhibitors and helps to shed light on the areas that require attention going forward.

Excitingly, the residence time of a RCI can be tuned over a wide range. For example, Taunton and coworkers reported a series of BTK inhibitors that show biochemical residence time ranging from minutes to days (Fig. 5A),[18] determined via a fluorescence competition assay against the target protein. Further, the authors show that the long-acting BTK inhibitors maintained their long residence time in live cells. Even 18 h after the inhibitor clearance from circulation, the inhibitor **9** was found to remain bound to BTK in live cells. As the same warhead structure, namely a α-cyanoacetamide, was used for all inhibitors shown in Fig. 5A, the residence time variation observed appears to originate from the peripheral substituents, which are not in direct conjugation with the warhead structure. These peripheral substituents presumably modulate the drug's residence time by engaging in additional interactions with the target protein. In a separate publication, the Taunton group introduced a different strategy to modulate the residence times of RCIs, which involves tuning the electronic properties of the warhead (Fig. 5B).[13] Specifically, various aryl groups were installed onto the α-carbon of the acrylonitrile warhead to give a panel of functional analogues of α-cyanoacrylamide. These analogues serve as Michael acceptors that conjugate with small molecule thiols or cysteine side chains in a reversible manner. While the thiol adduct of α-cyanoacrylamide was found to dissociate within a minute ($t_{1/2} < 1$ min), the heteroaryl substituted acrylonitriles gave $t_{1/2}$ values ranging from seconds to hours.[13] Interestingly, the rate of dissociation (inverse of $t_{1/2}$) of these analogues correlates nicely with the α-proton acidity of the thiol adduct, as expected for the postulated β-elimination mechanism for the dissociation. To further demonstrate the utility of this strategy, Taunton and coworkers synthesized a long-acting RSK2 inhibitor, which maintained RSK2 occupancy in live cells 24 h after treating cells with the drug. In contrast, under the same experimental conditions, little enzyme engagement was observed for a control molecule that carries a cyanoacrylamide warhead.

As illustrated in Fig. 1, another appealing feature of RCDs is the avoidance of undesired protein modifications. Undesired protein modifications encompass modifications of unintended proteins as well as denatured or degraded proteins. It is believed that protein fragments with chemical modifications could trigger immune response and hence should be avoided.[19] For denatured or degraded proteins, the loss of protein structure largely eliminates the noncovalent interactions between the protein and the drug, hence the reversibility of the covalent linkage is dictated by the intrinsic off-rate of the conjugation reaction. Excitingly, essentially all of the reversible covalent chemistries heretofore used for RCD design exhibit fast

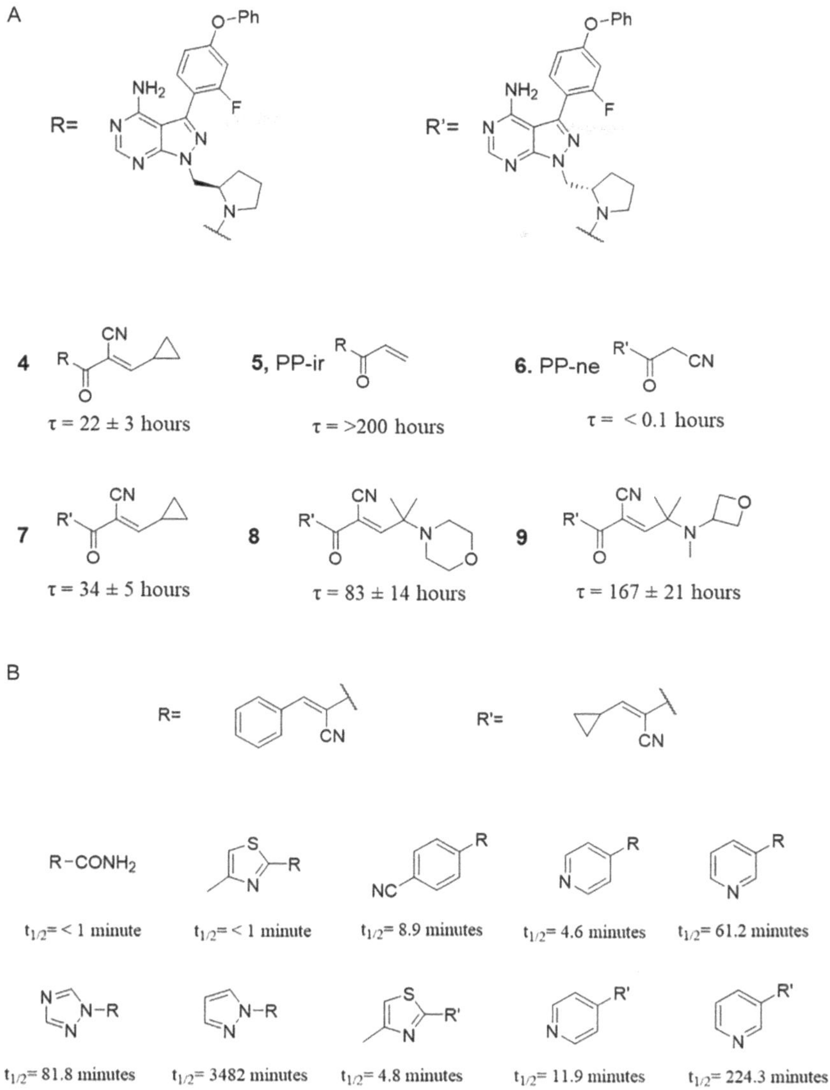

Fig. 5 Two distinct strategies for tuning the residence time of RCDs. (A) Tuning residence time by varying peripheral substituents. The compounds were numbered as they appeared in the original publication (Ref. 18), which reports the residence time (τ) for several RCDs that carry the same α-cyanoacrylamide warhead but varied peripheral substituents that are not in conjugation with the warhead. The τ values were determined via measurement of the drug dissociation from the protein over time. (B) Tuning the off-rate of the acrylonitrile warheads by manipulating the electronic properties. Cysteine adduct of the warheads were synthesized and examined for elimination half-life ($t_{1/2}$) values via ^{1}H NMR or LC-MS analysis.

dissociation kinetics. For example, the thiol-cyanoacrylamide/acrylate conjugation equilibrates in just several minutes[7] and the iminoboronate dissociation was found to be instantaneous under physiologic conditions.[15] We note, although intuitively making sense, it has been difficult to quantitatively correlate the dissociation kinetics and the potential immunogenicity or toxicity of RCDs.

5. Challenges and future perspectives

Although the field of RCD development is still very young, exciting progress and success have been achieved, yielding several clinically used drugs and drug candidates, as well as the creation of a growing number of reversible covalent reactions suitable for protein modification. Importantly, as seen in our discussions above, a conceptual framework, built on fundamental principles of chemistry thermodynamics and kinetics, is emerging for the rational design of RCDs. The framework provides useful parameters that guide future RCD design. Not surprisingly, the emerging field faces formidable challenges, which manifest in the still small repertoire of reversible covalent reactions suitable for biological use, as well as the difficulty of optimizing the combination of covalent warhead and noncovalent ligand that serves as a directing motif.

A significant challenge that must be addressed for future RCD development is the expansion of the tool kit available to researchers. Today there are a number of warheads known to operate via reversible covalent bond formation, however this number still pales in comparison to the number of irreversible warheads that have been developed, and so the development of new and novel chemistries that form reversible covalent bonds is an important task. While irreversible warhead design is mainly limited by the capabilities of organic chemistry, reversible covalent warhead design has a narrow set of requirements that must be met in order to form a covalent bond with the specific reactivity and stability needed to be reversible. Encouragingly, despite this added challenge, novel and powerful chemistries continue to emerge to allow expansion of the scope of RCDs. A recent example of this is the development of a novel chlorofluoroacetamide (CFA) warhead by Shindo et al. in their efforts to develop kinase inhibitors.[20] NS-062 (Fig. 6A, B) is an antiproliferative for nonsmall cell lung carcinoma that utilizes a CFA warhead. The warhead forms a reversible covalent bond with a reactive cysteine found in the ATP binding site of epidermal growth factor receptor (EGFR) via a standard S_N2 reaction substituting the Cl for the thiol. The reversibility of the reaction is achieved via hydrolysis of the F- and

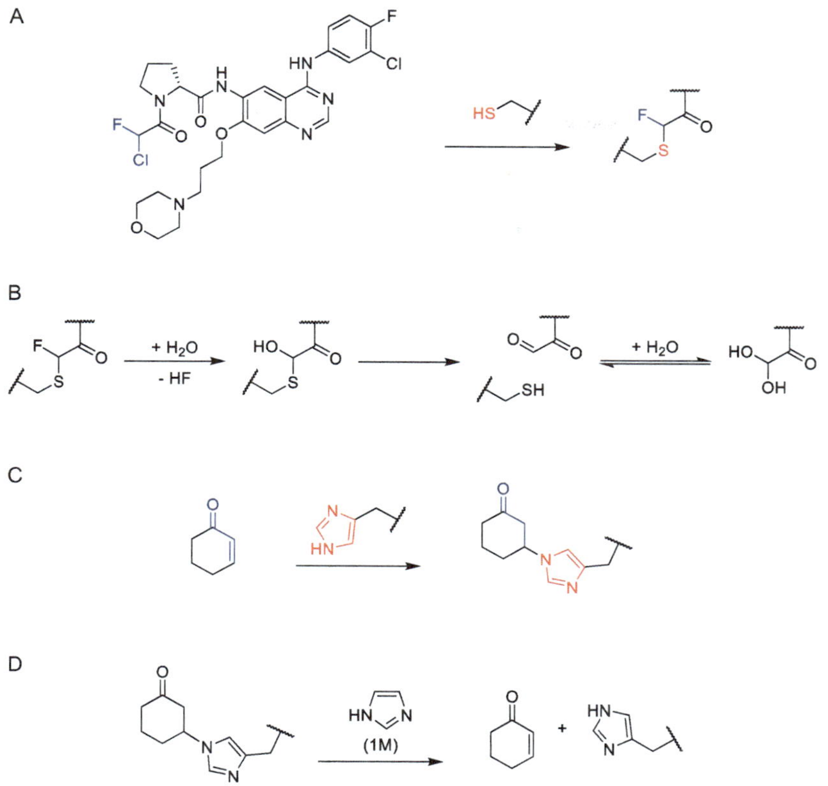

Fig. 6 Emerging reversible chemistries of proteins. (A) A unique reversible covalent modification of cysteines using chlorofluoroacetamides (CFAs). NS-062, which installs a CFA warhead, serves as highly effective antiproliferatives for nonsmall cell lung carcinoma as tested in animal models. (B) Proposed hydrolysis reaction of covalent adduct formed from a CFA warhead and cysteine. Unmodified cysteine is recovered but the warhead is not. (C) Histidine modification with 2-cyclohexenone developed for site-specific protein labeling. (D) Base activated elimination of covalent adduct of 2-cyclohexenone and histidine. Imidazole was found to be the most efficient base, yielding 93% recovery of unlabeled histidine.

S-geminally substituted adduct. NS-062 was compared to afatinib, a clinically approved irreversible inhibitor that targets EGFR in a similar manner. Both compounds had identical IC_{50} values of 0.19 μM when examined in the antiproliferative assay against erlotinib-resistant H1975 cells, however NS-062 had a significantly higher IC_{50} than afatinib against nontarget, EGFR-independent and wild-type cell lines, meaning that NS-062 is less toxic to these cell lines due to fewer off-target interactions. In terms of residence time, kinase inhibition by NS-062 was seen up to 8 h in EGFR-dependent cells after washout. This prolonged residence time is presumably due to the slow hydrolysis of the covalent adduct in the binding pocket of the enzyme. Oral administration (20 mg/kg) of afatinib and NS-062 were administered to mice, in which NS-062 had fewer instances of mortality and weight loss among the mice dosed meaning that NS-062 is less toxic. Tumor cell volume was monitored over a 25-day period and NS-062 showed comparable decrease in tumor growth suppression over afatinib.

Interestingly, the CFA warhead was also implemented to develop BTK inhibitors, which provided an opportunity to compare the CFA warhead to the cyanoacrylamide. An identical scaffold was modified with the CFA warhead or a α-cyanoacrylamide receptively. The resulting inhibitors (1 μM) were mixed with the target protein and incubated for 1 h. A competition assay with a competitive fluorescent probe was used to determine the occupancy rate of both inhibitors. The results show that, after 12 h, the CFA appended scaffold retained 82% occupancy while the cyanoacrylamide analogue gave 62%. The even longer residence time elicited by the CFA warhead highlights the power of novel chemistry in the development of RCDs.

Expanding the chemical toolbox also calls for developing reversible conjugation reactions of additional residues. As of now, only reversible covalent warheads of serine/threonine, cysteine and lysine have been explored in the RCD development (Fig. 2). Having suitable warheads that target additional residues such as histidine, aspartate/glutamate, and tyrosine will greatly expand the scope of proteins that can be targeted with a RCD. While it is nontrivial to develop reversible reactions for these residues, some encouraging developments have been seen in the pursuit of site-specific labeling chemistry for native proteins. A recent and relevant example of this is the work by Joshi and Rai studying β-enones, most notably 2-cyclohexenone (Fig. 6C, D). Joshi and Rai report rapid, nondisruptive, and selective single-site labeling of histidine residues in a number of proteins, confirmed via tryptic digestion and MS-MS analysis. The reaction yielded excellent site-selectivity from pH 5–7 and with up to a 250:1 ratio of 2-cyclohexenone to protein. Tryptic digestion and tandem MS analysis confirmed that even myoglobin, which contains 11 histidine and 19 lysine residues, was only singly labeled. Histidine residues are not traditionally targeted by RCDs, and so a reliable method that selectively targets histidine would greatly expand the number of potential proteins targetable with RCDs. The Michael-addition of 2-cyclohexenone with histidine acquires reversibility via induction by biocompatible small molecules: adding excess imidazole enables the reverse

Michael reaction and regeneration of unlabeled histidine through an elimination reaction. Note that this technology is not quite ready to be implemented into RCD design just yet, due to the high concentrations of imidazole needed for fast enough reverse reaction. However, it is exciting to speculate that, with further development and optimization, the reversible histidine modification could be achieved fully under physiologic conditions, which will open new doors for RCD design to target histidine residues.

In our earlier discussions, we attempted to draw similarities between reversible covalent interactions and noncovalent interactions such as hydrogen bond. It is important to note, however, most of the reversible covalent warheads available are significantly bulkier than hydrogen bonding-capable structural motifs. The steric bulk of the reversible covalent warheads can make it difficult for their installation onto a noncovalent scaffold as the bulky warhead may dramatically perturb the noncovalent ligand's binding geometry to the target protein. This problem is more evident for designing RCDs to target catalytic residues which typically reside in a well-crafted pocket. In contrast, it is potentially easier to target noncatalytic residues such as surface lysines or cysteines as the surface location of these residues would be more accommodating in terms of how the warhead is linked to the noncovalent scaffold. Excitingly, a joint publication by the Shoichet and Taunton groups showcases the possibility of using computational docking methods to assist the development of covalent inhibitors.[21] While docking methods have been routinely used in evaluating small molecule libraries to identify noncovalent inhibitors, covalent docking presents a unique challenge as both noncovalent and covalent interaction energies need to be properly counted. The Shoichet and Taunton collaboration created a docking program called DOCKovalen, which has been made publically available.[21] In this work, the authors created large libraries of readily accessible electrophiles, including α,β-unsaturated carbonyls, aldehydes, boronic acids, α-cyanoacrylamides, alkyl halides, carbamates, α-ketoamides and epoxides. For each ligand, the possible stereoisomers, protonation states, and conformations are pregenerated in respect to the covalent bond. The DOCKovalent program docks these pregenerated poses into the target protein and scores them using a physics-based energy function. This clever design avoids the computation of covalent bonding energies, which can be difficult to compute. The practicality of the DOCKovalent methodology was demonstrated with three enzymes including AmpC β-lactamase, p90 ribosomal protein S6 kinase-2 (RSK2) and JAK3 kinase. Multiple potent, reversible covalent inhibitors were found against all three targets as confirmed by biochemical and X-ray crystallography studies. It remains to be seen how widely applicable this methodology is. Conceivably, it can be expanded to include novel warheads whose chemistry is well characterized.

6. Conclusion

With this chapter, we seek to present an analysis of the current status and future perspectives of reversible covalent drugs, primarily from the chemist's perspective. Since the FDA approval of bortezomib in 2003, there have been at least five more RCDs that received FDA approval for clinical use. The early success has inspired increasing interest in RCD development within the past decades. While some of these early RCDs came through serendipitous discoveries, the following research within the past decade has revealed the possibility and promise of developing RCDs in rational and/or systematic ways. The accumulating literature data has allowed us to construct a conceptual framework that we believe will serve as a valuable guidance for future research in this area. In addition, we emphasize the remaining challenges, which manifest in the limited availability of reversible covalent conjugation reactions and the nontrivial optimization of the directing scaffold to direct warhead reactivities. With the growing interest in RCDs, we are optimistic that new exciting chemistries and methodologies will continue to emerge to deliver novel life-saving drugs.

Acknowledgments

We thank the National Science Foundation and the National Institutes of Health for supporting our work to develop novel reversible covalent chemistry to enable chemical biology and medicinal chemistry applications.

References

1. Singh, J.; Petter, R. C.; Baillie, T. A.; Whitty, A. The Resurgence of Covalent Drugs. *Nat. Rev. Drug Discov.* **2011**, *10*, 307–317.
2. Bandyopadhyay, A.; Gao, J. Targeting Biomolecules With Reversible Covalent Chemistry. *Curr. Opin. Chem. Biol.* **2016**, *34*, 110–116.
3. Gehringer, M.; Laufer, S. A. Emerging and Re-Emerging Warheads for Targeted Covalent Inhibitors: Applications in Medicinal Chemistry and Chemical Biology. *J. Med. Chem.* **2019**, *62*, 5673–5724.
4. Wang, Y.-H.; Zhang, F.; Diao, H.; Wu, R. Covalent Inhibition Mechanism of Antidiabetic Drugs—Vildagliptin vs Saxagliptin. *ACS Catal.* **2019**, *9*, 2292–2302.
5. O'Meara, J. A.; Lemke, C. T.; Godbout, C.; Kukolj, G.; Lagacé, L.; Moreau, B.; Thibeault, D.; White, P. W.; Llinàs-Brunet, M. Molecular Mechanism by Which a Potent Hepatitis C Virus NS3-NS4A Protease Inhibitor Overcomes Emergence of Resistance. *J. Biol. Chem.* **2013**, *288*, 5673–5681.
6. Greener, B. S.; Millan, D. S. Bortezomib (Velcade): A First-in-Class Proteasome Inhibitor. In *Modern Drug Synthesis*; Li, J. J., Johnson, D. S., Eds.; 2010; pp. 99–110.

7. Serafimova, I. M.; Pufall, M. A.; Krishnan, S.; Duda, K.; Cohen, M. S.; Maglathlin, R. L.; McFarland, J. M.; Miller, R. M.; Frodin, M.; Taunton, J. Reversible Targeting of Noncatalytic Cysteines With Chemically Tuned Electrophiles. *Nat. Chem. Biol.* **2012**, *8*, 471–476.
8. Fairhurst, R. A.; Knoepfel, T.; Buschmann, N.; Leblanc, C.; Mah, R.; Todorov, M.; Nimsgern, P.; Ripoche, S.; Niklaus, M.; Warin, N.; Luu, V. H.; Madoerin, M.; Wirth, J.; Graus-Porta, D.; Weiss, A.; Kiffe, M.; Wartmann, M.; Kinyamu-Akunda, J.; Sterker, D.; Stamm, C.; Adler, F.; Buhles, A.; Schadt, H.; Couttet, P.; Blank, J.; Galuba, I.; Trappe, J.; Voshol, J.; Ostermann, N.; Zou, C.; Berghausen, J.; Del Rio Espinola, A.; Jahnke, W.; Furet, P. Discovery of Roblitinib (FGF401) as a Reversible-Covalent Inhibitor of the Kinase Activity of Fibroblast Growth Factor Receptor 4. *J. Med. Chem.* **2020**,. https://doi.org/10.1021/acs.jmedchem.0c01019 (in press).
9. Benson, M. J.; Rodriguez, V.; von Schack, D.; Keegan, S.; Cook, T. A.; Edmonds, J.; Benoit, S.; Seth, N.; Du, S.; Messing, D.; Nickerson-Nutter, C. L.; Dunussi-Joannopoulos, K.; Rankin, A. L.; Ruzek, M.; Schnute, M. E.; Douhan, J. Modeling the Clinical Phenotype of BTK Inhibition in the Mature Murine Immune System. *J. Immunol.* **2014**, *193*, 185–197.
10. Oksenberg, D.; Dufu, K.; Patel, M. P.; Chuang, C.; Li, Z.; Xu, Q.; Silva-Garcia, A.; Zhou, C.; Hutchaleelaha, A.; Patskovska, L.; Patskovsky, Y.; Almo, S. C.; Sinha, U.; Metcalf, B. W.; Archer, D. R. GBT440 Increases Haemoglobin Oxygen Affinity, Reduces Sickling and Prolongs RBC Half-Life in a Murine Model of Sickle Cell Disease. *Brit. J. Haematol.* **2016**, *175*, 141–153.
11. Cambray, S.; Gao, J. Versatile Bioconjugation Chemistries of Ortho-Boronyl Aryl Ketones and Aldehydes. *Acc. Chem. Res.* **2018**, *51*, 2198–2206.
12. Akcay, G.; Belmonte, M. A.; Aquila, B.; Chuaqui, C.; Hird, A. W.; Lamb, M. L.; Rawlins, P. B.; Su, N.; Tentarelli, S.; Grimster, N. P.; Su, Q. Inhibition of Mcl-1 Through Covalent Modification of a Noncatalytic Lysine Side Chain. *Nat. Chem. Biol.* **2016**, *12*, 931–936.
13. Krishnan, S.; Miller, R. M.; Tian, B.; Mullins, R. D.; Jacobson, M. P.; Taunton, J. Design of Reversible, Cysteine-Targeted Michael Acceptors Guided by Kinetic and Computational Analysis. *J. Am. Chem. Soc.* **2014**, *136*, 12624–12630.
14. Crugeiras, J.; Rios, A.; Riveiros, E.; Richard, J. P. Substituent Effects on the Thermodynamic Stability of Imines Formed From Glycine and Aromatic Aldehydes: Implications for the Catalytic Activity of Pyridoxal-5′-phosphate. *J. Am. Chem. Soc.* **2009**, *131*, 15815–15824.
15. Bandyopadhyay, A.; Gao, J. Iminoboronate-Based Peptide Cyclization That Responds to pH, Oxidation and Small Molecule Modulators. *J. Am. Chem. Soc.* **2016**, *138*, 2098–2101.
16. Springsteen, G.; Wang, B. H. A Detailed Examination of Boronic Acid-Diol Complexation. *Tetrahedron* **2002**, *58*, 5291–5300.
17. Copeland, R. A. The Dynamics of Drug-Target Interactions: Drug-Target Residence Time and Its Impact on Efficacy and Safety. *Exp. Opin. Drug Discov.* **2010**, *5*, 305–310.
18. Bradshaw, J. M.; McFarland, J. M.; Paavilainen, V. O.; Bisconte, A.; Tam, D.; Phan, V. T.; Romanov, S.; Finkle, D.; Shu, J.; Patel, V.; Ton, T.; Li, X.; Loughhead, D. G.; Nunn, P. A.; Karr, D. E.; Gerritsen, M. E.; Funk, J. O.; Owens, T. D.; Verner, E.; Brameld, K. A.; Hill, R. J.; Goldstein, D. M.; Taunton, J. Prolonged and Tunable Residence Time Using Reversible Covalent Kinase Inhibitors. *Nat. Chem. Biol.* **2015**, *11*, 525–531.
19. Kuriakose, A.; Chirmule, N.; Nair, P. Immunogenicity of Biotherapeutics: Causes and Association With Posttranslational Modifications. *J. Immunol. Res.* **2016**, *2016*, 1298473.

20. Shindo, N.; Fuchida, H.; Sato, M.; Watari, K.; Shibata, T.; Kuwata, K.; Miura, C.; Okamoto, K.; Hatsuyama, Y.; Tokunaga, K.; Sakamoto, S.; Morimoto, S.; Abe, Y.; Shiroishi, M.; Caaveiro, J. M. M.; Ueda, T.; Tamura, T.; Matsunaga, N.; Nakao, T.; Koyanagi, S.; Ohdo, S.; Yamaguchi, Y.; Hamachi, I.; Ono, M.; Ojida, A. Selective and Reversible Modification of Kinase Cysteines with Chlorofluoroacetamides. *Nat. Chem. Biol.* **2019**, *15*, 250–258.
21. London, N.; Miller, R. M.; Krishnan, S.; Uchida, K.; Irwin, J. J.; Eidam, O.; Gibold, L.; Cimermancic, P.; Bonnet, R.; Shoichet, B. K.; Taunton, J. Covalent docking of large libraries for the discovery of chemical probes. *Nat. Chem. Biol.* **2014**, *10*, 1066–1072, The docking program can be found at http://covalent.docking.org/.

CHAPTER FOUR

Design of next-generation covalent inhibitors: Targeting residues beyond cysteine

Lyn H. Jones*
Dana-Farber Cancer Institute, Boston, MA, United States
*Corresponding author: e-mail address: lyn_jones@dfci.harvard.edu

Contents

1. Introduction

Cysteine is the most nucleophilic of the 20 canonical amino acid residues. Because of this, and due to its low abundance, particularly in small molecule binding sites in proteins, cysteine is often targeted in order to develop selective covalent protein inhibitors. Certain proteins contain a catalytic cysteine residue, such as cysteine proteases, and the heightened nucleophilicity of the amino acid is the result of facile deprotonation to the thiolate anion. In fact, there are several approved drugs that engage a nucleophilic cysteine in the active site of a protein (both catalytic and non-catalytic), and the rational design of such inhibitors is an area that has been reviewed extensively.[1–4] Additional to the opportunities of developing selective inhibitors, cysteine reactive covalent modulators possess longer pharmacodynamic (PD) duration than reversible binders. This is important when considering target proteins with long resynthesis rates—

Annual Reports in Medicinal Chemistry, Volume 56
ISSN 0065-7743
https://doi.org/10.1016/bs.armc.2020.10.001

in this scenario, a fast-acting covalent inhibitor does not need to possess sustained free drug levels to achieve efficacious inhibition.[1,5,6] In this way, systemic drug levels can be lowered to enhance safety margins (an effect that has been suggested for the BTK inhibitor ibrutinib, for example).[7]

Although the low frequency of cysteine enables the development of selective inhibitors, it of course limits the opportunities for creating covalent protein modulators when the residue is not available for templated addition chemistry.[8] Another challenge associated with irreversible cysteine-targeted inhibitors, especially those that engage non-catalytic residues in cancer targets, is the rapid emergence of cysteine-to-serine mutations, that retain enzymatic function while reducing drug potency and PD duration.[9,10] Therefore, there is a strong need to broaden amino acid target space beyond cysteine.[4,11,12]

In this chapter, advances in the development of reactive warheads that engage non-cysteine amino acid residues in protein binding sites will be described. Factors influencing the reactivity of the nucleophilic amino acid side chains (such as pK_a and the protein microenvironment) and the chemoselectivity of the electrophilic centers present in the targeted inhibitors are explored and discussed. Functionalized chemical probes that map reactive amino acid side chains across the proteome are featured throughout to illustrate the impact of chemical biology technologies in the drug discovery setting. Insights provided by chemoproteomic and structural biology analyses highlight opportunities for the prospective rational design of targeted covalent inhibitors, and the challenges that need to be overcome to achieve these goals are addressed.

2. Lysine (and *N*-terminal amine)

The ε-amino group present in the lysine side chain is on average highly basic (p$K_a \sim 10.6$) and fully protonated in proteins.[13] As a result, the nucleophilicity of lysine, particularly on the surface of a protein, is considerably tempered, but in the microenvironment of a binding site, the pK_a may be considerably lower, and its nucleophilicity significantly increased. Remarkably, lysine pK_a in proteins can range from 5 to 11.[13] Catalytic lysines possess low pK_as and are deprotonated at physiological pH, which may be due to a general desolvation effect in the hydrophobic microenvironment of the protein binding site, or neighboring residues directly facilitating specific deprotonation.[14–16] Non-catalytic, pK_a-perturbed lysines present in binding pockets also possess higher nucleophilicity and are

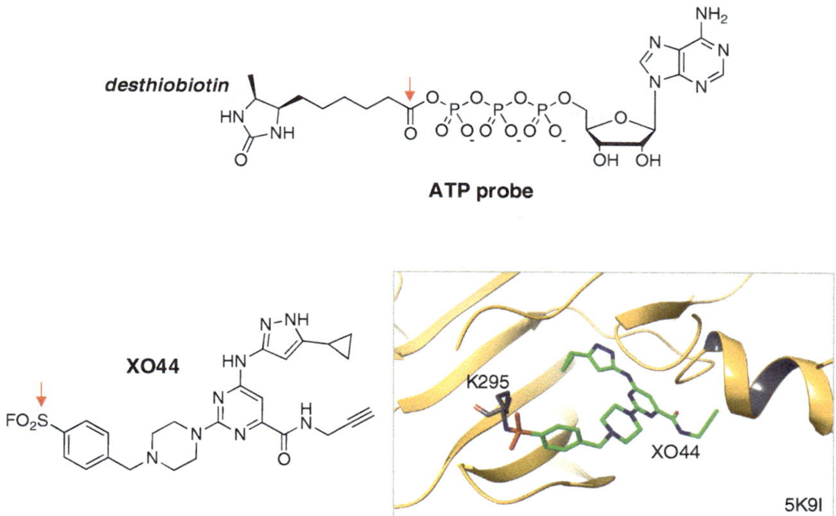

Fig. 1 Chemoproteomic kinase probes designed to target the conserved catalytic lysine residue in the ATP-site (sites of nucleophilic attack highlighted with red arrow). The crystal structure of the sulfonyl fluoride probe XO44 with the representative Src kinase is shown.

likewise targetable using various electrophilic species which are described below.[11,17,18]

An important example of lysine targeting in chemical biology is the use of reactive acyl phosphates that engage the catalytic and conserved lysine residue in the ATP-site of protein kinases.[19] These ADP- and ATP-desthiobiotinylated probes (Fig. 1) were designed to react with the conserved lysine in kinases, and following trypsinization and streptavidin-mediated enrichment, labeled peptides were identified and quantified using mass spectrometry (MS). By performing the reaction in the presence of a kinase inhibitor, that competes for the ATP-site, an assessment of its selectivity in a cellular context is revealed. This chemoproteomic technology has become a commonly applied platform in drug discovery because a more physiologically-relevant assessment of kinase selectivity is provided compared to in vitro biochemical kinase panels that utilize isolated recombinant proteins.[20]

In a related strategy, a clickable aryl sulfonyl fluoride-containing probe was developed to permit kinase selectivity determination in live cells.[21] Sulfonyl fluoride electrophiles are privileged warheads because they balance binding site-dependent amino acid reactivity with adequate aqueous stability suitable for chemical biology applications.[22] We incorporated a sulfonyl fluoride into a promiscuous kinase inhibitor template to target the conserved lysine residue in the ATP-site (a similar approach was employed previously using nucleoside-containing probes).[23,24] A terminal alkyne was used as a biologically silent reporter,[25] enabling the probe (called XO44, Fig. 1), to enter the cell and undergo copper(I)-catalyzed azide-alkyne cycloaddition (CuAAC)[26,27] click attachment of biotin for subsequent isolation and analysis of captured proteins. 133 endogenous kinases were identified using XO44 in Jurkat cells, an impressive list that could no doubt be expanded upon using additional lines and alternative sulfonyl fluoride probes.[21] XO44 was used in competition chemoproteomics experiments, and the anticancer drug dasatinib was found to be selective for SRC over the related LCK kinase in live cells (something not previously appreciated using alternative techniques). In fact, many non-kinases, including other nucleotide/nucleoside-binding proteins, were also isolated using XO44 which may highlight future opportunities for the development of lysine-targeted covalent inhibitors beyond the kinome.[28]

Sulfonyl fluoride chemistry is an example of sulfur(VI)-fluoride exchange (SuFEx) which has become an important addition to the chemical biology toolbox and will be highlighted throughout this review.[22,29] Additional to alignment of a suitably nucleophilic lysine with the sulfonyl fluoride that aids templated sulfonamide formation, the microenvironment of the binding site enhances site-selective reactivity through hydrogen-bonding interactions with the fluorine leaving group and oxygen atoms of the sulfonyl moiety.[30] Other examples of the rational design of lysine-targeting sulfonyl fluoride probes include covalent stabilizers of transthyretin which target a pK_a-perturbed lysine in the thyroxin binding site,[31] the structure-guided development of the first eIF4E covalent inhibitors with cellular activity[32] and the probe SF-erlotinib that supported the assessment of EGFR occupancy in live cells (Fig. 2).[33]

In order to canvass reactive lysines more broadly, a minimalized sulfotetrafluorophenyl pentynoate probe (STP-alkyne, Fig. 3) was developed for application in chemoproteomics experiments.[34] Treatment of cells with STP-alkyne yielded >4000 targeted lysine residues, and 310 that were deemed hyper-reactive. These lysines were enriched in enzyme active sites (catalytic residues) and ligand pockets that could be opportunities for the development of covalent inhibitors. To illustrate this, competition experiments using various activated ester fragments, chosen to capture quite a broad chemical space, led to the identification of 121 ligandable lysines in 113 proteins, many of which were deemed difficult-to-drug, including

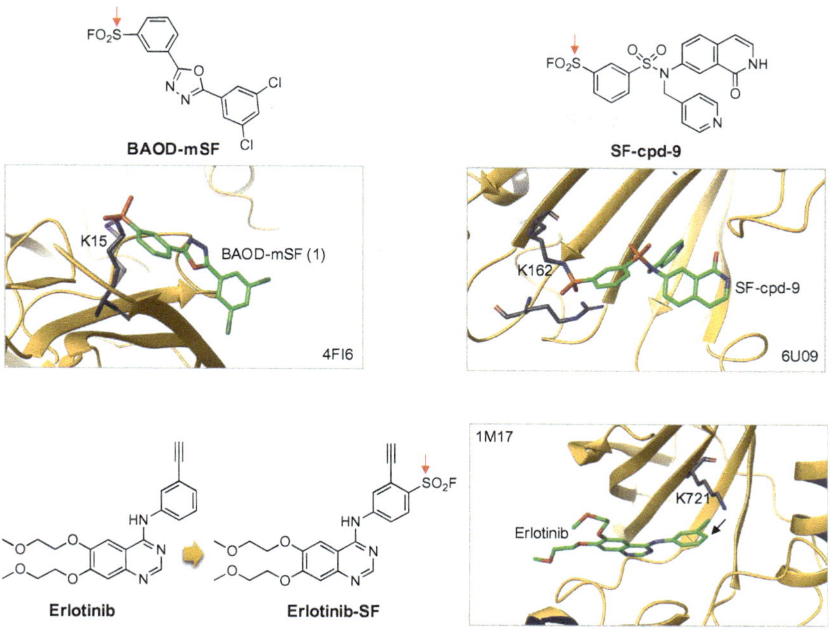

Fig. 2 Sulfonyl fluoride probes designed to engage reactive lysine residues in protein binding sites: BAOD-mSF (TTR), SF-cpd-9 (eIF4E) and erlotinib-SF (EGFR). Erlotinib-EGFR crystal structure showing site for incorporation of sulfonyl fluoride electrophile (black arrow).

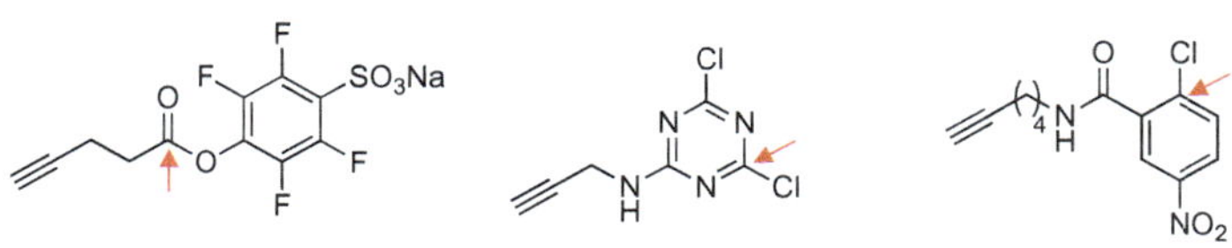

STP-alkyne **RB7** **RB2**

Fig. 3 Promiscuous chemoproteomic probes targeting lysine (STP-alkyne, RB7) and cysteine (RB2). Electrophilic sites are indicated with a red arrow.

transcription factors and scaffolding proteins. Structure-activity relationships (SARs) demonstrated that equilibrium binding of the fragment appeared to contribute to selective lysine labeling and optimization of these interactions would likely further improve potency and selectivity.

In another quantitative MS proteomics experiment, a clickable dichlorotriazine electrophilic probe RB7 was found to engage lysine residues selectively over other amino acid residues, while chloronitrobenzene RB2 was selective for cysteines (Fig. 3).[35] Protein hits from the RB7 probe include sites of lysine acetylation often exposed on the surface (such as glutamate dehydrogenase, Glud1) suggesting intrinsically high reactivity for these residues, and known binding site lysines such as Lys12 in the ATP-site of nucleoside diphosphate kinase A/B (NME1/2).

Although the reactive functionalized chemical probes such as those described above facilitate the chemoproteomic assessment of hyper-reactive lysines, their intrinsically high electrophilicity makes the development of drug-like compounds bearing these motifs challenging.[18] Warheads that possess attenuated reactivity enable the design of drug-like inhibitors with improved protein selectivity, and chemical and metabolic stability. Phenol esters appear to be ideal lysine-targeting warheads for drug design because potency can be finely tuned by making substitutions to the aromatic ring, and the intrinsic reactivity of the ester is intrinsically weaker than many electrophiles (such as the sulfonyl fluorides and acyl phosphates described above), so one would expect fewer instances of off-target labeling. A selective covalent inhibitor of the kinase PI3Kδ was developed recently that engaged the conserved catalytic lysine residue in the ATP pocket.[36] A series of phenol esters were designed based on an X-ray crystal structure of a reversible binding ligand with the protein. The potency of these covalent derivatives broadly correlated with the electron-withdrawing nature of the substituent although highly reactive electrophiles suffered from selectivity issues. However, detailed time course experiments revealed that changes in potency arose from variations in reversible binding interactions (K_I) of the different esters, not the rate constant of inactivation (k_{inact}). In fact, electron-withdrawing groups on phenol esters reduce the electrophilicity of the carbonyl carbon atom potentially through a π-polarization mechanism, but acyl substitution is enhanced through ground-state destabilization of the reacting ester.[37] Further detailed mechanistic studies of the protein labeling event are required to help delineate these complex effects. Nevertheless, through a relatively simple SAR study with PI3Kδ, the *para*-fluorophenol ester PI-cpd-4 (Fig. 4A) was found to possess the right potency/selectivity balance and covalent adduct formation was confirmed by MS and structural biology. Importantly, the compound was shown to possess a long duration of action (>48 h) in cell wash-out experiments although no in vivo activity was reported. This strategy will likely become particularly useful for anticancer targets due to the immutable nature of the catalytic residue. The *p*-fluorophenol ester warhead was also incorporated into a stilbene TTR ligand to create the lysine-targeting agent A1 (and a related thioester A2),

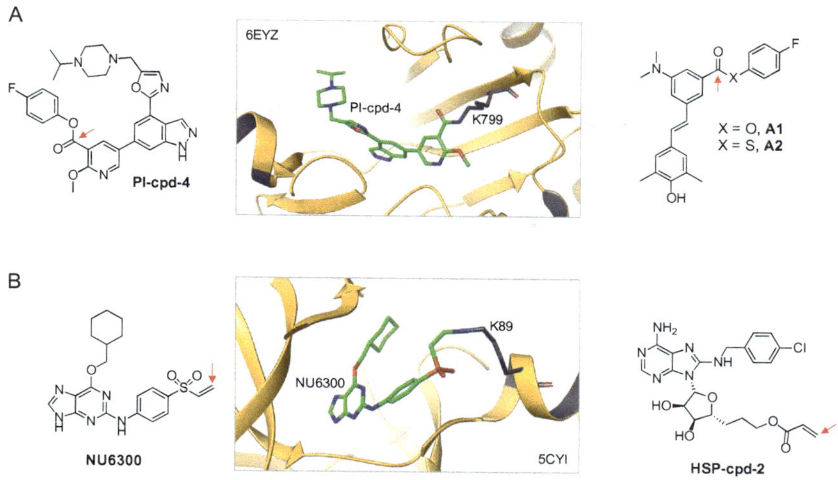

Fig. 4 (A) Fluorophenol esters designed to label reactive lysine residues in PI3Kδ (PI-cpd-4) and TTR (A1, A2). (B) Serendipitous Michael addition modifications of non-catalytic lysines by vinyl sulfone NU6300 (CDK2) and acrylate HSP-cpd-2 (Hsp72).

suggesting these electrophiles could find broader utility, providing the required geometry for nucleophilic attack can be satisfied (Fig. 4A).[38,39]

Non-catalytic lysines are also targetable in the ATP-site as illustrated by the fortuitous discovery that a synthetic intermediate en route to reversible CDK2 inhibitors, the vinyl sulfone NU6300 (Fig. 4B), engages Lys89 in the kinase site.[40] It is conceivable that improvements in potency could be achieved by optimizing reversible binding interactions and related lysines in other kinases could be rationally targeted using a similar strategy.[11] Another serendipitous example of lysine labeling was the Michael addition of Lys56 in the heat shock protein Hsp72 into the acrylate present in the benzyl adenosine HSP-cpd-2 (Fig. 4B) that was originally designed to engage a neighboring cysteine residue in the ligand binding pocket.[41]

Although sulfonyl fluoride reactivity is ideally suited to the creation of chemical biology probes, it is generally considered too high for drug development where higher intrinsic stability is required to avoid off-target protein labeling and poor pharmacokinetics.[32,42–44] The electrophilicity of the fluorosulfate warhead is significantly attenuated compared to a sulfonyl fluoride because the additional oxygen atom reduces the intrinsic reactivity of the sulfur.[45] The improved chemical and metabolic stability of the fluorosulfate electrophile makes it suitable for covalent drug design (also

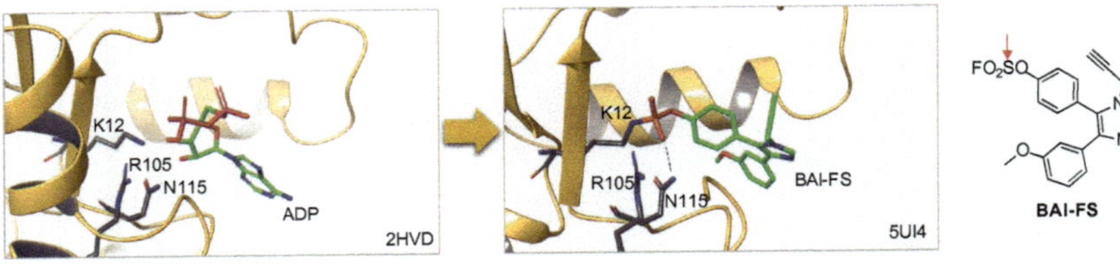

Fig. 5 The active site of NME1 possesses a catalytic lysine residue that reacts with the fluorosulfate electrophile present in BAI-FS. Arg105 may reduce the pK_a of Lys12, and the proximal Asn115 undergoes a hydrogen bond with a sulfonyl oxygen atom.

see serine targeting section below)[42] and several binding site lysines have been shown to be targetable.[30,46–50] Building on advances in chemoproteomics and appreciating the need to develop more attractive starting points for design, a recent strategy termed "inverse drug discovery" used drug-like fluorosulfate fragments to canvass reactive residues in cells.[48] BAI-FS (Fig. 5) engaged Lys12 in NME1 the same residue found to react with RB7 above. The amino acid appears to be a catalytic residue involved in DNA cleavage[51] and thus intrinsically reactive, confirmed by the unbiased chemoproteomics experiments described here (a neighboring arginine residue likely lowers the lysine pK_a, and a proximal asparagine side chain also makes a hydrogen bond to a sulfonyl oxygen atom).[48] These studies could be used to elaborate a selective covalent chemical probe to understand the emerging role of NME1 in cancer metastasis.[52]

Schiff base formation between lysines (or *N*-terminal amines) and aldehydes or ketones also affords opportunities for site-selective covalent inhibitor development. Motivated by bioconjugation chemistry using iminoboronates,[53] an inhibitor of the anti-apoptotic protein Mcl-1 was developed whereby the boronic acid stabilized the imine adduct formed with a non-catalytic lysine residue (Fig. 6).[54] The *N*-terminal amine of hemoglobin is also reactive (pK_a 6.9) and undergoes Schiff base adduct formation with glucose followed by an Amadori rearrangement to yield HbA_{1c}, a reliable biomarker of elevated glucose levels in the blood (Fig. 6). Discovery of this chemistry led to identification of a series of synthetic and natural product salicylaldehydes which form related Schiff base adducts with the *N*-terminus of sickle hemoglobin that increase its affinity for oxygen and reduce polymerization.[55] The *ortho*-hydroxy moiety stabilizes the imine adduct in a similar manner to that for the iminoboronic acids. Subsequent optimization of salicylaldehyde compounds furnished GBT440 (voxelotor, Fig. 6) which is approved for the treatment of sickle cell disease.[56]

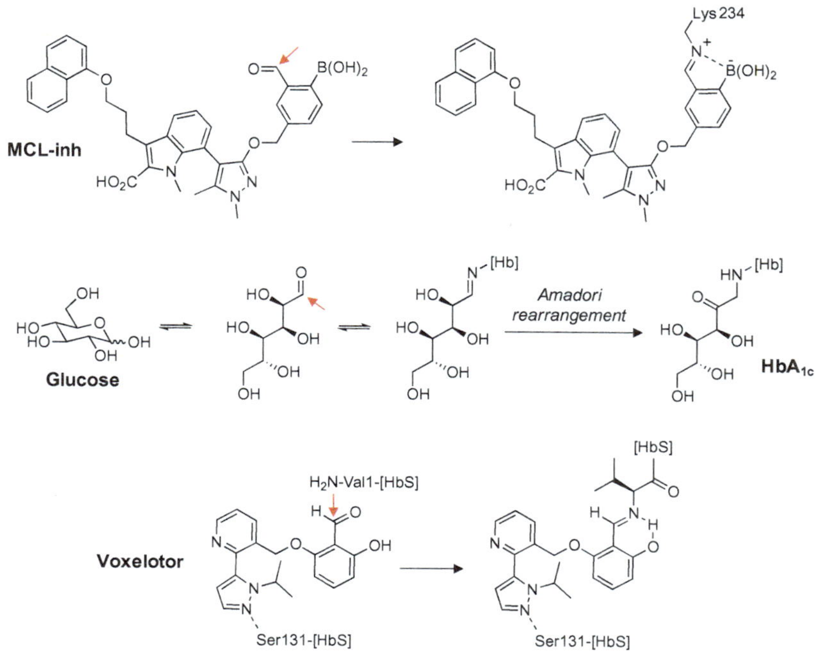

Fig. 6 Schiff base adducts formed between aldehydes and Lys234 in Mcl-1 or the *N*-terminal amine (Val1) in hemoglobin.

Other natural products may serve as inspiration for developing selective covalent modulators. For instance, the PI3K/mTOR inhibitors wortmannin and neolymphostin engage the conserved lysine through Michael addition chemistry (Fig. 7).[57,58] Advances in total synthesis and synthetic biology could be used to create novel and optimized covalent inhibitors based on the lymphostin and wortmannin scaffolds and although there have been some efforts in this area, more research is warranted to further exploit these opportunities.[59–62] Equally, lysine-targeting electrophilic warheads discovered in natural products, such as the vinylogous ester in neolymphostin, could be incorporated into the design of synthetic irreversible inhibitors.

3. Tyrosine

In recent years, the rational design of tyrosine-targeted irreversible inhibitors and chemical probes has been made possible through innovations in SuFEx chemical biology.[22] Unbiased chemoproteomic studies using sulfonyl fluoride-containing probes have identified tyrosine as a preferred

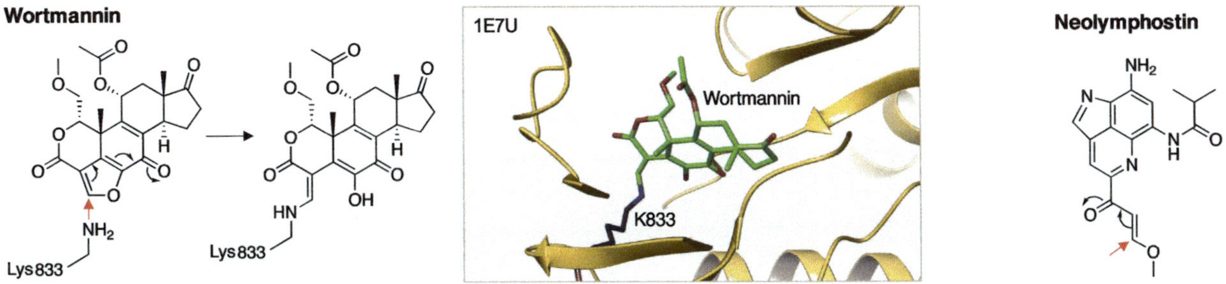

Fig. 7 Michael addition chemistry of natural products wortmannin and neolymphostin that modify the conserved lysine in kinases PI3K/mTOR (wortmannin-PI3Kγ crystal structure shown).

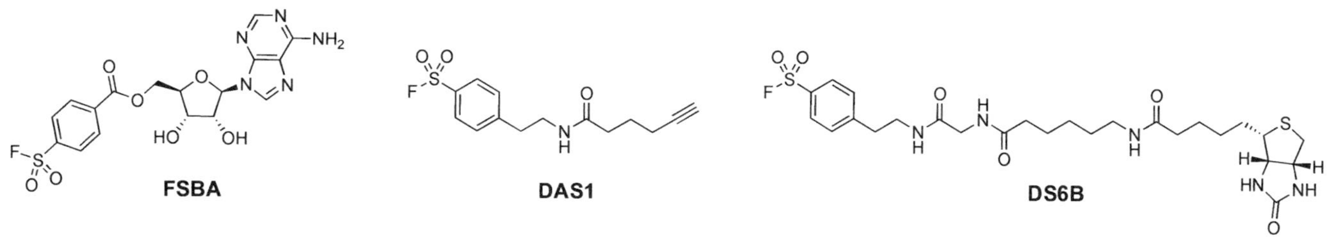

FSBA DAS1 DS6B

Fig. 8 Sulfonyl fluoride chemoproteomic probes that map nucleophilic residues in proteins (tyrosine residues in particular).

target of these electrophiles across the proteome. For example, treatment of Jurkat cells with the purine nucleotide affinity probe 5′-fluorosulfonylbenzoyl 5′-adenosine (FSBA, Fig. 8) identified 185 labeled sites using MS/MS analysis, with 67% being tyrosine and 33% lysine.[63] Unsurprisingly, many of the labeled proteins contain a nucleotide-binding site, similar to XO44 described above, and so these probes allow for the physiologically-relevant assessment of nucleotide-binding site occupancy in cells.[23] Simpler aryl sulfonyl fluoride probes DAS1 and DS6B bearing click and biotin reporters (Fig. 8) similarly labeled tyrosines in chemoproteomics experiments, including functionally important residues in the binding sites of glutathione S-transferase (GST) enzymes.[64]

An analysis of proteomics experiments using such probes and cross-reference with structural information in the Protein Data Bank (PDB) revealed that in the microenvironment of a protein binding site, most reactive tyrosines are proximal to basic residues (Lys, Arg, His) that may increase phenol-OH reactivity by creating an electropositive field that lowers pK_a.[13,16,30,65–67] Additionally, serine protease inhibitors containing sulfonyl fluorides (thus targeting the catalytic serine, see below), that are commonly used in cell lysate preparations to prevent degradation of proteins-of-interest, have been found to fortuitously engage similarly reactive tyrosine residues, as confirmed by crystallography.[22]

Inspired by work showing that sulfonyl fluoride probes such as FSBA fortuitously engage tyrosine residues[23,63,68,69] we rationally designed inhibitors and functionalized probes to target specific tyrosines in a protein binding site for the first time. A series of diaminoquinazolines (DAQs) had been found previously to be effective in phenotypic models of spinal muscular atrophy (SMA), and a protein microarray binding assay using a radiometric DAQ probe identified the mRNA decapping scavenger enzyme (DcpS) as a putative target of these compounds.[70,71] In order to validate DcpS as the relevant target in human primary cells, and to understand the levels of target engagement required to drive functional pharmacology in live cells, a covalent functionalized probe was required.[72] From cocrystal structures, it was noticeable that a targetable cysteine residue was not available in the binding site, but tyrosine side chains were found to be proximal to the DAQ scaffold.[70] Additionally, neighboring basic residues (lysine and histidine) likely reduced the phenol pK_a thus enhancing Tyr reactivity.[13,30] Sulfonyl fluoride isomers SF-o1, SF-m1 and SF-p1 were developed to specifically target these residues and were confirmed as covalent inhibitors of DcpS by intact MS (Fig. 9).[65] Peptide mapping and crystallography confirmed the desired

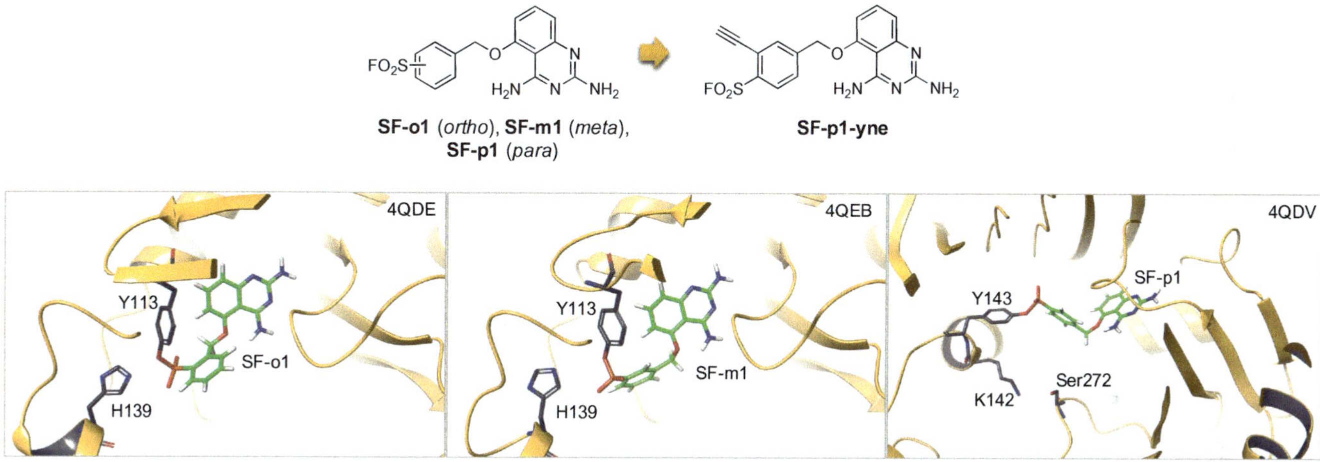

Fig. 9 Sulfonyl fluoride covalent inhibitors of DcpS that were designed to engage different tyrosine residues in the protein active site. Proximal basic residues likely reduce the phenol pK_a enhancing nucleophilicity. SF-p1-yne was used in target validation experiments as an occupancy biomarker probe in living human primary cells. PDB 4QDV illustrates the proximity of Ser272 that is modified by the fluorosulfate congener FS-p1 (see Section 5).

labeling of Tyr113 and Tyr143. Structure-based design was used to incorporate a terminal alkyne click handle to occupy a small pocket in the binding site that furnished the covalent chemical biology probe SF-p1-yne (Fig. 9). Human peripheral mononuclear cells (PBMCs) were treated with SF-p1-yne, then lysed, click conjugated to biotin to enable isolation, and analysis by western blot confirmed DcpS target engagement by the probe. Pre-incubation of cells with the advanced DAQ candidate compound RG3039 (a reversible inhibitor) competed labeling of DcpS by SF-p1-yne in a dose-related manner, thus providing an assessment of the in cell occupancy of the enzyme by the inhibitor.[73] This study confirmed reduced occupancy levels in live cells due to lysosomotropism that led to the design of DcpS inhibitors with considerably improved drug-like properties which demonstrated efficacy in SMA mouse models.[74,75]

Since this work, sulfonyl fluoride probes that engage binding site tyrosines have found considerable utility in chemical biology and drug discovery. Some recent examples of targeted proteins include IL-17A,[76] SRPK1,[77] Bcl-xL,[78] BCL6,[79] A_1-AR,[80] A_3-AR,[81] and RalA GTPase (and their respective probes are shown in Fig. 10).[82] Notably, these studies illustrate the effectiveness of sulfonyl fluoride labeling chemistry in developing inhibitors of protein-protein interactions, and indeed, tyrosine residues are significantly enriched at such protein interfaces.[83]

Fluorosulfate SuFEx motifs mirror the reactivity of sulfonyl fluoride in that they also appear to prefer labeling of tyrosine and lysine residues.[45] Unbiased chemoproteomic profiling of fragments bearing the fluorosulfate electrophile using the concept of inverse drug discovery described above identified reactive tyrosine residues. AQ-FS-yne was found to label Tyr229 (proximal to Lys57, and thus possibly pK_a-perturbed) in the binding site of GSTO1 (confirmed by a crystal structure with the inhibitor lacking the alkyne, Fig. 11A).[48] In a similar manner, the clickable probe FS-PEG-yne engaged analogous tyrosines flanked by neighboring arginine residues in the lipid-binding proteins CRABP2, FABP4 and FABP5 (Fig. 11B).[47]

In order to modify the leaving group of the electrophile, which is always fluorine in SuFEx chemistry, sulfur-triazole exchange (SuTEx) was recently developed to fine-tune the reactivity and selectivity of the sulfonyl center.[84] Regioisomeric probes HHS-465 and HHS-475 (Fig. 11C) collectively modified 8259 tyrosine residues in multiple human cell lines, assessed using MS proteomics. 1368 tyrosines were uniquely labeled by HHS-465 and 1595 by HHS-475, and many of these were annotated as phosphosites, enabling the development of a chemical strategy to delineate tyrosine

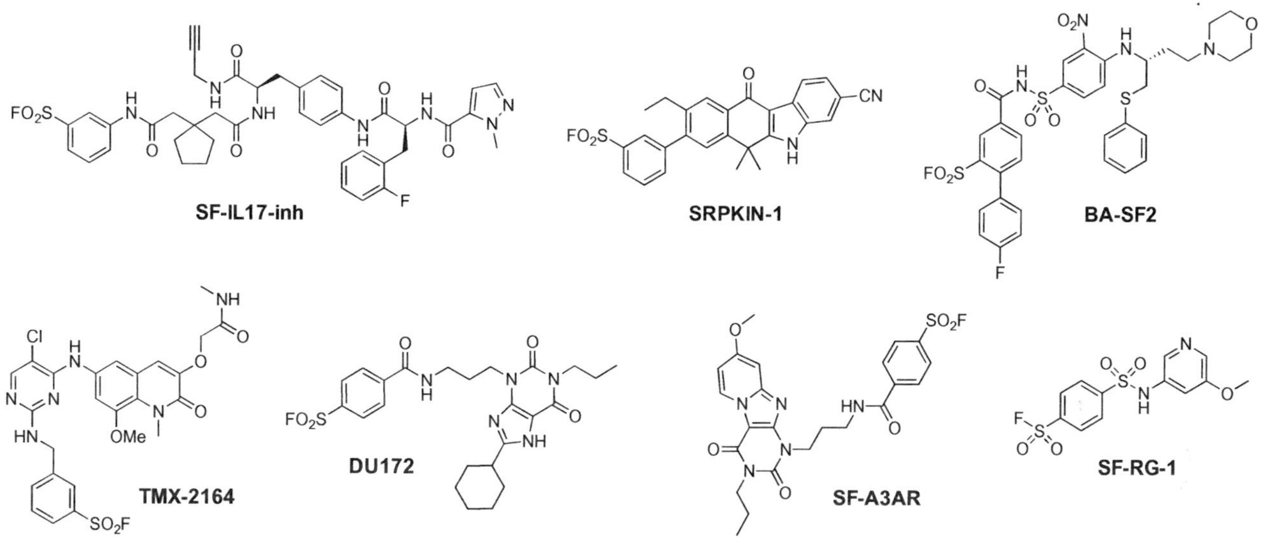

Fig. 10 Some recently reported tyrosine-targeting covalent inhibitor probes using the sulfonyl fluoride warhead: SF-IL17-inh (IL-17A), SRPKIN-1 (SRPK1), BA-SF2 (Bcl-xL), TMX-2164 (BCL6), DU172 (A_1-AR), SF-A3AR (A_3-AR), and SF-RG-1 (RalA GTPase).

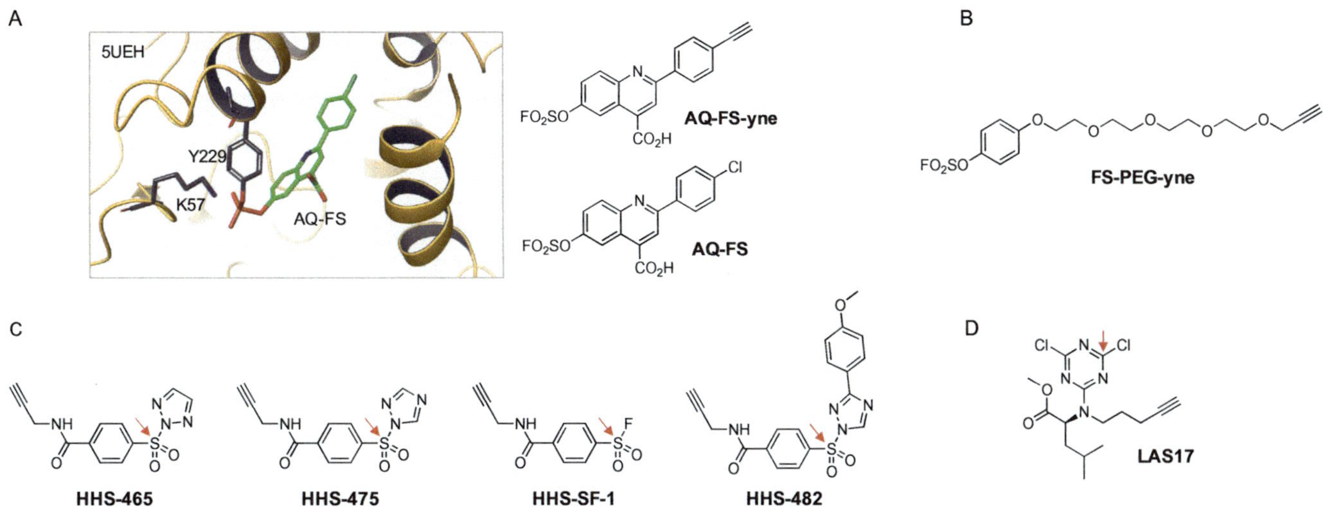

Fig. 11 Clickable chemical probes that label tyrosine residues in protein binding sites. (A) Fluorosulfates AQ-FS-yne and AQ-FS modify Tyr229 in GSTO1. (B) Minimalized fluorosulfate probe FS-PEG-yne labels tyrosines in lipid-binding proteins. (C) Sulfur-triazole exchange chemoproteomic probes, and the sulfonyl fluoride congener HHS-SF-1. (D) Dichlorotriazine LAS17 engages Tyr108 in the binding site of GSTP1.

phosphorylations in signaling pathways. 127 hyper-reactive tyrosine sites were identified and these appear to be enriched in enzyme domains and some had already been reported to be reactive residues.[84] HHS-465, HHS-475 and the sulfonyl fluoride congener HHS-SF-1 (Fig. 11C) preferably labeled tyrosine over lysine in cell proteomes (ratio Y/K = 2.5, 2.8, and 2.3, respectively) but interestingly, minor elaboration of the triazole scaffold led to significantly higher chemoselectivity (HHS-482 Y/K ~ 5). Related work showed that a small SuTEx fragment library based on the HHS-482 lead structure engaged 305 tyrosine sites on 213 distinct proteins with clear SAR.[85] These results suggest that SuTEx could be harnessed in the creation of site-specific tyrosine-targeting drugs through tuning of intrinsic reactivity and optimization of binding interactions. Intriguingly, the minimalized sulfo-triazole probes HHS-465 and HHS-475 demonstrate prominent labeling of nucleotide-binding domains, as shown for the elaborated kinase/purine nucleotide SuFEx probes XO44 and FSBA. Considering that tyrosine residues in DcpS were also readily targeted using sulfonyl fluoride probes, then SuTEx and SuFEx chemistry appear to be well-suited to the development of covalent inhibitors of proteins involved in RNA biology, and mRNA translation and metabolism in particular, since many possess reactive tyrosine residues.[65,84–87]

As discussed above, dichlorotriazines like RB7 predominantly engage lysine residues in cells.[35] However, in a library of 20 clickable dichlorotriazines, several were found to selectively label a low molecular weight protein using an in-gel fluorescence assay.[88] The protein was subsequently identified using LAS17 (Fig. 11D) and MS proteomics as GSTP1, and peptide mapping showed labeling of Tyr108 in the active site, that had been engaged previously by sulfonyl fluoride probes.[64] Evidently, a small change in the binding interactions of a dichlorotriazine probe can switch reactivity of the electrophile from lysine (RB7) to tyrosine (LAS17) depending on the vicinity of the reactive amino acid. Further exploration of halogenated aromatic heterocycles may yield additional selective tyrosine or lysine preferring electrophiles in other proteins.

4. Histidine

Although FSBA prefers tyrosine and lysine residues, there are some reports of the probe engaging histidine in protein active sites. In mitochondrial F_1-ATPase, a nucleotide-driven motor, both Tyr368 and His427 were labeled by FSBA with equal efficiency in a mutually exclusive manner.[89] At higher pH, that presumably facilitates deprotonation of the phenol and increases reactivity, tyrosine labeling is predominant, which switches to histidine at lower pH. FSBA also modified a functionally important histidine residue that binds the γ-phosphate of ATP in the active site of *Salmonella typhimurium* 5-phosphoribosyl-α-1-pyrophosphate (PRPP) synthetase.[90] Recently, genetic code expansion was used to incorporate a fluorosulfate tyrosine amino acid into PD-1 that labeled a histidine residue on PD-L1 highlighting an alternative protein-based targeted covalent inhibition strategy.[91]

Fumagillin is an antibiotic natural product approved for the treatment of intestinal microsporidiosis in immunocompromised patients that was shown previously to bind metalloprotease methionine aminopeptidase 2 (MetAP-2) (Fig. 12).[92] Fumagillin contains an electrophilic spiroepoxide moiety that was found to label a key histidine residue (His231) in the enzyme active site, causing irreversible inhibition.[93,94] An early study inspired by fumagillin chemical biology explored the targets of clickable spiroepoxides, but only a single protein was identified using gel-based proteomics (phosphoglycerate mutase PGAM1) which was modified on a lysine residue in the active site.[95,96] It would be interesting to further apply global quantitative MS proteomics of fumagillin-related probes to understand the amino acid preferences and SAR of the spiroepoxide electrophile in cells. Unanticipated histidine alkylation was also observed for the aliphatic bromide PBRM (Fig. 12) that was the result of standard SAR exploration of inhibitors of 17β-hydroxysteroid dehydrogenase type 1 (17β-HSD1).[97,98] To demonstrate the importance of site-specific labeling and intrinsically weak electrophilicity of the aliphatic bromide, PBRM did not react with histidine or imidazole even when used in large excess, and it exhibited excellent selectivity in a broad off-target panel.[99] However, the only approved drug possessing an aliphatic bromide is the anticancer compound pipobroman, a likely DNA alkylating agent, suggesting the development of drugs bearing this electrophile will be problematic.

Histidines have also been modified through Michael addition chemistry. Human serum albumin (HSA) is involved in prostaglandin metabolism and is known to bind and actively carry Δ^{12}-PGJ_2 (a natural PGD_2 metabolite, Fig. 13) in the bloodstream. Unexpectedly, the prostaglandin was found to modify His146 in a fatty acid binding site of HSA to form the unusual histidine adduct.[100] A high-throughput screen of inhibitors of the oncogene isocitrate dehydrogenase 1 (IDH1) led to the fortuitous discovery of a cyanoenone hit CNE1 (Fig. 13) that covalently modified an active site

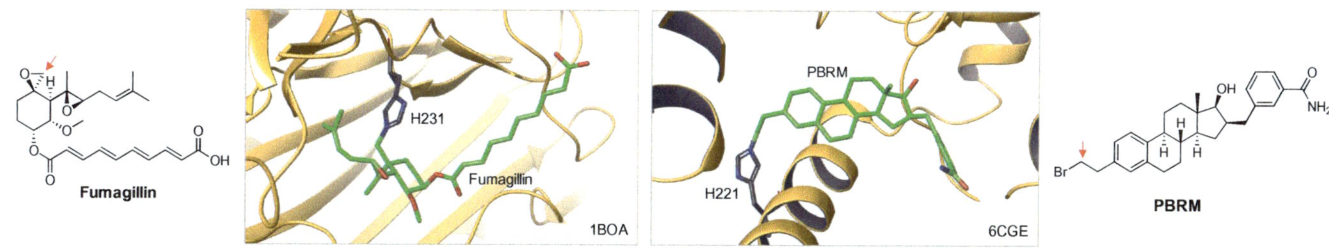

Fig. 12 Alkylation of active site histidine residues by the spiroepoxide in fumagillin and the aliphatic bromide of PBRM (irreversible inhibitors of MetAP-2 and 17β-HSD1, respectively).

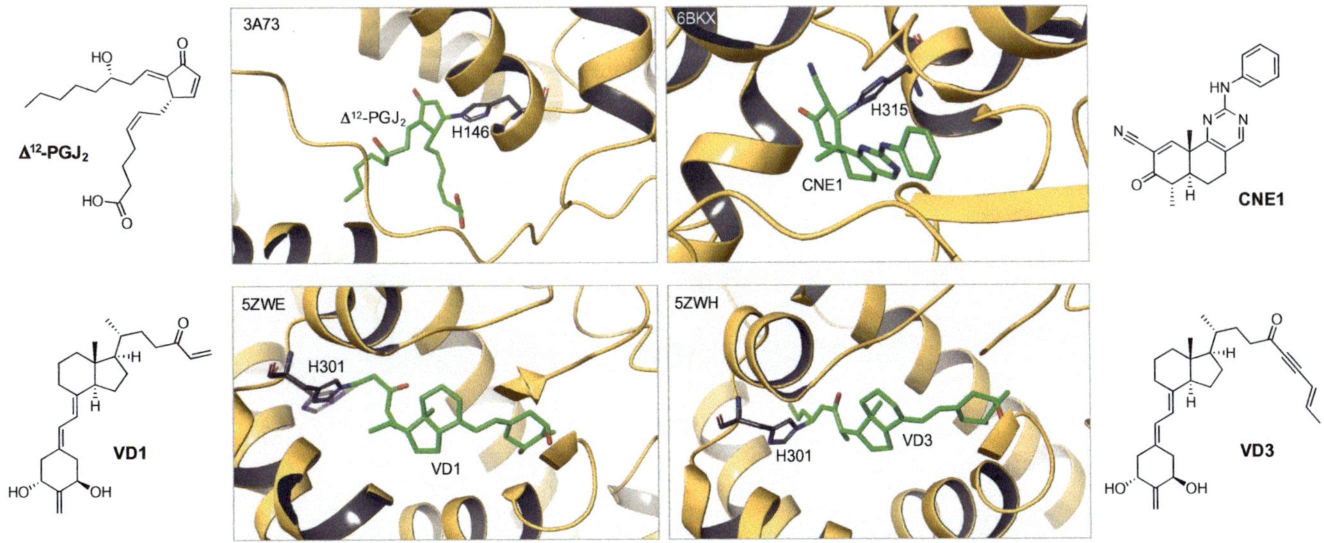

Fig. 13 Michael addition modification of active site histidine residues: Δ^{12}-PGJ_2 (HSA); CNE1 (IDH1); VD1 and VD3 (VDR).

histidine of the enzyme in a reversible manner.[101] Structure-guided optimization of binding interactions delivered a compound with improved IDH1 potency that exhibited a dose-dependent decrease in reductive glutaminolysis in a pharmacodynamic cell-based assay. Similarly, cyanoacrylamide kinase inhibitors were designed previously to form reversible covalent bonds with cysteine residues whereby the increased α-proton acidity imparted by the presence of the nitrile group drives β-elimination,[102,103] a mechanism that may play a role in CNE1 reversibility.

Rational structure-based drug design was used to create a series of vitamin D receptor unsaturated ketones based on the active form of vitamin D_3 (1α,25-dihydroxyvitamin D_3).[104] These compounds engaged a histidine residue in the ligand binding pocket through a 1,4-conjugate addition reaction and were full agonists of the receptor. Rates of addition with VDR protein were assessed using intact MS and the reaction with VD1 was complete after 90 min, while VD3 was incomplete after 6 h (Fig. 13). These differences appear to be due to the different geometry and more rigid nature of the ynone VD3 electrophile, which appears to be suboptimal for attack by the nucleophilic histidine imidazole.

KRAS (G12C) is an exciting anticancer target and several covalent inhibitors that engage the reactive cysteine residue have been reported recently (such as AMG 510, Fig. 14).[105–107] An alternative approach appears to focus on the development of an irreversible inhibitor targeting His95 in the cryptic KRAS pocket, but details have yet to be released.[108] Noticeably, a tyrosine residue in the same site also seems targetable (Tyr96), although this has not been investigated (Fig. 14).

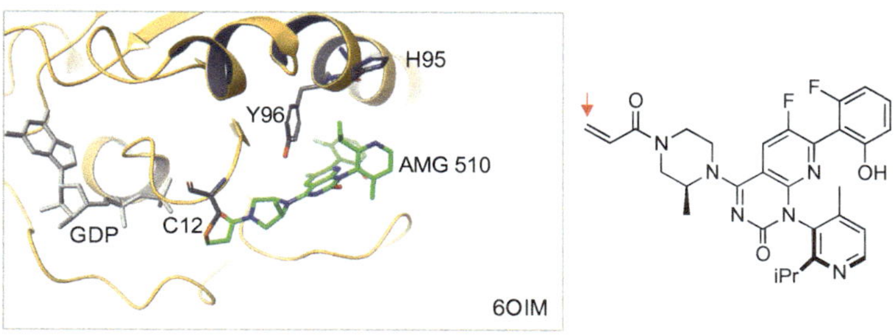

AMG 510

Fig. 14 Crystal structure of cysteine-targeting covalent inhibitor AMG 510 with $KRAS^{G12C}$ showing the proximity of His95 and Y96 to the ligand.[105]

5. Serine, threonine

Covalent modification of catalytic serine and threonine amino acids has been reviewed extensively,[109,110] and so for illustrative purposes the area will be briefly summarized here. Serine hydrolases contain a highly nucleophilic serine residue that participates in a catalytic dyad (Ser-Asp/Ser-Lys) or triad (Ser-His-Asp/Ser-Ser-Lys) which mediates enzyme acylation and is subsequently hydrolyzed to release the catalytic serine.[111] Several clinically approved drugs modify the reactive serine in these enzymes, shown in Fig. 15A. β-Lactam antibiotics inhibit bacterial transpeptidases and β-lactamases,[112] and the β-lactone tetrahydrolipstatin, an anti-obesity drug, covalently modifies the active site serine of gastric and pancreatic lipases.[113] Beyond the strained lactam and lactones, electrophilic centers present in serine hydrolase drugs include carbamate (rivastigmine),[114] α-amino nitrile (saxagliptin),[115] and phenol ester (sivelestat).[116] Covalent proteasome inhibitor drugs react with the catalytic *N*-terminal threonine residue and include the epoxy ketone carfilzomib,[117] bortezomib that utilizes a boronate warhead[118] and marizomib that bears a chloroalkyl β-lactone[119] (Fig. 15B illustrates their fascinating mechanisms of inhibition).

As for kinases described above, reactive chemical reporters have been designed to engage the conserved catalytic residue leading to visualization and identification of labeled proteins allowing binding site occupancy and inhibitor selectivity to be determined. This is the principle behind activity-based protein profiling (ABPP), and in the serine hydrolase area, activity-based probes often utilize the promiscuous fluorophosphonate affinity label.[120,121] Comparative ABPP across different biological samples has influenced target discovery and fundamental biological insights by identifying changes in activity (not necessarily expression) of serine hydrolases involved in cancer, neurology and other diseases.[111,122,123] Changes in activity measured by ABPP has also led to advances in biomarker development with potential clinical applications.[124] Using competition proteomics, serine hydrolase selectivity of many inhibitors has been assessed in whole proteomes leading to the design and development of high-quality chemical probes and drug candidates.[125–127]

Sulfonyl fluoride warheads, as described above, are used increasingly to target reactive lysine and tyrosine residues in protein binding sites. However, some of the earliest chemical biology studies in this area demonstrated labeling of catalytic serines residues in proteases using simple reagents such as

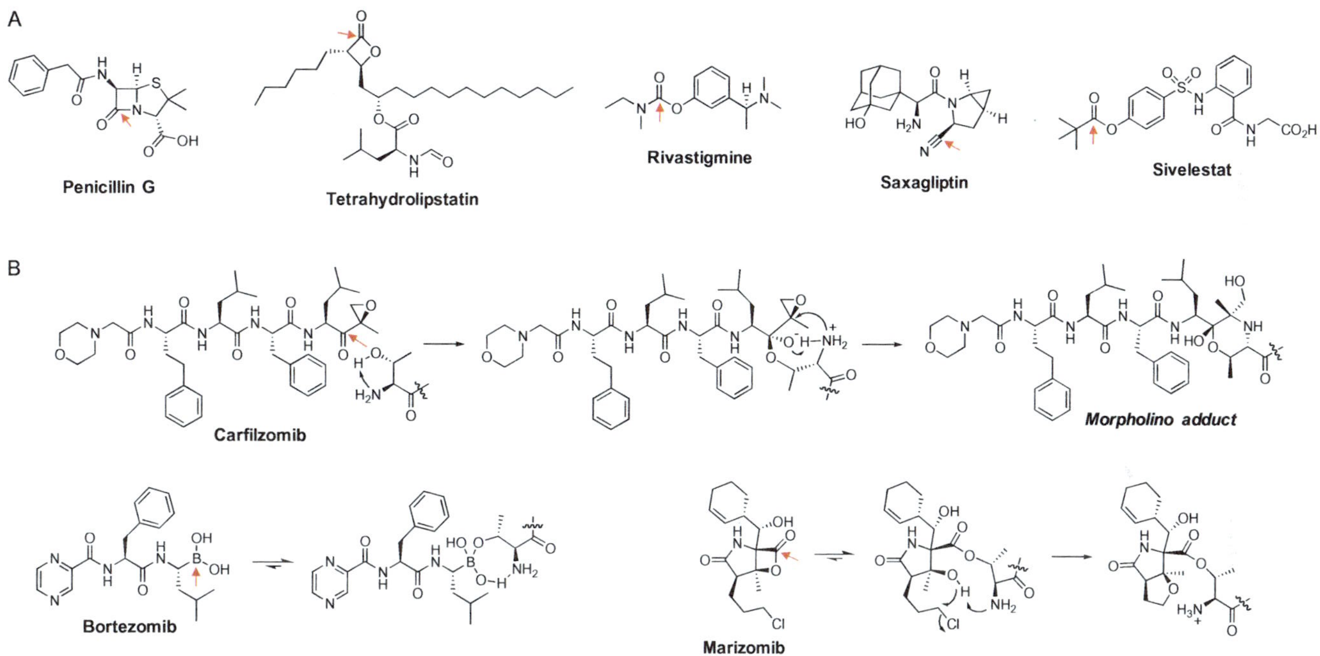

Fig. 15 (A) Examples of targeted covalent inhibitor drugs that engage the catalytic serine residue in serine hydrolase enzymes; (B) targeted covalent inhibitor drugs that react with the catalytic threonine of the proteasome (sites of nucleophilic attack highlighted with red arrow).

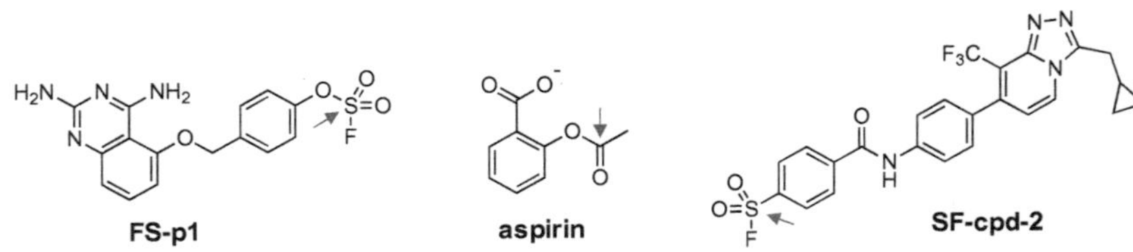

Fig. 16 Covalent modifiers of non-catalytic serine and threonine residues: FS-p1 (Ser272 in DcpS), aspirin (Ser530 in COX1/2), and SF-cpd-2 (Thr791 in $mGlu_2$). Red arrows indicate electrophilic centers.

phenylmethylsulfonyl fluoride.[128] Since then many examples of sulfonyl fluoride probes targeting serine hydrolases and peptidic inhibitors of the proteasome have been reported.[22,129,130]

The fluorosulfate congener FS-p1 (Fig. 16) of the sulfonyl fluoride SF-p1 (Fig. 9) that labels a reactive tyrosine in the DcpS functional site was found to serendipitously engage a non-catalytic serine residue in a different location of the same pocket.[42] The microenvironment of Ser272 is highly electropositive which may reduce the pK_a of the hydroxyl group thus facilitating enzyme labeling. Although the rate of serine modification was slow, further improvements in reversible binding interactions and optimal positioning of the electrophilic center may enhance irreversible inhibition efficacy. Importantly, the high chemical and metabolic stability of FS-p1 confirms the suitability of the fluorosulfate warhead for targeted covalent drug design.[42] The only other reported example of non-catalytic serine modification is the acetylation of cyclooxygenase enzymes COX1 and COX2 by aspirin (Fig. 16).[131] The carboxylate in aspirin forms a salt bridge with Arg120 and hydrogen bonding between Tyr385 and the acetyl group facilitates transacylation to Ser530 near the COX active site, resulting in enzyme inactivation.[132,133] Importantly, the carboxylate also acts as a general base facilitating serine deprotonation. Recently, the likely position of labeling of the class C GPCR $mGlu_2$ by a positive allosteric modulator bearing a sulfonyl fluoride electrophile was determined to be a threonine residue using computational modeling and receptor mutagenesis (Fig. 16).[134]

6. Others (glutamate, aspartate, methionine)

The carboxylic acid side chains of glutamate and aspartate are fully deprotonated at physiological pH, but the carboxylate anion is a poor nucleophile, making site-selective modification challenging. A recent study reported the labeling of Glu88 in the binding site of PDE6δ through the structure-based design of an inhibitor bearing the isoxazolium salt Woodward's reagent K (WRK).[135] The WRK warhead had been used to label carboxylates many years previously[136] and the reaction proceeds via the highly reactive ketenimine intermediate (Fig. 17A). A label-free cellular thermal shift experiment using MS proteomics as a read-out[137] identified a small number of proteins engaged by the covalent PDE6δ inhibitor in cell lysate illustrating surprising selectivity across the proteome. The tandem reaction mechanism, potentially initiated by amino acid-mediated rearrangement of the WRK motif, and templated nucleophilic attack, presumably aids context-dependent reactivity that could be harnessed in the development of other selective covalent modulators.

A clickable photoreactive 2,5-disubstituted tetrazole Tet-yne (Fig. 17B) was used to map aspartate and glutamate residues in the bacterial proteome.[138] UV light activation of tetrazoles yields highly reactive nitrilimines that label many amino acid side chains, including carboxylates.[139] In *Staphylococcus aureus* Tet-yne labeled aspartate and glutamate selectively as demonstrated by MS proteomics (nearly 3000 residues modified). In competition chemoproteomic experiments using Tet-yne, proteins engaged by the isoxazolium fragments Iso-1 and Iso-2 were identified in *S. aureus*. Nitrilimine formation without the need for UV light was achieved using the hydrazonyl chloride HC-1 that eliminates HCl to reveal the electrophile. Similarly, Tet-yne was used to determine the selectivity of HC-1 across the *S. aureus* proteome. A related chemoproteomic platform was developed recently using a clickable 2*H*-azirine probe AZ-9 (Fig. 17C) that chemoselectively labeled carboxylates, although far fewer than Tet-yne, in the breast cancer cell line MCF-7.[140]

Methionine is weakly nucleophilic, unlike hyper-reactive cysteine residues that possess lower pK_a facilitating thiolate anion formation. To address this, sulfur redox chemistry was harnessed to create oxaziridine-based reagents for effective methionine bioconjugation, an approach termed redox-activated chemical tagging (ReACT).[141] In this work the clickable oxaziridine Ox4 (Fig. 18A) was used to map 235 hyper-reactive methionines in cells, and several were known to be functionally relevant in the binding sites of proteins. Further changes to the oxaziridine substituents may enable fine-tuning of protein reactivity, although its intrinsically high electrophilicity will make the development of targeted covalent inhibitors challenging. The rational targeting of methionine in the binding site of

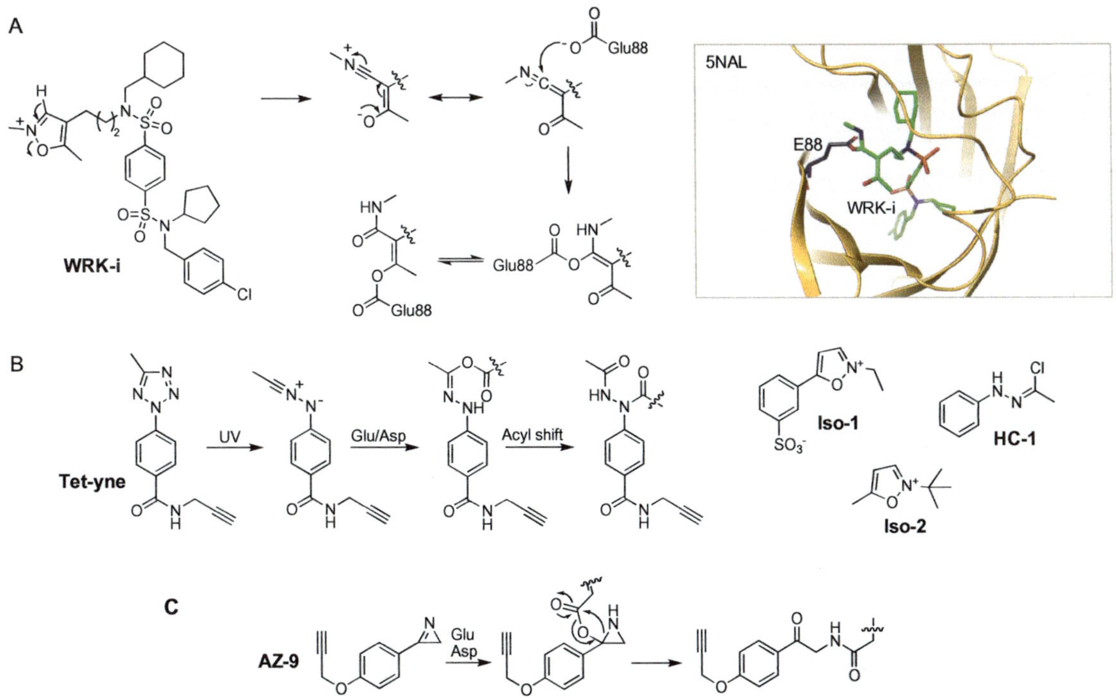

Fig. 17 (A) WRK-i contains the isoxazolium warhead that was designed to label Glu88 in the PDE6δ binding pocket. The reaction proceeds via the ketenimine intermediate which undergoes nucleophilic attack by the carboxylate side chain. (B) 2,5-Substituted tetrazole Tet-yne photoaffinity labels Glu and Asp residues across the proteome and was used to map binding sites of the fragments Iso-1, Iso-2 and HC-1. (C) 2*H*-Azirine AZ-9 was used to map reactive Glu and Asp residues in cells.

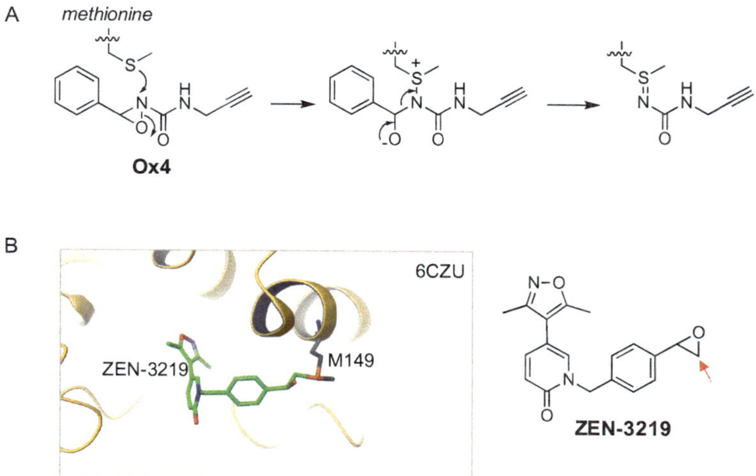

Fig. 18 Labeling of methionine using (A) oxaziridine and (B) epoxide reactive warheads.

BRD4 was achieved recently using an epoxide warhead leading to the development of an irreversible inhibitor (ZEN-3219, Fig. 18B) that forms a sulfonium ion adduct with the amino acid residue.[142]

7. Conclusions and future opportunities

Cysteine targeting has dominated rational inhibitor drug discovery and development, but the residue is rarely present in protein binding sites. This review has highlighted the prospects for rationally targeting amino acid side chains beyond cysteine in a site-selective manner using covalent small molecule inhibitors. The review has also revealed a number of opportunities that are further elucidated and explored below:

- *Chemoproteomic profiling* using clickable electrophilic fragments has unearthed targetable residues beyond cysteine in numerous proteins. Further profiling of additional warheads and broader diversity in the fragment binding element are required to broaden the addressable proteome and expand chemogenomic space.[28] Clickable probes also enable the development of binding site occupancy biomarkers that establish drug-target engagement and selectivity in cells through competition proteomics experiments.[21,34,72] Moreover, clickable fragments could be used as activity-based probes to reveal new context-dependent targetable sites that may emerge in mutant or fusion proteins, following cellular activation or in various disease states.

- *Natural products* continue to serve an important role in covalent drug discovery and several have been highlighted in this review. Further chemoproteomic profiling of reactive natural products that are active in phenotypic screens will identify not only new therapeutic targets but also provide starting points for medicinal chemistry optimization.[143] These studies also match intrinsically biologically relevant electrophilic warheads with functional nucleophilic sites in proteins.
- Covalent protein modulators harnessing neighboring group participation and *tandem reactions* (e.g., aspirin and marizomib) could inspire prospective design strategies that incorporate related mechanisms to enhance site-selective labeling.
- *Structure-based analysis* of chemoproteomic studies and adducted protein crystal structures provide fundamental insights into the microenvironment of reactive amino acids that will influence the rational design of covalent inhibitors that target specific residues.[30,144] A key factor is the perturbation of the pK_a of the nucleophilic side chain that can significantly enhance its reactivity—accurate measurement and calculation of pK_a will thus become increasingly important.[13,16,145] Computational methods such as *covalent docking* and molecular dynamics simulations also facilitate the design of selective inhibitors.[32]
- The prospects for site-selective sulfonyl-exchange labeling of reactive tyrosine or lysine residues in nucleotide-binding domains appears to be particularly promising from recent chemoproteomic studies. Further directed drug discovery follow-up may yield covalent inhibitors of *RNA-binding proteins* (RBPs)[87] that are emerging anticancer targets.[146] For example, selective inhibitors of two different mRNA cap-binding proteins important in cancer (eIF4E[147] and DcpS[148]) were rationally designed to engage a lysine and tyrosine residue in their respective pockets.[32,65]
- Through the application of *SuFEx and SuTEx* chemistry, optimization of on-target pharmacology and proteome selectivity through fine-tuning of electrophilicity and equilibrium binding interactions, while maintaining adequate pharmacokinetics (including high chemical and metabolic stability) would appear to be readily achievable.[22,29,84]
- *Non-catalytic histidine, serine and threonine* side chains seem to be intrinsically more difficult to target than tyrosine or lysine. However, several fortuitous discoveries have shown that numerous electrophiles may engage such residues providing they are suitably reactive (pK_a-perturbed) and reversible ligand binding interactions are favorable. Concerted efforts

targeting these amino acids will likely yield covalent inhibitors that may subsequently lead to rational drug design strategies.

- As for cysteine-targeting covalent kinase inhibitors, the best-suited targets could be those that possess long resynthesis rates, thus enhancing *pharmacodynamic duration* and lessening the need for sustained high concentrations of the covalent drug in vivo so reducing potential off-target modification.[7] Additionally, modification of unique residues in a binding pocket may deliver highly selective inhibitors, a strategy often applied in cysteine-targeted therapeutics development.
- In contrast to the last point, numerous emerging examples illustrate the potential to target catalytic and conserved/essential residues while retaining selectivity across the gene family or related class of enzymes. This becomes particularly important in anti-infectives and anticancer drug development since theoretically these residues will be less likely to mutate than non-catalytic amino acids. Studies should now assess the *mutational profiles* of these covalent inhibitors and compare the propensities for emergence of resistance with cysteine-targeting agents, particularly those targeting non-catalytic residues where this is a major issue.
- Proteolysis targeting chimeras (*PROTACs*) are heterobifunctional small molecules that collocate target proteins with E3 ubiquitin ligases resulting in their ubiquitination and subsequent proteasomal degradation. Some examples have reported effective protein depletion (which is often associated with superior pharmacology compared to inhibitors) using covalent target protein or ligase warheads that engage reactive cysteine residues in binding sites.[149–152] It will be interesting to create and assess the performance of non-cysteine-targeted PROTACs in the future.
- Amelioration of the reactivity of highly electrophilic warheads used in *bioconjugation*, and optimization of reversible binding interactions may deliver more discerning site-selective covalent inhibitors. Conversely, advances in the design of targeted covalent inhibitors may reveal warhead chemistries that are suitable for incorporation into bioconjugation strategies.

The applications of covalent targeting beyond cysteine are too numerous to detail here, but hopefully the examples described above provide some insight as to the future possibilities in this space. Clearly, a multidisciplinary effort is required to make advances in the area, and the breakthroughs that are required to realize its full potential will rely heavily on innovations being made in synthetic and computational chemistry. Moreover, medicinal chemical biology, which continues to evolve as a discipline in the drug discovery setting, will likely lead the charge to deliver a new generation of targeted covalent inhibitors and chemical probes.

Competing interests

L.H.J. receives research funding from Deerfield.

https://orcid.org/0000-0002-8388-5865

References

1. Liu, Q.; Sabnis, Y.; Zhao, Z.; Zhang, T.; Buhrlage, S. J.; Jones, L. H.; Gray, N. S. Developing Irreversible Inhibitors of the Protein Kinase Cysteinome. *Chem. Biol.* **2013**, *20* (2), 146–159.
2. Zhao, Z.; Liu, Q.; Bliven, S.; Xie, L.; Bourne, P. E. Determining Cysteines Available for Covalent Inhibition Across the Human Kinome. *J. Med. Chem.* **2017**, *60* (7), 2879–2889.
3. Ábrányi-Balogh, P.; Petri, L.; Imre, T.; Szijj, P.; Scarpino, A.; Hrast, M.; Mitrović, A.; Fonovič, U. P.; Németh, K.; Barreteau, H.; Roper, D. I.; Horváti, K.; Ferenczy, G. G.; Kos, J.; Ilaš, J.; Gobec, S.; Keserű, G. M. A Road Map for Prioritizing Warheads for Cysteine Targeting Covalent Inhibitors. *Eur. J. Med. Chem.* **2018**, *160*, 94–107.
4. Gehringer, M.; Laufer, S. A. Emerging and Re-Emerging Warheads for Targeted Covalent Inhibitors: Applications in Medicinal Chemistry and Chemical Biology. *J. Med. Chem.* **2019**, *62* (12), 5673–5724.
5. Strelow, J. M. A Perspective on the Kinetics of Covalent and Irreversible Inhibition. *SLAS Discov.* **2017**, *22* (1), 3–20.
6. Daryaee, F.; Zhang, Z.; Gogarty, K. R.; Li, Y.; Merino, J.; Fisher, S. L.; Tonge, P. J. A Quantitative Mechanistic PK/PD Model Directly Connects Btk Target Engagement and In Vivo Efficacy. *Chem. Sci.* **2017**, *8* (5), 3434–3443.
7. Honigberg, L. A.; Smith, A. M.; Sirisawad, M.; Verner, E.; Loury, D.; Chang, B.; Li, S.; Pan, Z.; Thamm, D. H.; Miller, R. A.; Buggy, J. J. The Bruton Tyrosine Kinase Inhibitor PCI-32765 Blocks B-Cell Activation and Is Efficacious in Models of Autoimmune Disease and B-Cell Malignancy. *Proc. Natl. Acad. Sci. U. S. A.* **2010**, *107* (29), 13075–13080.
8. Khazanov, N. A.; Carlson, H. A. Exploring the Composition of Protein-Ligand Binding Sites on a Large Scale. *PLoS Comput. Biol.* **2013**, *9* (11), e1003321.
9. Woyach, J. A.; Furman, R. R.; Liu, T. M.; Ozer, H. G.; Zapatka, M.; Ruppert, A. S.; Xue, L.; Li, D. H.; Steggerda, S. M.; Versele, M.; Dave, S. S.; Zhang, J.; Yilmaz, A. S.; Jaglowski, S. M.; Blum, K. A.; Lozanski, A.; Lozanski, G.; James, D. F.; Barrientos, J. C.; Lichter, P.; Stilgenbauer, S.; Buggy, J. J.; Chang, B. Y.; Johnson, A. J.; Byrd, J. C. Resistance Mechanisms for the Bruton's Tyrosine Kinase Inhibitor Ibrutinib. *N. Engl. J. Med.* **2014**, *370* (24), 2286–2294.
10. Niederst, M. J.; Hu, H.; Mulvey, H. E.; Lockerman, E. L.; Garcia, A. R.; Piotrowska, Z.; Sequist, L. V.; Engelman, J. A. The Allelic Context of the C797S Mutation Acquired Upon Treatment With Third-Generation EGFR Inhibitors Impacts Sensitivity to Subsequent Treatment Strategies. *Clin. Cancer Res.* **2015**, *21* (17), 3924–3933.
11. Jones, L. H. Reactive Chemical Probes: Beyond the Kinase Cysteinome. *Angew. Chem. Int. Ed. Engl.* **2018**, *57*, 9220–9223.
12. Mukherjee, H.; Grimster, N. P. Beyond Cysteine: Recent Developments in the Area of Targeted Covalent Inhibition. *Curr. Opin. Chem. Biol.* **2018**, *44*, 30–38.

13. Pahari, S.; Sun, L.; Alexov, E. PKAD: A Database of Experimentally Measured pKa Values of Ionizable Groups in Proteins. *Database (Oxford)* **2019**, *2019*, baz024.
14. Ishikita, H. Origin of the pKa Shift of the Catalytic Lysine in Acetoacetate Decarboxylase. *FEBS Lett.* **2010**, *584* (15), 3464–3468.
15. Isom, D. G.; Castañeda, C. A.; Cannon, B. R.; García-Moreno, B. Large Shifts in pKa Values of Lysine Residues Buried Inside a Protein. *Proc. Natl. Acad. Sci. U. S. A.* **2011**, *108* (13), 5260–5265.
16. Harris, T. K.; Turner, G. J. Structural Basis of Perturbed pKa Values of Catalytic Groups in Enzyme Active Sites. *IUBMB Life* **2002**, *53* (2), 85–98.
17. Cuesta, A.; Taunton, J. Lysine-Targeted Inhibitors and Chemoproteomic Probes. *Annu. Rev. Biochem.* **2019**, *88*, 365–381.
18. Pettinger, J.; Jones, K.; Cheeseman, M. D. Lysine-Targeting Covalent Inhibitors. *Angew. Chem. Int Ed. Engl.* **2017**, *56* (48), 15200–15209.
19. Patricelli, M. P.; Szardenings, A. K.; Liyanage, M.; Nomanbhoy, T. K.; Wu, M.; Weissig, H.; Aban, A.; Chun, D.; Tanner, S.; Kozarich, J. W. Functional Interrogation of the Kinome Using Nucleotide Acyl Phosphates. *Biochemistry* **2007**, *46* (2), 350–358.
20. Patricelli, M. P.; Nomanbhoy, T. K.; Wu, J.; Brown, H.; Zhou, D.; Zhang, J.; Jagannathan, S.; Aban, A.; Okerberg, E.; Herring, C.; Nordin, B.; Weissig, H.; Yang, Q.; Lee, J. D.; Gray, N. S.; Kozarich, J. W. In Situ Kinase Profiling Reveals Functionally Relevant Properties of Native Kinases. *Chem. Biol.* **2011**, *18* (6), 699–710.
21. Zhao, Q.; Ouyang, X.; Wan, X.; Gajiwala, K. S.; Kath, J. C.; Jones, L. H.; Burlingame, A. L.; Taunton, J. Broad-Spectrum Kinase Profiling in Live Cells With Lysine-Targeted Sulfonyl Fluoride Probes. *J. Am. Chem. Soc.* **2017**, *139* (2), 680–685.
22. Narayanan, A.; Jones, L. H. Sulfonyl Fluorides as Privileged Warheads in Chemical Biology. *Chem. Sci.* **2015**, *6*, 2650–2659.
23. Colman, R. F. Affinity Labeling of Purine Nucleotide Sites in Proteins. *Annu. Rev. Biochem.* **1983**, *52*, 67–91.
24. Gushwa, N. N.; Kang, S.; Chen, J.; Taunton, J. Selective Targeting of Distinct Active Site Nucleophiles by Irreversible SRC-Family Kinase Inhibitors. *J. Am. Chem. Soc.* **2012**, *134* (50), 20214–20217.
25. Jones, L. H. Cell Permeable Affinity- and Activity-Based Probes. *Future Med. Chem.* **2015**, *7* (16), 2131–2141.
26. Rostovtsev, V. V.; Green, L. G.; Fokin, V. V.; Sharpless, K. B. A Stepwise Huisgen Cycloaddition Process: Copper(I)-Catalyzed Regioselective "Ligation" of Azides and Terminal Alkynes. *Angew. Chem. Int. Ed. Engl.* **2002**, *41* (14), 2596–2599.
27. Tornøe, C. W.; Christensen, C.; Meldal, M. Peptidotriazoles on Solid Phase: [1,2,3]-Triazoles by Regiospecific Copper(I)-Catalyzed 1,3-Dipolar Cycloadditions of Terminal Alkynes to Azides. *J. Org. Chem.* **2002**, *67* (9), 3057–3064.
28. Jones, L. H. Expanding Chemogenomic Space Using Chemoproteomics. *Bioorg. Med. Chem.* **2019**, *27* (15), 3451–3453.
29. Dong, J.; Krasnova, L.; Finn, M. G.; Sharpless, K. B. Sulfur(VI) Fluoride Exchange (SuFEx): Another Good Reaction for Click Chemistry. *Angew. Chem. Int. Ed. Engl.* **2014**, *53* (36), 9430–9448.
30. Jones, L. H.; Kelly, J. W. Structure-Based Design and Analysis of SuFEx Chemical Probes. *RSC Med. Chem.* **2020**, *11*, 10–17.
31. Grimster, N. P.; Connelly, S.; Baranczak, A.; Dong, J.; Krasnova, L. B.; Sharpless, K. B.; Powers, E. T.; Wilson, I. A.; Kelly, J. W. Aromatic Sulfonyl Fluorides Covalently Kinetically Stabilize Transthyretin to Prevent Amyloidogenesis While Affording a Fluorescent Conjugate. *J. Am. Chem. Soc.* **2013**, *135* (15), 5656–5668.
32. Wan, X.; Yang, T.; Cuesta, A.; Pang, X.; Balius, T. E.; Irwin, J. J.; Shoichet, B. K.; Taunton, J. Discovery of Lysine-Targeted eIF4E Inhibitors Through Covalent Docking. *J. Am. Chem. Soc.* **2020**, *142* (11), 4960–4964.

33. Fadeyi, O.; Parikh, M. D.; Chen, M. Z.; Kyne, R. E.; Taylor, A. P.; O'Doherty, I.; Kaiser, S. E.; Barbas, S.; Niessen, S.; Shi, M.; Weinrich, S. L.; Kath, J. C.; Jones, L. H.; Robinson, R. P. Chemoselective Preparation of Clickable Aryl Sulfonyl Fluoride Monomers: A Toolbox of Highly Functionalized Intermediates for Chemical Biology Probe Synthesis. *ChemBioChem* **2016**, *17* (20), 1925–1930.
34. Hacker, S. M.; Backus, K. M.; Lazear, M. R.; Forli, S.; Correia, B. E.; Cravatt, B. F. Global Profiling of Lysine Reactivity and Ligandability in the Human Proteome. *Nat. Chem.* **2017**, *9* (12), 1181–1190.
35. Shannon, D. A.; Banerjee, R.; Webster, E. R.; Bak, D. W.; Wang, C.; Weerapana, E. Investigating the Proteome Reactivity and Selectivity of Aryl Halides. *J. Am. Chem. Soc.* **2014**, *136* (9), 3330–3333.
36. Dalton, S. E.; Dittus, L.; Thomas, D. A.; Convery, M. A.; Nunes, J.; Bush, J. T.; Evans, J. P.; Werner, T.; Bantscheff, M.; Murphy, J. A.; Campos, S. Selectively Targeting the Kinome-Conserved Lysine of PI3Kδ as a General Approach to Covalent Kinase Inhibition. *J. Am. Chem. Soc.* **2018**, *140* (3), 932–939.
37. Neuvonen, H.; Neuvonen, K.; Koch, A.; Kleinpeter, E.; Pasanen, P. Electron-Withdrawing Substituents Decrease the Electrophilicity of the Carbonyl Carbon. An Investigation With the Aid of (13)C NMR Chemical Shifts, Nu(C[Double Bond]O) Frequency Values, Charge Densities, and Isodesmic Reactions to Interpret Substituent Effects on Reactivity. *J. Org. Chem.* **2002**, *67* (20), 6995–7003.
38. Choi, S.; Ong, D. S.; Kelly, J. W. A Stilbene That Binds Selectively to Transthyretin in Cells and Remains Dark Until It Undergoes a Chemoselective Reaction to Create a Bright Blue Fluorescent Conjugate. *J. Am. Chem. Soc.* **2010**, *132* (45), 16043–16051.
39. Pettinger, J.; Carter, M.; Jones, K.; Cheeseman, M. D. Kinetic Optimization of Lysine-Targeting Covalent Inhibitors of HSP72. *J. Med. Chem.* **2019**, *62* (24), 11383–11398.
40. Anscombe, E.; Meschini, E.; Mora-Vidal, R.; Martin, M. P.; Staunton, D.; Geitmann, M.; Danielson, U. H.; Stanley, W. A.; Wang, L. Z.; Reuillon, T.; Golding, B. T.; Cano, C.; Newell, D. R.; Noble, M. E. M.; Wedge, S. R.; Endicott, J. A.; Griffin, R. J. Identification and Characterization of an Irreversible Inhibitor of CDK2. *Chem. Biol.* **2015**, *22*, 1159–1164.
41. Pettinger, J.; Le Bihan, Y. V.; Widya, M.; van Montfort, R. L.; Jones, K.; Cheeseman, M. D. An Irreversible Inhibitor of HSP72 That Unexpectedly Targets Lysine-56. *Angew. Chem. Int. Ed. Engl.* **2017**, *56* (13), 3536–3540.
42. Fadeyi, O. O.; Hoth, L. R.; Choi, C.; Feng, X.; Gopalsamy, A.; Hett, E. C.; Kyne, R. E.; Robinson, R. P.; Jones, L. H. Covalent Enzyme Inhibition Through Fluorosulfate Modification of a Noncatalytic Serine Residue. *ACS Chem. Biol.* **2017**, *12* (8), 2015–2020.
43. Gambini, L.; Baggio, C.; Udompholkul, P.; Jossart, J.; Salem, A. F.; Perry, J. J. P.; Pellecchia, M. Covalent Inhibitors of Protein-Protein Interactions Targeting Lysine, Tyrosine, or Histidine Residues. *J. Med. Chem.* **2019**, *62* (11), 5616–5627.
44. Mukherjee, H.; Debreczeni, J.; Breed, J.; Tentarelli, S.; Aquila, B.; Dowling, J. E.; Whitty, A.; Grimster, N. P. A Study of the Reactivity of S(VI)-F Containing Warheads With Nucleophilic Amino-Acid Side Chains Under Physiological Conditions. *Org. Biomol. Chem.* **2017**, *15* (45), 9685–9695.
45. Jones, L. H. Emerging Utility of Fluorosulfate Chemical Probes. *ACS Med. Chem. Lett.* **2018**, *9* (7), 584–586.
46. Baranczak, A.; Liu, Y.; Connelly, S.; Du, W. G.; Greiner, E. R.; Genereux, J. C.; Wiseman, R. L.; Eisele, Y. S.; Bradbury, N. C.; Dong, J.; Noodleman, L.; Sharpless, K. B.; Wilson, I. A.; Encalada, S. E.; Kelly, J. W. A Fluorogenic Aryl Fluorosulfate for Intraorganellar Transthyretin Imaging in Living Cells and in Caenorhabditis elegans. *J. Am. Chem. Soc.* **2015**, *137* (23), 7404–7414.

47. Chen, W.; Dong, J.; Plate, L.; Mortenson, D. E.; Brighty, G. J.; Li, S.; Liu, Y.; Galmozzi, A.; Lee, P. S.; Hulce, J. J.; Cravatt, B. F.; Saez, E.; Powers, E. T.; Wilson, I. A.; Sharpless, K. B.; Kelly, J. W. Arylfluorosulfates Inactivate Intracellular Lipid Binding Protein(s) Through Chemoselective SuFEx Reaction With a Binding Site Tyr Residue. *J. Am. Chem. Soc.* **2016**, *138* (23), 7353–7364.
48. Mortenson, D. E.; Brighty, G. J.; Plate, L.; Bare, G.; Chen, W.; Li, S.; Wang, H.; Cravatt, B. F.; Forli, S.; Powers, E. T.; Sharpless, K. B.; Wilson, I. A.; Kelly, J. W. "Inverse Drug Discovery" Strategy to Identify Proteins That Are Targeted by Latent Electrophiles as Exemplified by Aryl Fluorosulfates. *J. Am. Chem. Soc.* **2018**, *140* (1), 200–210.
49. Baggio, C.; Udompholkul, P.; Gambini, L.; Salem, A. F.; Jossart, J.; Perry, J. J. P.; Pellecchia, M. Aryl-Fluorosulfate-Based Lysine Covalent Pan-Inhibitors of Apoptosis Protein (IAP) Antagonists With Cellular Efficacy. *J. Med. Chem.* **2019**, *62* (20), 9188–9200.
50. Martín-Gago, P.; Olsen, C. A. Arylfluorosulfate-Based Electrophiles for Covalent Protein Labeling: A New Addition to the Arsenal. *Angew. Chem. Int. Ed. Engl.* **2019**, *58* (4), 957–966.
51. Leonard, M. K.; McCorkle, J. R.; Snyder, D. E.; Novak, M.; Zhang, Q.; Shetty, A. C.; Mahurkar, A. A.; Kaetzel, D. M. Identification of a Gene Expression Signature Associated With the Metastasis Suppressor Function of NME1: Prognostic Value in Human Melanoma. *Lab. Invest.* **2018**, *98* (3), 327–338.
52. Wang, Y.; Leonard, M. K.; Snyder, D. E.; Fisher, M. L.; Eckert, R. L.; Kaetzel, D. M. NME1 Drives Expansion of Melanoma Cells With Enhanced Tumor Growth and Metastatic Properties. *Mol. Cancer Res.* **2019**, *17* (8), 1665–1674.
53. Cal, P. M.; Vicente, J. B.; Pires, E.; Coelho, A. V.; Veiros, L. F.; Cordeiro, C.; Gois, P. M. Iminoboronates: A New Strategy for Reversible Protein Modification. *J. Am. Chem. Soc.* **2012**, *134* (24), 10299–10305.
54. Akçay, G.; Belmonte, M. A.; Aquila, B.; Chuaqui, C.; Hird, A. W.; Lamb, M. L.; Rawlins, P. B.; Su, N.; Tentarelli, S.; Grimster, N. P.; Su, Q. Inhibition of Mcl-1 Through Covalent Modification of a Noncatalytic Lysine Side Chain. *Nat. Chem. Biol.* **2016**, *12* (11), 931–936.
55. Zaugg, R. H.; Walder, J. A.; Klotz, I. M. Schiff Base Adducts of Hemoglobin. Modifications That Inhibit Erythrocyte Sickling. *J. Biol. Chem.* **1977**, *252* (23), 8542–8548.
56. Metcalf, B.; Chuang, C.; Dufu, K.; Patel, M. P.; Silva-Garcia, A.; Johnson, C.; Lu, Q.; Partridge, J. R.; Patskovska, L.; Patskovsky, Y.; Almo, S. C.; Jacobson, M. P.; Hua, L.; Xu, Q.; Gwaltney, S. L.; Yee, C.; Harris, J.; Morgan, B. P.; James, J.; Xu, D.; Hutchaleelaha, A.; Paulvannan, K.; Oksenberg, D.; Li, Z. Discovery of GBT440, an Orally Bioavailable R-State Stabilizer of Sickle Cell Hemoglobin. *ACS Med. Chem. Lett.* **2017**, *8* (3), 321–326.
57. Castro-Falcón, G.; Seiler, G. S.; Demir, Ö.; Rathinaswamy, M. K.; Hamelin, D.; Hoffmann, R. M.; Makowski, S. L.; Letzel, A. C.; Field, S. J.; Burke, J. E.; Amaro, R. E.; Hughes, C. C. Neolymphostin A Is a Covalent Phosphoinositide 3-Kinase (PI3K)/Mammalian Target of Rapamycin (mTOR) Dual Inhibitor That Employs an Unusual Electrophilic Vinylogous Ester. *J. Med. Chem.* **2018**, *61* (23), 10463–10472.
58. Brunn, G. J.; Williams, J.; Sabers, C.; Wiederrecht, G.; Lawrence, J. C.; Abraham, R. T. Direct Inhibition of the Signaling Functions of the Mammalian Target of Rapamycin by the Phosphoinositide 3-Kinase Inhibitors, Wortmannin and LY294002. *EMBO J.* **1996**, *15* (19), 5256–5267.
59. Seiler, G. S.; Hughes, C. C. Progress Toward the Total Synthesis of Lymphostins: Preparation of a Functionalized Tetrahydropyrrolo[4,3,2-*de*]Quinoline and Unusual Oxidative Dimerization. *J. Org. Chem.* **2019**, *84* (14), 9339–9343.

60. Miyanaga, A.; Janso, J. E.; McDonald, L.; He, M.; Liu, H.; Barbieri, L.; Eustáquio, A. S.; Fielding, E. N.; Carter, G. T.; Jensen, P. R.; Feng, X.; Leighton, M.; Koehn, F. E.; Moore, B. S. Discovery and Assembly-Line Biosynthesis of the Lymphostin Pyrroloquinoline Alkaloid Family of mTOR Inhibitors in Salinispora Bacteria. *J. Am. Chem. Soc.* **2011**, *133* (34), 13311–13313.
61. Wipf, P.; Halter, R. J. Chemistry and Biology of Wortmannin. *Org. Biomol. Chem.* **2005**, *3* (11), 2053–2061.
62. Howes, A. L.; Chiang, G. G.; Lang, E. S.; Ho, C. B.; Powis, G.; Vuori, K.; Abraham, R. T. The Phosphatidylinositol 3-Kinase Inhibitor, PX-866, Is a Potent Inhibitor of Cancer Cell Motility and Growth in Three-Dimensional Cultures. *Mol. Cancer Ther.* **2007**, *6* (9), 2505–2514.
63. Hanoulle, X.; van Damme, J.; Staes, A.; Martens, L.; Goethals, M.; Vandekerckhove, J.; Gevaert, K. A New Functional, Chemical Proteomics Technology to Identify Purine Nucleotide Binding Sites in Complex Proteomes. *J. Proteome Res.* **2006**, *5*, 3438–3445.
64. Gu, C.; Shannon, D. A.; Colby, T.; Wang, Z.; Shabab, M.; Kumari, S.; Villamor, J. G.; McLaughlin, C. J.; Weerapana, E.; Kaiser, M.; Cravatt, B. F.; van der Hoorn, R. A. Chemical Proteomics With Sulfonyl Fluoride Probes Reveals Selective Labeling of Functional Tyrosines in Glutathione Transferases. *Chem. Biol.* **2013**, *20* (4), 541–548.
65. Hett, E. C.; Xu, H.; Geoghegan, K. F.; Gopalsamy, A.; Kyne, R. E.; Menard, C. A.; Narayanan, A.; Parikh, M. D.; Liu, S.; Roberts, L.; Robinson, R. P.; Tones, M. A.; Jones, L. H. Rational Targeting of Active-Site Tyrosine Residues Using Sulfonyl Fluoride Probes. *ACS Chem. Biol.* **2015**, *10* (4), 1094–1098.
66. Jones, L. H. Understanding the Chemically-Reactive Proteome. *Mol. BioSyst.* **2016**, *12* (6), 1728–1730.
67. Pettigrew, N. E.; Moyer-Myers, M.; Colman, R. F. Affinity Labeling of Pig Lung Glutathione S-Transferase Pi by 4-(Fluorosulfonyl)Benzoic Acid. *Arch. Biochem. Biophys.* **1999**, *364* (1), 107–114.
68. Esch, F. S.; Allison, W. S. Identification of a Tyrosine Residue at a Nucleotide Binding Site in the Beta Subunit of the Mitochondrial ATPase With P-Fluorosulfonyl[14C]-Benzoyl-5′-Adenosine. *J. Biol. Chem.* **1978**, *253* (17), 6100–6106.
69. Manvar, D.; Singh, K.; Pandey, V. N. Affinity Labeling of Hepatitis C Virus Replicase With a Nucleotide Analogue: Identification of Binding Site. *Biochemistry* **2013**, *52* (2), 432–444.
70. Singh, J.; Salcius, M.; Liu, S. W.; Staker, B. L.; Mishra, R.; Thurmond, J.; Michaud, G.; Mattoon, D. R.; Printen, J.; Christensen, J.; Bjornsson, J. M.; Pollok, B. A.; Kiledjian, M.; Stewart, L.; Jarecki, J.; Gurney, M. E. DcpS as a Therapeutic Target for Spinal Muscular Atrophy. *ACS Chem. Biol.* **2008**, *3* (11), 711–722.
71. Thurmond, J.; Butchbach, M. E.; Palomo, M.; Pease, B.; Rao, M.; Bedell, L.; Keyvan, M.; Pai, G.; Mishra, R.; Haraldsson, M.; Andresson, T.; Bragason, G.; Thosteinsdottir, M.; Bjornsson, J. M.; Coovert, D. D.; Burghes, A. H.; Gurney, M. E.; Singh, J. Synthesis and Biological Evaluation of Novel 2,4-Diaminoquinazoline Derivatives as SMN2 Promoter Activators for the Potential Treatment of Spinal Muscular Atrophy. *J. Med. Chem.* **2008**, *51* (3), 449–469.
72. Jones, L. H.; Xu, H.; Fadeyi, O. O. Quantifying Drug-Target Engagement in Live Cells Using Sulfonyl Fluoride Chemical Probes. *Methods Enzymol.* **2019**, *622*, 201–220.
73. Xu, H.; Gopalsamy, A.; Hett, E. C.; Salter, S.; Aulabaugh, A.; Kyne, R. E.; Pierce, B.; Jones, L. H. Cellular Thermal Shift and Clickable Chemical Probe Assays for the Determination of Drug-Target Engagement in Live Cells. *Org. Biomol. Chem.* **2016**, *14* (26), 6179–6183.

74. Gopalsamy, A.; Narayanan, A.; Liu, S.; Parikh, M. D.; Kyne, R. E.; Fadeyi, O.; Tones, M. A.; Cherry, J. J.; Nabhan, J. F.; LaRosa, G.; Petersen, D. N.; Menard, C.; Foley, T. L.; Noell, S.; Ren, Y.; Loria, P. M.; Maglich-Goodwin, J.; Rong, H.; Jones, L. H. Design of Potent mRNA Decapping Scavenger Enzyme (DcpS) Inhibitors With Improved Physicochemical Properties to Investigate the Mechanism of Therapeutic Benefit in Spinal Muscular Atrophy (SMA). *J. Med. Chem.* **2017**, *60* (7), 3094–3108.
75. Cherry, J. J.; DiDonato, C. J.; Androphy, E. J.; Calo, A.; Potter, K.; Custer, S. K.; Du, S.; Foley, T. L.; Gopalsamy, A.; Reedich, E. J.; Gordo, S. M.; Gordon, W.; Hosea, N.; Jones, L. H.; Krizay, D. K.; LaRosa, G.; Li, H.; Mathur, S.; Menard, C. A.; Patel, P.; Ramos-Zayas, R.; Rietz, A.; Rong, H.; Zhang, B.; Tones, M. A. In Vitro and In Vivo Effects of 2,4 Diaminoquinazoline Inhibitors of the Decapping Scavenger Enzyme DcpS: Context-Specific Modulation of SMN Transcript Levels. *PLoS One* **2017**, *12* (9), e0185079.
76. Liu, S.; Dakin, L. A.; Xing, L.; Withka, J. M.; Sahasrabudhe, P. V.; Li, W.; Banker, M. E.; Balbo, P.; Shanker, S.; Chrunyk, B. A.; Guo, Z.; Chen, J. M.; Young, J. A.; Bai, G.; Starr, J. T.; Wright, S. W.; Bussenius, J.; Tan, S.; Gopalsamy, A.; Lefker, B. A.; Vincent, F.; Jones, L. H.; Xu, H.; Hoth, L. R.; Geoghegan, K. F.; Qiu, X.; Bunnage, M. E.; Thorarensen, A. Binding Site Elucidation and Structure Guided Design of Macrocyclic IL-17A Antagonists. *Sci. Rep.* **2016**, *6*, 30859.
77. Hatcher, J. M.; Wu, G.; Zeng, C.; Zhu, J.; Meng, F.; Patel, S.; Wang, W.; Ficarro, S. B.; Leggett, A. L.; Powell, C. E.; Marto, J. A.; Zhang, K.; Ki Ngo, J. C.; Fu, X. D.; Zhang, T.; Gray, N. S. SRPKIN-1: A Covalent SRPK1/2 Inhibitor That Potently Converts VEGF From Pro-Angiogenic to Anti-Angiogenic Isoform. *Cell. Chem. Biol.* **2018**, *25* (4), 460–470.e6.
78. Mukherjee, H.; Su, N.; Belmonte, M. A.; Hargreaves, D.; Patel, J.; Tentarelli, S.; Aquila, B.; Grimster, N. P. Discovery and Optimization of Covalent Bcl-xL Antagonists. *Bioorg. Med. Chem. Lett.* **2019**, *29* (23), 126682.
79. Teng, M.; Ficarro, S. B.; Yoon, H.; Che, J.; Zhou, J.; Fischer, E. S.; Marto, J. A.; Zhang, T.; Gray, N. S. Rationally Designed Covalent BCL6 Inhibitor That Targets a Tyrosine Residue in the Homodimer Interface. *ACS Med. Chem. Lett.* **2020**, *11* (6), 1269–1273.
80. Glukhova, A.; Thal, D. M.; Nguyen, A. T.; Vecchio, E. A.; Jörg, M.; Scammells, P. J.; May, L. T.; Sexton, P. M.; Christopoulos, A. Structure of the Adenosine A_1 Receptor Reveals the Basis for Subtype Selectivity. *Cell* **2017**, *168* (5), 867–877.e13.
81. Yang, X.; van Veldhoven, J. P. D.; Offringa, J.; Kuiper, B. J.; Lenselink, E. B.; Heitman, L. H.; van der Es, D.; IJzerman, A. P. Development of Covalent Ligands for G Protein-Coupled Receptors: A Case for the Human Adenosine A. *J. Med. Chem.* **2019**, *62* (7), 3539–3552.
82. Bum-Erdene, K.; Liu, D.; Gonzalez-Gutierrez, G.; Ghozayel, M. K.; Xu, D.; Meroueh, S. O. Small-Molecule Covalent Bond Formation at Tyrosine Creates a Binding Site and Inhibits Activation of Ral GTPases. *Proc. Natl. Acad. Sci. U. S. A.* **2020**, *117* (13), 7131–7139.
83. Jones, L. H.; Narayanan, A.; Hett, E. C. Understanding and Applying Tyrosine Biochemical Diversity. *Mol. BioSyst.* **2014**, *10*, 952–969.
84. Hahm, H. S.; Toroitich, E. K.; Borne, A. L.; Brulet, J. W.; Libby, A. H.; Yuan, K.; Ware, T. B.; McCloud, R. L.; Ciancone, A. M.; Hsu, K. L. Global Targeting of Functional Tyrosines Using Sulfur-Triazole Exchange Chemistry. *Nat. Chem. Biol.* **2020**, *16* (2), 150–159.
85. Brulet, J. W.; Borne, A. L.; Yuan, K.; Libby, A. H.; Hsu, K. L. Liganding Functional Tyrosine Sites on Proteins Using Sulfur-Triazole Exchange Chemistry. *J. Am. Chem. Soc.* **2020**, *142* (18), 8270–8280.

86. Castello, A.; Fischer, B.; Eichelbaum, K.; Horos, R.; Beckmann, B. M.; Strein, C.; Davey, N. E.; Humphreys, D. T.; Preiss, T.; Steinmetz, L. M.; Krijgsveld, J.; Hentze, M. W. Insights into RNA Biology From an Atlas of Mammalian mRNA-Binding Proteins. *Cell* **2012**, *149* (6), 1393–1406.
87. Castello, A.; Fischer, B.; Frese, C. K.; Horos, R.; Alleaume, A. M.; Foehr, S.; Curk, T.; Krijgsveld, J.; Hentze, M. W. Comprehensive Identification of RNA-Binding Domains in Human Cells. *Mol. Cell* **2016**, *63* (4), 696–710.
88. Crawford, L. A.; Weerapana, E. A Tyrosine-Reactive Irreversible Inhibitor for Glutathione S-Transferase Pi (GSTP1). *Mol. BioSyst.* **2016**, *12* (6), 1768–1771.
89. Bullough, D. A.; Allison, W. S. Three Copies of the Beta Subunit Must be Modified to Achieve Complete Inactivation of the Bovine Mitochondrial F1-ATPase by 5′-P-Fluorosulfonylbenzoyladenosine. *J. Biol. Chem.* **1986**, *261* (13), 5722–5730.
90. Harlow, K. W.; Switzer, R. L. Chemical Modification of Salmonella typhimurium Phosphoribosylpyrophosphate Synthetase With 5′-(P-Fluorosulfonylbenzoyl)Adenosine. Identification of an Active Site Histidine. *J. Biol. Chem.* **1990**, *265* (10), 5487–5493.
91. Li, Q.; Chen, Q.; Klauser, P. C.; Li, M.; Zheng, F.; Wang, N.; Li, X.; Zhang, Q.; Fu, X.; Wang, Q.; Xu, Y.; Wang, L. Developing Covalent Protein Drugs via Proximity-Enabled Reactive Therapeutics. *Cell* **2020**, *182* (1), 85–97.e16.
92. Griffith, E. C.; Su, Z.; Turk, B. E.; Chen, S.; Chang, Y. H.; Wu, Z.; Biemann, K.; Liu, J. O. Methionine Aminopeptidase (Type 2) Is the Common Target for Angiogenesis Inhibitors AGM-1470 and Ovalicin. *Chem. Biol.* **1997**, *4* (6), 461–471.
93. Griffith, E. C.; Su, Z.; Niwayama, S.; Ramsay, C. A.; Chang, Y. H.; Liu, J. O. Molecular Recognition of Angiogenesis Inhibitors Fumagillin and Ovalicin by Methionine Aminopeptidase 2. *Proc. Natl. Acad. Sci. U. S. A.* **1998**, *95* (26), 15183–15188.
94. Liu, S.; Widom, J.; Kemp, C. W.; Crews, C. M.; Clardy, J. Structure of Human Methionine Aminopeptidase-2 Complexed With Fumagillin. *Science* **1998**, *282* (5392), 1324–1327.
95. Evans, M. J.; Saghatelian, A.; Sorensen, E. J.; Cravatt, B. F. Target Discovery in Small-Molecule Cell-Based Screens by In Situ Proteome Reactivity Profiling. *Nat. Biotechnol.* **2005**, *23* (10), 1303–1307.
96. Evans, M. J.; Morris, G. M.; Wu, J.; Olson, A. J.; Sorensen, E. J.; Cravatt, B. F. Mechanistic and Structural Requirements for Active Site Labeling of Phosphoglycerate Mutase by Spiroepoxides. *Mol. BioSyst.* **2007**, *3* (7), 495–506.
97. Maltais, R.; Ayan, D.; Trottier, A.; Barbeau, X.; Lagüe, P.; Bouchard, J. E.; Poirier, D. Discovery of a Non-estrogenic Irreversible Inhibitor of 17β-Hydroxysteroid Dehydrogenase Type 1 From 3-Substituted-16β-(M-Carbamoylbenzyl)-Estradiol Derivatives. *J. Med. Chem.* **2014**, *57* (1), 204–222.
98. Li, T.; Maltais, R.; Poirier, D.; Lin, S. X. Combined Biophysical Chemistry Reveals a New Covalent Inhibitor With a Low-Reactivity Alkyl Halide. *J. Phys. Chem. Lett.* **2018**, *9* (18), 5275–5280.
99. Trottier, A.; Maltais, R.; Ayan, D.; Barbeau, X.; Roy, J.; Perreault, M.; Poulin, R.; Lagüe, P.; Poirier, D. Insight into the Mode of Action and Selectivity of PBRM, a Covalent Steroidal Inhibitor of 17β-Hydroxysteroid Dehydrogenase Type 1. *Biochem. Pharmacol.* **2017**, *144*, 149–161.
100. Yamaguchi, S.; Aldini, G.; Ito, S.; Morishita, N.; Shibata, T.; Vistoli, G.; Carini, M.; Uchida, K. Delta12-Prostaglandin J2 as a Product and Ligand of Human Serum Albumin: Formation of an Unusual Covalent Adduct at His146. *J. Am. Chem. Soc.* **2010**, *132* (2), 824–832.
101. Jakob, C. G.; Upadhyay, A. K.; Donner, P. L.; Nicholl, E.; Addo, S. N.; Qiu, W.; Ling, C.; Gopalakrishnan, S. M.; Torrent, M.; Cepa, S. P.; Shanley, J.; Shoemaker, A. R.; Sun, C. C.; Vasudevan, A.; Woller, K. R.; Shotwell, J. B.;

Shaw, B.; Bian, Z.; Hutti, J. E. Novel Modes of Inhibition of Wild-Type Isocitrate Dehydrogenase 1 (IDH1): Direct Covalent Modification of His315. *J. Med. Chem.* **2018**, *61* (15), 6647–6657.

102. Bradshaw, J. M.; McFarland, J. M.; Paavilainen, V. O.; Bisconte, A.; Tam, D.; Phan, V. T.; Romanov, S.; Finkle, D.; Shu, J.; Patel, V.; Ton, T.; Li, X.; Loughhead, D. G.; Nunn, P. A.; Karr, D. E.; Gerritsen, M. E.; Funk, J. O.; Owens, T. D.; Verner, E.; Brameld, K. A.; Hill, R. J.; Goldstein, D. M.; Taunton, J. Prolonged and Tunable Residence Time Using Reversible Covalent Kinase Inhibitors. *Nat. Chem. Biol.* **2015**, *11* (7), 525–531.
103. Serafimova, I. M.; Pufall, M. A.; Krishnan, S.; Duda, K.; Cohen, M. S.; Maglathlin, R. L.; McFarland, J. M.; Miller, R. M.; Frödin, M.; Taunton, J. Reversible Targeting of Noncatalytic Cysteines With Chemically Tuned Electrophiles. *Nat. Chem. Biol.* **2012**, *8* (5), 471–476.
104. Yoshizawa, M.; Itoh, T.; Hori, T.; Kato, A.; Anami, Y.; Yoshimoto, N.; Yamamoto, K. Identification of the Histidine Residue in Vitamin D Receptor That Covalently Binds to Electrophilic Ligands. *J. Med. Chem.* **2018**, *61* (14), 6339–6349.
105. Lanman, B. A.; Allen, J. R.; Allen, J. G.; Amegadzie, A. K.; Ashton, K. S.; Booker, S. K.; Chen, J. J.; Chen, N.; Frohn, M. J.; Goodman, G.; Kopecky, D. J.; Liu, L.; Lopez, P.; Low, J. D.; Ma, V.; Minatti, A. E.; Nguyen, T. T.; Nishimura, N.; Pickrell, A. J.; Reed, A. B.; Shin, Y.; Siegmund, A. C.; Tamayo, N. A.; Tegley, C. M.; Walton, M. C.; Wang, H. L.; Wurz, R. P.; Xue, M.; Yang, K. C.; Achanta, P.; Bartberger, M. D.; Canon, J.; Hollis, L. S.; McCarter, J. D.; Mohr, C.; Rex, K.; Saiki, A. Y.; San Miguel, T.; Volak, L. P.; Wang, K. H.; Whittington, D. A.; Zech, S. G.; Lipford, J. R.; Cee, V. J. Discovery of a Covalent Inhibitor of KRAS. *J. Med. Chem.* **2020**, *63* (1), 52–65.
106. Fell, J. B.; Fischer, J. P.; Baer, B. R.; Blake, J. F.; Bouhana, K.; Briere, D. M.; Brown, K. D.; Burgess, L. E.; Burns, A. C.; Burkard, M. R.; Chiang, H.; Chicarelli, M. J.; Cook, A. W.; Gaudino, J. J.; Hallin, J.; Hanson, L.; Hartley, D. P.; Hicken, E. J.; Hingorani, G. P.; Hinklin, R. J.; Mejia, M. J.; Olson, P.; Otten, J. N.; Rhodes, S. P.; Rodriguez, M. E.; Savechenkov, P.; Smith, D. J.; Sudhakar, N.; Sullivan, F. X.; Tang, T. P.; Vigers, G. P.; Wollenberg, L.; Christensen, J. G.; Marx, M. A. Identification of the Clinical Development Candidate MRTX849, A Covalent $KRAS^{G12C}$ Inhibitor for the Treatment of Cancer. *J. Med. Chem.* **2020**, *63*, 6679–6693.
107. Zeng, M.; Lu, J.; Li, L.; Feru, F.; Quan, C.; Gero, T. W.; Ficarro, S. B.; Xiong, Y.; Ambrogio, C.; Paranal, R. M.; Catalano, M.; Shao, J.; Wong, K. K.; Marto, J. A.; Fischer, E. S.; Jänne, P. A.; Scott, D. A.; Westover, K. D.; Gray, N. S. Potent and Selective Covalent Quinazoline Inhibitors of KRAS G12C. *Cell Chem. Biol.* **2017**, *24* (8), 1005–1016.e3.
108. https://www.cancer.gov/research/key-initiatives/ras/screens-assays/drug-discovery.
109. Powers, J. C.; Asgian, J. L.; Ekici, O. D.; James, K. E. Irreversible Inhibitors of Serine, Cysteine, and Threonine Proteases. *Chem. Rev.* **2002**, *102* (12), 4639–4750.
110. Bachovchin, D. A.; Cravatt, B. F. The Pharmacological Landscape and Therapeutic Potential of Serine Hydrolases. *Nat. Rev. Drug Discov.* **2012**, *11* (1), 52–68.
111. Long, J. Z.; Cravatt, B. F. The Metabolic Serine Hydrolases and Their Functions in Mammalian Physiology and Disease. *Chem. Rev.* **2011**, *111* (10), 6022–6063.
112. Silver, L. L. Multi-Targeting by Monotherapeutic Antibacterials. *Nat. Rev. Drug Discov.* **2007**, *6* (1), 41–55.
113. Hadváry, P.; Sidler, W.; Meister, W.; Vetter, W.; Wolfer, H. The Lipase Inhibitor Tetrahydrolipstatin Binds Covalently to the Putative Active Site Serine of Pancreatic Lipase. *J. Biol. Chem.* **1991**, *266* (4), 2021–2027.

114. Bar-On, P.; Millard, C. B.; Harel, M.; Dvir, H.; Enz, A.; Sussman, J. L.; Silman, I. Kinetic and Structural Studies on the Interaction of Cholinesterases With the Anti-Alzheimer Drug Rivastigmine. *Biochemistry* **2002**, *41* (11), 3555–3564.
115. Augeri, D. J.; Robl, J. A.; Betebenner, D. A.; Magnin, D. R.; Khanna, A.; Robertson, J. G.; Wang, A.; Simpkins, L. M.; Taunk, P.; Huang, Q.; Han, S. P.; Abboa-Offei, B.; Cap, M.; Xin, L.; Tao, L.; Tozzo, E.; Welzel, G. E.; Egan, D. M.; Marcinkeviciene, J.; Chang, S. Y.; Biller, S. A.; Kirby, M. S.; Parker, R. A.; Hamann, L. G. Discovery and Preclinical Profile of Saxagliptin (BMS-477118): A Highly Potent, Long-Acting, Orally Active Dipeptidyl Peptidase IV Inhibitor for the Treatment of Type 2 Diabetes. *J. Med. Chem.* **2005**, *48* (15), 5025–5037.
116. Nakayama, Y.; Odagaki, Y.; Fujita, S.; Matsuoka, S.; Hamanaka, N.; Nakai, H.; Toda, M. Clarification of Mechanism of Human Sputum Elastase Inhibition by a New Inhibitor, ONO-5046, Using Electrospray Ionization Mass Spectrometry. *Bioorg. Med. Chem. Lett.* **2002**, *12* (17), 2349–2353.
117. Kim, K. B.; Crews, C. M. From Epoxomicin to Carfilzomib: Chemistry, Biology, and Medical Outcomes. *Nat. Prod. Rep.* **2013**, *30* (5), 600–604.
118. Adams, J. Development of the Proteasome Inhibitor PS-341. *Oncologist* **2002**, 7 (1), 9–16.
119. Potts, B. C.; Albitar, M. X.; Anderson, K. C.; Baritaki, S.; Berkers, C.; Bonavida, B.; Chandra, J.; Chauhan, D.; Cusack, J. C.; Fenical, W.; Ghobrial, I. M.; Groll, M.; Jensen, P. R.; Lam, K. S.; Lloyd, G. K.; McBride, W.; McConkey, D. J.; Miller, C. P.; Neuteboom, S. T.; Oki, Y.; Ovaa, H.; Pajonk, F.; Richardson, P. G.; Roccaro, A. M.; Sloss, C. M.; Spear, M. A.; Valashi, E.; Younes, A.; Palladino, M. A. Marizomib, a Proteasome Inhibitor for All Seasons: Preclinical Profile and a Framework for Clinical Trials. *Curr. Cancer Drug Targets* **2011**, *11* (3), 254–284.
120. Cravatt, B. F.; Wright, A. T.; Kozarich, J. W. Activity-Based Protein Profiling: From Enzyme Chemistry to Proteomic Chemistry. *Annu. Rev. Biochem.* **2008**, 77, 383–414.
121. Liu, Y.; Patricelli, M. P.; Cravatt, B. F. Activity-Based Protein Profiling: The Serine Hydrolases. *Proc. Natl. Acad. Sci. U. S. A.* **1999**, *96* (26), 14694–14699.
122. Nomura, D. K.; Dix, M. M.; Cravatt, B. F. Activity-Based Protein Profiling for Biochemical Pathway Discovery in Cancer. *Nat. Rev. Cancer* **2010**, *10* (9), 630–638.
123. Viader, A.; Ogasawara, D.; Joslyn, C. M.; Sanchez-Alavez, M.; Mori, S.; Nguyen, W.; Conti, B.; Cravatt, B. F. A Chemical Proteomic Atlas of Brain Serine Hydrolases Identifies Cell Type-Specific Pathways Regulating Neuroinflammation. *eLife* **2016**, *5*, e12345.
124. Jones, L. H.; Neubert, H. Clinical Chemoproteomics-Opportunities and Obstacles. *Sci. Transl. Med.* **2017**, *9* (386), eaaf7951.
125. Cisar, J. S.; Weber, O. D.; Clapper, J. R.; Blankman, J. L.; Henry, C. L.; Simon, G. M.; Alexander, J. P.; Jones, T. K.; Ezekowitz, R. A. B.; O'Neill, G. P.; Grice, C. A. Identification of ABX-1431, a Selective Inhibitor of Monoacylglycerol Lipase and Clinical Candidate for Treatment of Neurological Disorders. *J. Med. Chem.* **2018**, *61* (20), 9062–9084.
126. Suciu, R. M.; Cognetta, A. B.; Potter, Z. E.; Cravatt, B. F. Selective Irreversible Inhibitors of the Wnt-Deacylating Enzyme NOTUM Developed by Activity-Based Protein Profiling. *ACS Med. Chem. Lett.* **2018**, *9* (6), 563–568.
127. Lentz, C. S.; Sheldon, J. R.; Crawford, L. A.; Cooper, R.; Garland, M.; Amieva, M. R.; Weerapana, E.; Skaar, E. P.; Bogyo, M. Identification of a S. aureus Virulence Factor by Activity-Based Protein Profiling (ABPP). *Nat. Chem. Biol.* **2018**, *14* (6), 609–617.

128. Gold, A. M.; Fahrney, D. Sulfonyl Fluorides as Inhibitors of Esterases. II. Formation and Reactions of Phenylmethanesulfonyl Alpha-Chymotrypsin. *Biochemistry* **1964**, *3*, 783–791.
129. Brouwer, A. J.; Jonker, A.; Werkhoven, P.; Kuo, E.; Li, N.; Gallastegui, N.; Kemmink, J.; Florea, B. I.; Groll, M.; Overkleeft, H. S.; Liskamp, R. M. Peptido Sulfonyl Fluorides as New Powerful Proteasome Inhibitors. *J. Med. Chem.* **2012**, *55* (24), 10995–11003.
130. Zheng, Q.; Woehl, J. L.; Kitamura, S.; Santos-Martins, D.; Smedley, C. J.; Li, G.; Forli, S.; Moses, J. E.; Wolan, D. W.; Sharpless, K. B. SuFEx-Enabled, Agnostic Discovery of Covalent Inhibitors of Human Neutrophil Elastase. *Proc. Natl. Acad. Sci. U. S. A.* **2019**, *116* (38), 18808–18814.
131. Roth, G. J.; Machuga, E. T.; Ozols, J. Isolation and Covalent Structure of the Aspirin-Modified, Active-Site Region of Prostaglandin Synthetase. *Biochemistry* **1983**, *22* (20), 4672–4675.
132. Tosco, P.; Lazzarato, L. Mechanistic Insights Into Cyclooxygenase Irreversible Inactivation by Aspirin. *ChemMedChem* **2009**, *4* (6), 939–945.
133. Lei, J.; Zhou, Y.; Xie, D.; Zhang, Y. Mechanistic Insights Into a Classic Wonder Drug—Aspirin. *J. Am. Chem. Soc.* **2015**, *137* (1), 70–73.
134. Doornbos, M. L. J.; Wang, X.; Vermond, S. C.; Peeters, L.; Pérez-Benito, L.; Trabanco, A. A.; Lavreysen, H.; Cid, J. M.; Heitman, L. H.; Tresadern, G.; IJzerman, A. P. Covalent Allosteric Probe for the Metabotropic Glutamate Receptor 2: Design, Synthesis, and Pharmacological Characterization. *J. Med. Chem.* **2019**, *62* (1), 223–233.
135. Martín-Gago, P.; Fansa, E. K.; Winzker, M.; Murarka, S.; Janning, P.; Schultz-Fademrecht, C.; Baumann, M.; Wittinghofer, A.; Waldmann, H. Covalent Protein Labeling at Glutamic Acids. *Cell Chem. Biol.* **2017**, *24* (5), 589–597.e5.
136. Bodlaender, P.; Feinstein, G.; Shaw, E. The Use of Isoxazolium Salts for Carboxyl Group Modification in Proteins. Trypsin. *Biochemistry* **1969**, *8* (12), 4941–4949.
137. Savitski, M. M.; Reinhard, F. B.; Franken, H.; Werner, T.; Savitski, M. F.; Eberhard, D.; Martinez Molina, D.; Jafari, R.; Dovega, R. B.; Klaeger, S.; Kuster, B.; Nordlund, P.; Bantscheff, M.; Drewes, G. Tracking Cancer Drugs in Living Cells by Thermal Profiling of the Proteome. *Science* **2014**, *346*, 1255784.
138. Bach, K.; Beerkens, B. L. H.; Zanon, P. R. A.; Hacker, S. M. Light-Activatable, 2,5-Disubstituted Tetrazoles for the Proteome-Wide Profiling of Aspartates and Glutamates in Living Bacteria. *ACS Cent. Sci.* **2020**, *6* (4), 546–554.
139. Zhao, S.; Dai, J.; Hu, M.; Liu, C.; Meng, R.; Liu, X.; Wang, C.; Luo, T. Photo-Induced Coupling Reactions of Tetrazoles With Carboxylic Acids in Aqueous Solution: Application in Protein Labelling. *Chem. Commun.* **2016**, *52* (25), 4702–4705.
140. Ma, N.; Hu, J.; Zhang, Z. M.; Liu, W.; Huang, M.; Fan, Y.; Yin, X.; Wang, J.; Ding, K.; Ye, W.; Li, Z. *2H*-Azirine-Based Reagents for Chemoselective Bioconjugation at Carboxyl Residues Inside Live Cells. *J. Am. Chem. Soc.* **2020**, *142* (13), 6051–6059.
141. Lin, S.; Yang, X.; Jia, S.; Weeks, A. M.; Hornsby, M.; Lee, P. S.; Nichiporuk, R. V.; Iavarone, A. T.; Wells, J. A.; Toste, F. D.; Chang, C. J. Redox-Based Reagents for Chemoselective Methionine Bioconjugation. *Science* **2017**, *355* (6325), 597–602.
142. Kharenko, O. A.; Patel, R. G.; Brown, S. D.; Calosing, C.; White, A.; Lakshminarasimhan, D.; Suto, R. K.; Duffy, B. C.; Kitchen, D. B.; McLure, K. G.; Hansen, H. C.; van der Horst, E. H.; Young, P. R. Design and Characterization of Novel Covalent Bromodomain and Extra-Terminal Domain (BET) Inhibitors Targeting a Methionine. *J. Med. Chem.* **2018**, *61* (18), 8202–8211.

143. Wright, M. H.; Sieber, S. A. Chemical Proteomics Approaches for Identifying the Cellular Targets of Natural Products. *Nat. Prod. Rep.* **2016**, *33* (5), 681–708.
144. Lonsdale, R.; Ward, R. A. Structure-Based Design of Targeted Covalent Inhibitors. *Chem. Soc. Rev.* **2018**, *47* (11), 3816–3830.
145. Liu, R.; Yue, Z.; Tsai, C. C.; Shen, J. Assessing Lysine and Cysteine Reactivities for Designing Targeted Covalent Kinase Inhibitors. *J. Am. Chem. Soc.* **2019**, *141* (16), 6553–6560.
146. Mohibi, S.; Chen, X.; Zhang, J. Cancer the 'RBP'eutics-RNA-Binding Proteins as Therapeutic Targets for Cancer. *Pharmacol. Ther.* **2019**, *203*, 107390.
147. Siddiqui, N.; Sonenberg, N. Signalling to eIF4E in Cancer. *Biochem. Soc. Trans.* **2015**, *43* (5), 763–772.
148. Yamauchi, T.; Masuda, T.; Canver, M. C.; Seiler, M.; Semba, Y.; Shboul, M.; Al-Raqad, M.; Maeda, M.; Schoonenberg, V. A. C.; Cole, M. A.; Macias-Trevino, C.; Ishikawa, Y.; Yao, Q.; Nakano, M.; Arai, F.; Orkin, S. H.; Reversade, B.; Buonamici, S.; Pinello, L.; Akashi, K.; Bauer, D. E.; Maeda, T. Genome-Wide CRISPR-Cas9 Screen Identifies Leukemia-Specific Dependence on a Pre-mRNA Metabolic Pathway Regulated by DCPS. *Cancer Cell* **2018**, *33* (3), 386–400.e5.
149. Gabizon, R.; Shraga, A.; Gehrtz, P.; Livnah, E.; Shorer, Y.; Gurwicz, N.; Avram, L.; Unger, T.; Aharoni, H.; Albeck, S.; Brandis, A.; Shulman, Z.; Katz, B. Z.; Herishanu, Y.; London, N. Efficient Targeted Degradation via Reversible and Irreversible Covalent PROTACs. *J. Am. Chem. Soc.* **2020**, *142* (27), 11734–11742.
150. Xue, G.; Chen, J.; Liu, L.; Zhou, D.; Zuo, Y.; Fu, T.; Pan, Z. Protein Degradation Through Covalent Inhibitor-Based PROTACs. *Chem. Commun.* **2020**, *56* (10), 1521–1524.
151. Zhang, X.; Crowley, V. M.; Wucherpfennig, T. G.; Dix, M. M.; Cravatt, B. F. Electrophilic PROTACs That Degrade Nuclear Proteins by Engaging DCAF16. *Nat. Chem. Biol.* **2019**, *15* (7), 737–746.
152. Spradlin, J. N.; Hu, X.; Ward, C. C.; Brittain, S. M.; Jones, M. D.; Ou, L.; To, M.; Proudfoot, A.; Ornelas, E.; Woldegiorgis, M.; Olzmann, J. A.; Bussiere, D. E.; Thomas, J. R.; Tallarico, J. A.; McKenna, J. M.; Schirle, M.; Maimone, T. J.; Nomura, D. K. Harnessing the Anti-Cancer Natural Product Nimbolide for Targeted Protein Degradation. *Nat. Chem. Biol.* **2019**, *15* (7), 747–755.

CHAPTER FIVE

Boron and covalent inhibition

Andrew G. Leach*
Division of Pharmacy and Optometry, University of Manchester, Manchester, United Kingdom
*Corresponding author: e-mail address: andrew.leach@manchester.ac.uk

Contents

Annual Reports in Medicinal Chemistry, Volume 56
ISSN 0065-7743
https://doi.org/10.1016/bs.armc.2020.10.002

Boron has proved itself to be a very useful addition to the field of medicinal chemistry. The first drug to contain boron approved by the FDA was bortezomib in 2003.[1] Subsequently, four further boron-containing compounds have been approved (Fig. 1) and several more are undergoing clinical trials.[2–8] The story of the discovery and design of these compounds has been described elsewhere, and many of the structure–activity relationships that have been established have been reviewed.[2,3,9,10] Some general comments can be made about these drugs that set the context for this chapter. Bortezomib inhibits the proteolytic behavior of the proteasome and is administered as weekly injections whereas Ixazomib has the same target but is administered orally. Tavaborole is a broad spectrum anti-fungal, usually administered topically. Crisaborole is a topical treatment for eczema (atopic dermatitis). Vaborbactam in combination with meripenem is an anti-bacterial; it works by block of β-lactamase activity. The six-membered ring boronic ester in vaborbactam is likely important for selectivity as it is too large to be accommodated in some active sites.[3] Notable failures for boron compounds in clinical studies are also enlightening and include AN3365, which failed due to the development of resistance, talabostat, PHX1766 and delanzomib that failed due to lack of efficacy.[3] Given the indications for which compounds are approved and their routes of administration, and the reasons for failure of others it is highly likely that challenges will involve potency, pharmacokinetics and safety and therefore these are areas that are focused on in this chapter.

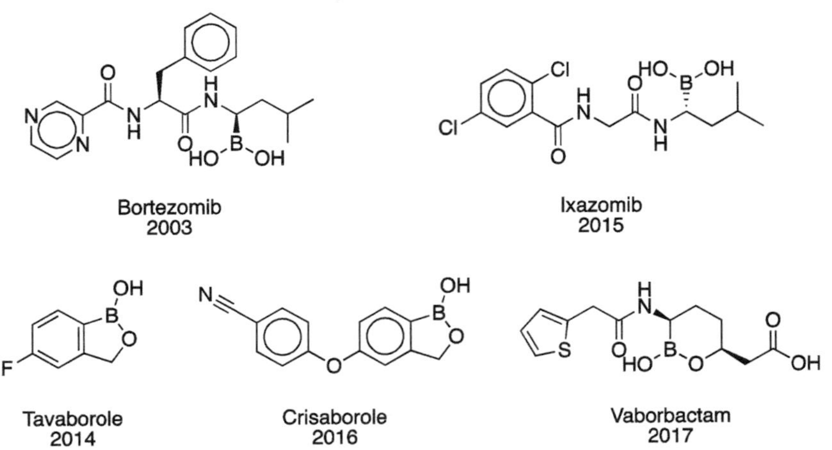

Fig. 1 Approved drugs that contain boron and the year of their first approval by the FDA.

1. General design principles—Covalent and non-covalent approaches

In Hansch and co-workers' compilation of aromatic substituent constants, the boronic acid group has a π value of -0.55, a σ_m of -0.01, σ_p of $+0.12$ and molar refractivity of 11.04.[11] The boronic acid group is polar (similar in magnitude to CN or COMe), weakly electron-withdrawing as a para-substituent (similar to an alkyne group) and similar in terms of steric bulk to COMe or NO_2. This is a rather unusual combination of properties and suggests that regardless of its versatility as a covalent recognition element, the boronic acid group might be able to manipulate physical properties such as logD and pK_a in ways that could be beneficial.

Given the small decrease in logP expected when adding a boronic acid group to a molecule, if the binding were non-covalent there should usually be a small decrease in the strength of binding, in line with the general expectation that lipophilicity drives binding.[12] Those occasions when adding a boronic acid group gives an increase in potency likely indicates that covalent binding has occurred. In order to probe this, all of the molecules in the ChEMBL database that differ only by the addition of a boronic acid group and which have been measured for activity against the same target have been identified; this is a matched molecular pair analysis.[13–16] In the plot in Fig. 2, the IC_{50} for the unsubstituted compound is plotted against that for the corresponding boronic acid (imagine replacing the boronic acid in each of the molecules shown with hydrogen). The solid line that is plotted is $y=x$ and it can be seen that many of the pairs of compounds are clustered around this line (with a tendency to be just below it, as would be expected for a small decrease in logP). However, there are a group of compounds that have surprising increases in potency and this is consistent with these involving covalent interactions. The factors that influence this covalent binding and the types of protein that are amenable are discussed in greater detail in the following sections describing the considerations implied in designing high potency boron-containing covalent compounds.

In non-covalent applications, boron has recently seen frequent use as part of large carborane clusters. These can be replacements for hydrophobic groups such as phenyl rings,[17–21] steroid frameworks,[19,22–25] and adamantanes.[18,19,26] The unusual shape and size of these boron-containing groups could be valuable, but substantial lipophilicity dominates their effect[27–29]; one example sees

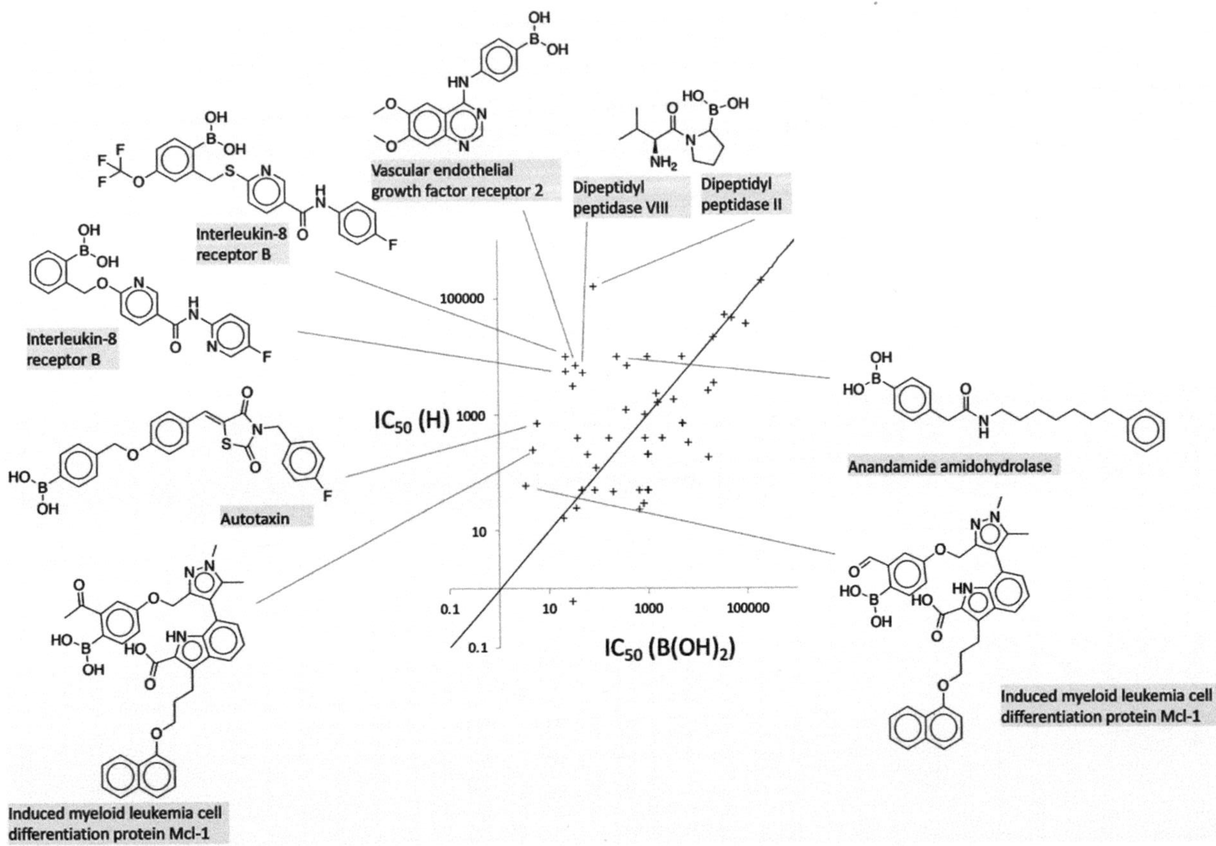

Fig. 2 Matched pairs for the addition of a boronic acid group. The IC_{50} for the unsubstituted compound is shown on the y-axis and that for the boronic acid on the x-axis. The scales are both logarithmic.

the effect of exchanging an adamantane for a carborane framework increasing logP by 0.2.[19] Some of the physical property challenges involved in working with these compounds have been explored.[30] An interesting application of an anionic carborane group has seen it employed as a protonophore that can carry protons through membranes thanks to its neutral and hydrophobic character when protonated.[31] Boron paired with nitrogen can be a replacement for the C=C functionality, particularly in aromatic rings.[29] By changing the electronic properties of such rings, their metabolic liability can be significantly altered and the lipophilicity reduced.[32] Hall and co-workers have highlighted that these B—N containing heterocycles can be unstable in water and have elaborated tactics to include stabilizing groups (such as phenolates that can form an interaction with boron to make a boronate). These intramolecular interactions yield groups that are stable in water and are likely to be a rich seam for future progress—with boron possibly as covalent as well as a non-covalent binding element.[33]

2. Designing for efficacy

A number of the drug discovery programs that have led to approved medicines or clinical candidates have been reviewed and presented previously and will not be described further in this chapter.[2,9] These descriptions generally focus on the efforts to optimize the potency of the compounds—their binding to a primary biological target. When proposing that a molecule binds to a protein in a covalent fashion, one of the most frequently employed methods to validate such a claim is to solve the structure of the protein in complex with the covalent modifier (mass spectrometry and kinetic studies could also be used but are challenging for the reversible covalent complexation that is common with boron). Therefore, a key source concerning the applicability of boron-based covalent compounds is the protein databank (PDB).[34] This repository contains the structures of most published protein–ligand complexes that have been solved by a range of methods (predominantly X-ray crystallography but also NMR and cryo-electron microscopy). In order to probe this valuable resource, all structures that feature a boron-containing ligand were found. The corresponding protein–ligand structures were scrutinized manually to ensure that they involved covalent interaction. This yielded a set of 231 structures. Analysis of the enzyme classes to which these boron compounds bind (Table 1) makes it clear that covalent boron inhibitors have been studied structurally in great detail and frequently for two dominant groups of enzymes: the serine endo peptidases (E.C. class 3.4.21, 65 structures) and the

Table 1 The E.C. classification of each of the PDB structures that involve covalent bonds to boron.

E.C. class	Class	Number	Species
1.3.1.9	Enoyl-[acyl-carrier-protein] reductase	5	*Eschericia coli* (4), *Brassica napus* (1)
2.3.2.2, 3.4.19.13, 3.4.19.14	γ-glutamyltransferase/glutathione γ-glutamate hydrolase/leukotriene-C_4 hydrolase	1	*Homo sapiens* (1)
2.4.1.129	Peptidoglycan glycosyltransferase	11	*Streptococcus pneumoniae* (11)
3.1.1.X	Carboxylic Ester Hydrolases	7	*Lucilia cuprina* (7)
3.1.1.84	Monoterpene ε-lactone hydrolase	1	*Rhodococcus* MB1 (1)
3.1.4.39	Alkylglycerophosphoethanolamine phosphodiesterase	1	*Rattus norvegicus* (1)
3.4.14.X	Dipeptidyl-peptidases and tripeptidyl-peptidases	2	*Porphyromonas gingivalis* (2)
3.4.14.5	Dipeptidyl-peptidase IV	2	*Sus scrofa* (1), *Homo sapiens* (1)
3.4.16.4	Srine-type D-Ala-D-Ala carboxypeptidase	11	*Eschericia coli* (1), *Actinomadura SP* (10)
3.4.21.X	Serine endopeptidases	1	*Homo sapiens* (1)
3.4.21.1	Chymotrypsin	4	*Bos taurus* (4)
3.4.21.4	Trypsin	10	*Sus scrofa* (6), *Bos taurus* (4)
3.4.21.5	Thrombin	8	*Homo sapiens* (8)
3.4.21.12	α-Lytic endopeptidase	23	*Lysobacter enzymogenes* (23)
3.4.21.27	Coagulation factor XIa	4	*Homo sapiens* (4)
3.4.21.36	Pancreatic elastase	1	*Sus scrofa* (1)
3.4.21.53	Endopeptidase La	4	*Meiothermus taiwanensis* (4)
3.4.21.61	Kexin	2	*Saccharomyces cerevisiae* (2)
3.4.21.62	Subtilisin	4	*Bacillus licheniformis* (4)
3.4.21.64	Peptidase K	1	*Engyodontium album* (1)

Table 1 The E.C. classification of each of the PDB structures that involve covalent bonds to boron.—cont'd

E.C. class	Class	Number	Species
3.4.21.92	Endopeptidase Clp	3	*Staphylococcus aureus* (2), *Thermus thermophilus* (1)
3.4.25.1	Proteasome endopeptidase complex	25	*Saccharomyces cerevisiae* (21), *Homo sapiens* (3), *Mycobacterium tuberculosis* (1)
3.5.1.97	Acyl-homoserine-lactone acylase	5	*Pseudomonas aeruginosa* (5)
3.5.2.6	β-lactamase	85	*Eschericia coli* (55), *Acinetobacter baumannii* (12), *Klebsiella pneumoniae* (9), *Pseudomonas aeruginosa* (6), *Enterobacter cloacae* (2), *Stenotrophomonas maltophilia* (1)
3.5.3.1	Arginase	2	*Rattus norvegicus* (1), *Homo sapiens* (1)
6.1.1.4	Leucine-tRNA ligase	1	*Thermus thermophilus* (1)
Not classified		7	

β-lactamases (E.C. class 3.5.2.6, 85 structures). It is also clear that while hydrolases (E.C. class 3.X) dominate the structures, boron has also found utility in the oxidoreductase (E.C. class 1.X), transferase (E.C. class 2.X) and ligase (E.C. class 6.X) areas. The species to which the protein in each complex belongs is also revealing, with most being bacterial proteins; mammalian proteins account for only 17% of the structures. A very major application of boron in medicinal chemistry thus far is in anti-infectives despite the first approved drug being an oncology medicine. As far as is possible, the illustrations of structures provided alongside each section that follows are of the structure with the best resolution that is relevant. The only exceptions is when the best resolution structure involves complicating factors (for instance, co-located glycerol).

Turning to the chemistry involved in the formation of covalent complexes with boron compounds, the set of structures has been inspected to identify key recognition elements. The most important feature is the type of residue with which boron forms covalent bonds. In the structures available, boron is almost exclusively (225/231, 97% of structures) bonded through oxygen atoms to either the protein or a co-factor. Even in those structures that feature a bond to the nitrogen of lysine (4 structures, all in the D-Ala-D-Ala carboxypeptidase class) or histidine (2 structures, both of the chymotrypsin class) there are also boron oxygen bonds present (these are classified as multiple in Fig. 3). It is therefore clear that up to this point in time, boron has been a selective recognition element for hydroxyl-containing residues. The type of residue through which the covalent bond occurs is summarized in Fig. 3 and shows that serine is the most important, accounting for over 80% of the known structures. There is one enzyme class that dominates the structures that bind to boron through threonine, the proteasome endopeptidase complexes, which account for 25 out of 27 structures. Of the oxidoreductases with available structures, all are enoyl reductases that feature a boronic ester with the 2′-hydroxy in the NAD co-factor while there is one ligase structure that features a boronate complex with the diol motif of tRNA. The final covalent recognition type is bonds to metal, that are formed by a boronate complex and are present in only two structures but represent an interesting further class with the potential to be sensitive to pH. Each of these different types of bonding is explored further and subdivided in order to identify and understand the particular environments that are liable to form covalent bonds to boron.

One feature that is present across all of the structures and is likely important for the successful application of boron in medicinal chemistry is that the

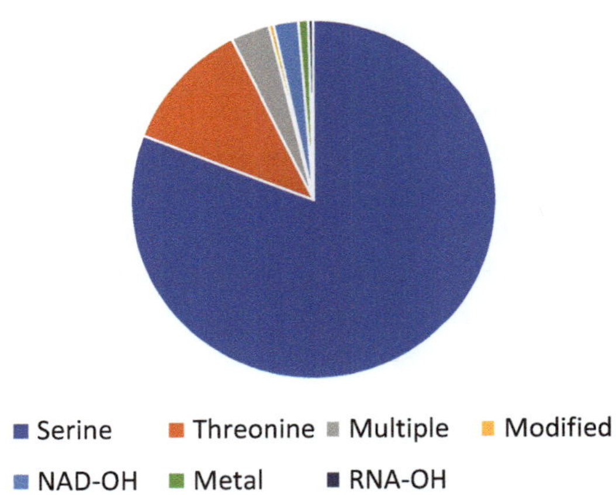

Fig. 3 Distribution of the residue types through which boron forms covalent interactions in protein–ligand complexes.

B-OH or B-OR functionality is able to form a set of hydrogen bonding interactions with a network of water molecules. The flexibility of boron to interact in a range of binding motifs has also been highlighted as a tool that may permit boron-containing compounds to achieve broad inhibition across the classes of β-lactamases permitting many of the β-lactam antibiotics to regain utility.[35] While class B β-lactamases are mono- or di-zinc hydrolases, classes A, C and D have a nucleophilic serine. The geometry at boron in the bound state can be sp^2 or sp^3 and this flexibility might be important. The challenge of achieving desired selectivity patterns among the β-lactamase classes is further complicated by the varying kinetics of forming covalent intermediates.

A general design concept for boron-containing compounds is that boronate complexes with their anionic character and tetrahedral disposition of substituents (usually including two or more bonds to oxygen) are a good mimic for the tetrahedral transition state (and intermediate) involved in the attack of a nucleophile on the amide bond in peptides or proteins. This analogy likely underpins the disproportionate use of boron in the hydrolase class of enzymes that includes the enzymes that cleave amide bonds (including those in β-lactams). A recent study of the mechanism of action of the zika virus NS2B/NS3 serine protease provides a visualization of the geometry of the rate limiting transition state for this enzyme (Fig. 4). The corresponding tetrahedral intermediate (TI) is quite similar in geometry to the transition state. The effectiveness of enzymes generally depends on them stabilizing the transition state of their reaction such that the rate-limiting TS should be structure best stabilized by the protein.[36,37] The structure of the protease transition state is compared to the high resolution (0.9 Å) geometry of a boronate complex of the alpha-lytic protease of Lysobacter enzymogenes in Fig. 4. The comparison reveals that although there are some shared features, notably, the histidine that

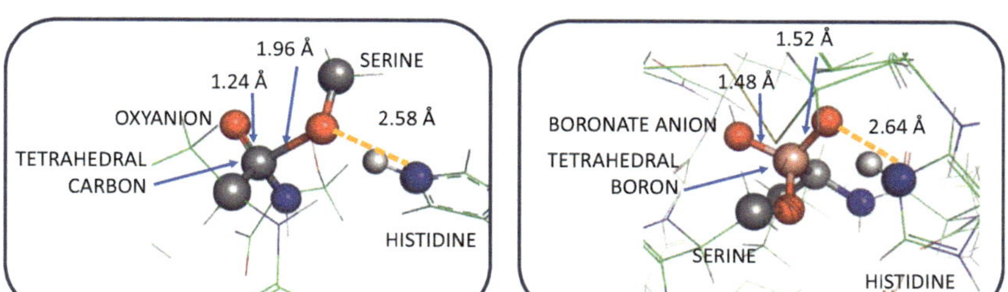

Fig. 4 (Left) Key atoms involved in the calculated rate-limiting transition state for the zika virus NS2b/NS3 serine protease and (right) The analogous atoms in the complex of a serine-bound boronate inhibitor in an alpha-lytic protease.

deprotonates the serine in the transition state exchanges this interaction for a hydrogen bond to the tetrahedral boron. There are also differences. The geometry around the tetrahedral center is generally a little tighter in the boronate (which features full bonds) than in the transition state (with partial bonds). The serine has moved away from its position in the transition state and the network of hydrogen bonds has changed. The B-OH group benefits from the same interactions that stabilize the oxyanion but the OH has the potential to present an unfavorable hydrogen bond donor in this region of the active site. The amide group of the peptide that is cleaved via the transition state is not mimicked directly by the boron-containing compound because a carbon atom intervenes between the tetrahedral center and the nitrogen atom. Although the tetrahedral intermediate is an alluring comparator for a boron-based inhibitor, the limitations of this should be considered and it may be that the most important feature is that these enzymes are able to present an anionic, activated alcohol. This is explored further in the following sections in which the various types of environment where boron forms covalent bonds are explored.

As it seems certain that in most cases interaction of oxygen with boron is preceded by deprotonation of the oxygen atom, the pK_a of different types of boron is a useful design consideration. Those with a lower pK_a have a higher affinity for hydroxide and therefore for deprotonated biological alkoxides (Fig. 5).[38,39] It is likely that the enhanced Lewis acidity of the benzoboroxole arises from ring strain in the five-membered ring that leads to a preference for an sp^3 hybridized atom in the ring.

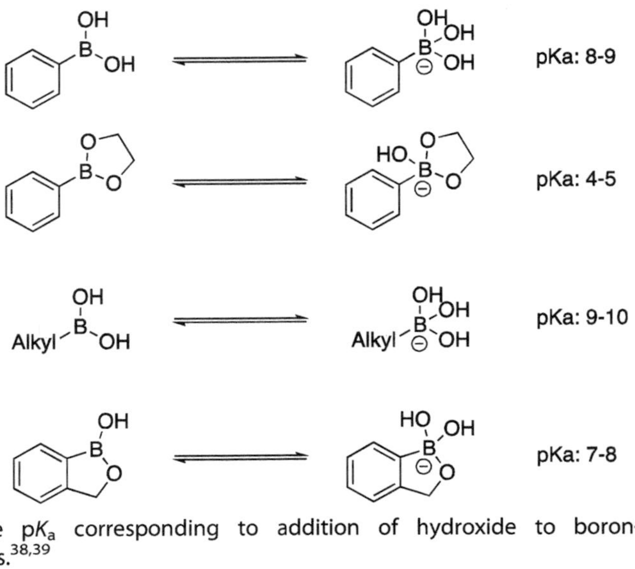

Fig. 5 The pK_a corresponding to addition of hydroxide to boron-containing compounds.[38,39]

2.1 Boron as a bioisostere

Viral proteases are a common target for boron-containing compounds and one strategy that can be considered is to view boron as an alternative to the α-ketoamide functionality.[3] It has also been proposed that boronic acids could be viewed as bioisosteres for aldehydes, nitriles or very electron-poor amides, all of which could potentially be reactive (Fig. 6).[3] A further equivalent that has been proposed is that aromatic boronic acids might be good replacements for phenols although, as noted in the stability section, it is also possible that they are pro-drugs that release the phenol itself.

2.2 Binding to serine

As mentioned above, serine is the most common residue that binds to boron. In this section the residues flanking the serine and the types of boron-containing compounds that bind to serine are considered.

2.2.1 Flanking histidine

The most common residue flanking serine is histidine and enzymes in the available structures that feature this fall into the protease or esterase classes (EC 3.1 and 3.4) and account for 72 of the 187 complexes with serine (39%). This reflects the role of histidine in the catalytic machinery of the enzymes, which tend to involve a nucleophilic serine that requires activation by deprotonation (Fig. 7). In the catalytic process, this deprotonation can either be the rate-limiting step leading to the formation of a high-energy serine alkoxide or, more likely, a concerted deprotonation-addition. The details and geometric requirements may be important for the covalent boron-containing inhibitors which can hijack the same catalytic machinery to form a boron–oxygen bond. The analogy is illustrated in Fig. 7B.

The set of interactions made by the boron and its immediately attached atoms has been examined for each of the 72 structures that have the boron linked to a serine flanked by histidine. Most commonly, the ligands in this class are amino acid mimics in which a boronic acid replaces the C-terminal carboxylic acid. These can be direct mimics of natural amino acids but are more commonly other variants. A representative of the interactions for these

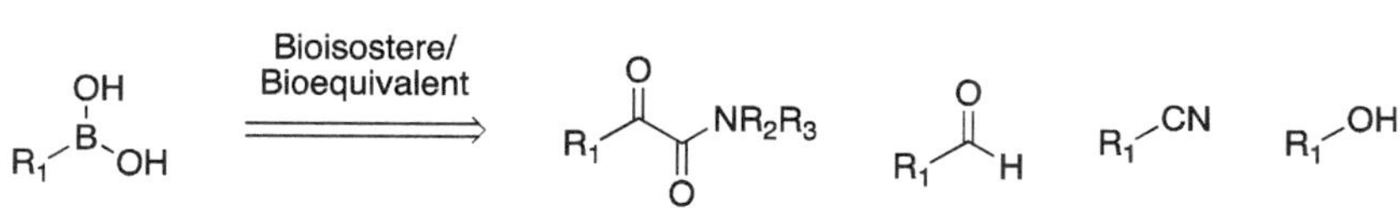

Fig. 6 Bioisosteric replacements with boronic acids.

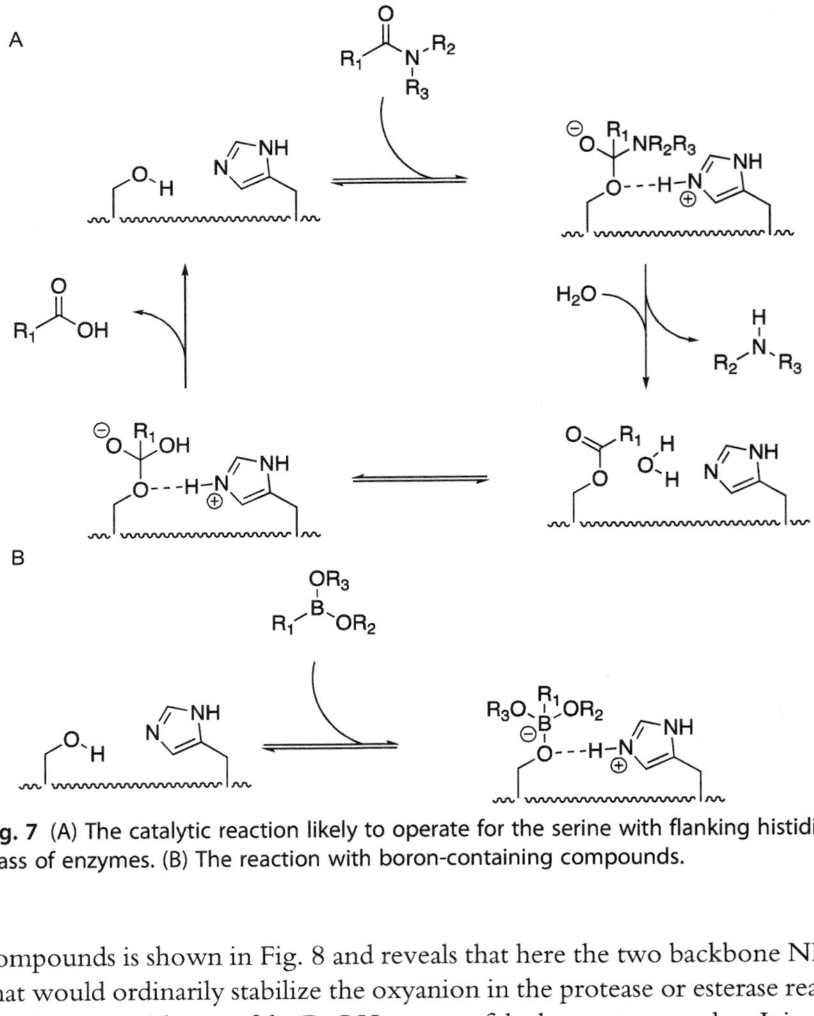

Fig. 7 (A) The catalytic reaction likely to operate for the serine with flanking histidine class of enzymes. (B) The reaction with boron-containing compounds.

compounds is shown in Fig. 8 and reveals that here the two backbone NHs that would ordinarily stabilize the oxyanion in the protease or esterase reaction interact with one of the B-OH groups of the boronate complex. It is not clear what the consequence of this might be for the orientation of the OH but previous studies have revealed the importance of intramolecular hydrogen bonding in boronic acids and boronate so it could be envisaged that the OH is oriented toward the other B-OH.[40,41] This group in turn can act as a hydrogen bond acceptor to the histidine ring NH and can form a hydrogen bond interaction with a water molecule. As discussed above, this second B-OH group has taken the place of the serine OH in the protease transition state and so this group may be surface exposed because the enzyme has not evolved to stabilize such a group. By remaining in contact with water, significant interactions with the enzyme are not essential.

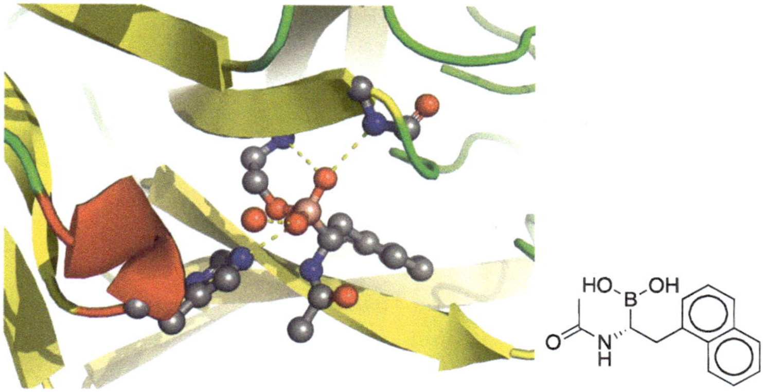

Fig. 8 A boronic acid bound to γ-chymotrypsin (PDB 3vgc).

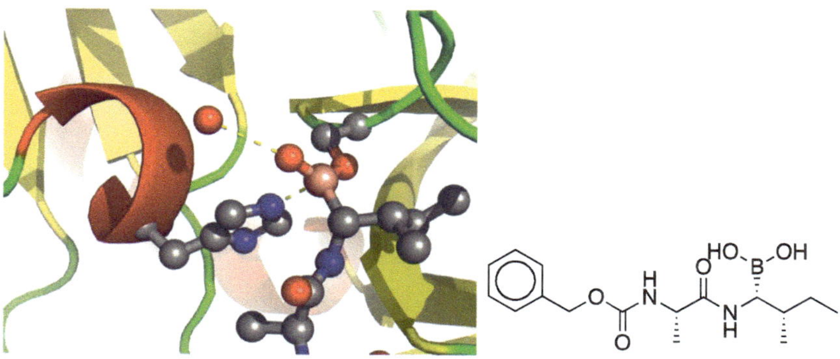

Fig. 9 The boronic ester complex of a boronic acid with porcine pancreatic elastase (PDB 5est).

A small number (7 out of 72) of boronic acids in this group form boronic esters rather than boronate complexes.[42–45] A representative example is shown in Fig. 9 of a peptide mimic bound to porcine pancreatic elastase. The formation of the boronic ester requires the initial formation of the boronate complex followed by loss of one of the ligands, likely an OH group that will remove a proton as it departs. This is analogous to proceeding further through the catalytic cycle for the protease activity, shown in Fig. 7A, where the acylated serine is formed. Although it is not clear whether the origin of the different behavior for certain boron-inhibitor/enzyme combinations is kinetic or thermodynamic, the driving force could plausibly be the formation of a histidine–boron Lewis acid-Lewis base interaction. The boronic ester that forms with serine is a stronger Lewis acid than the precursor boronic acid (Fig. 5). As with the boronate, the boronic ester forms a hydrogen bond with a water molecule but because it lacks the other oxygen and is not anionic, it does not form any other hydrogen bonds.

An alternative mode of hydrogen bonding is with the sidechain amide NH of an asparagine (or potentially glutamine) and this features in 7 of the 72 complexes with a flanking histidine. An example is shown in Fig. 10 of the complex of a boronic acid with subtilisin Carlsberg serine protease. Hydrogen bonding with the sidechain of a tyrosine is also possible and this is present in 5 out of 72 of these complexes. An example is shown in Fig. 11 of a complex with bacterial cocaine esterase. This involves the two boronate oxygens protruding into the binding site and making hydrogen bonds with only one backbone NH as well as the tyrosine OH. This could be a hydrogen bond in either direction (i.e., either Tyr-OH···OH-B or B-OH···OH-Tyr). Given that the boronate oxygen is likely to be a superior hydrogen bond acceptor, and weaker hydrogen bond donor due to the adjacent negative charge, it is likely that the Tyr-OH···OH-B is relevant in this case.

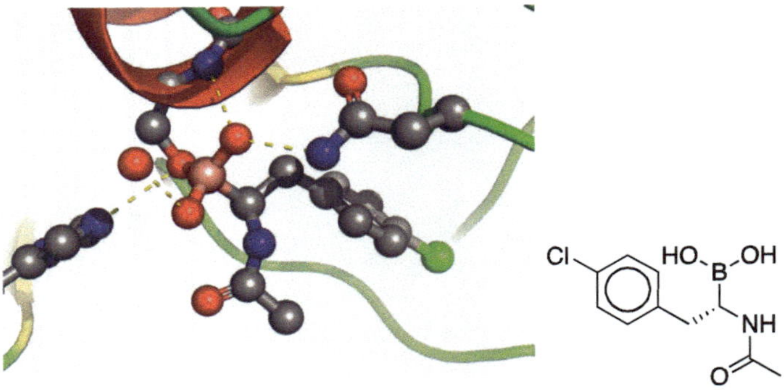

Fig. 10 The boronate complex of a boronic acid with subtilisin Carlsberg serine protease (PDB 1avt).

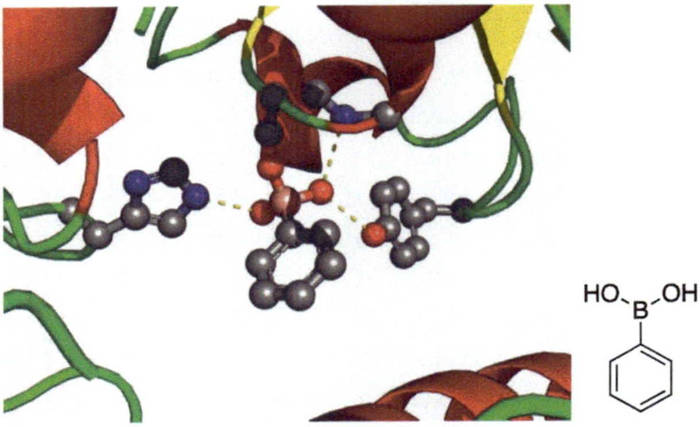

Fig. 11 The complex of phenyl boronic acid with bacterial cocaine esterase (PDB 1ju3).

In addition to the complexes of boronic acids, the structures in this class also include some borate esters $B(OR)_3$ and cyclic boronic esters (Fig. 12). As noted above, the boronic esters have enhanced Lewis acidity compared to the equivalent boronic acid but both boronic and borate esters are potentially unstable with regards to hydrolysis.

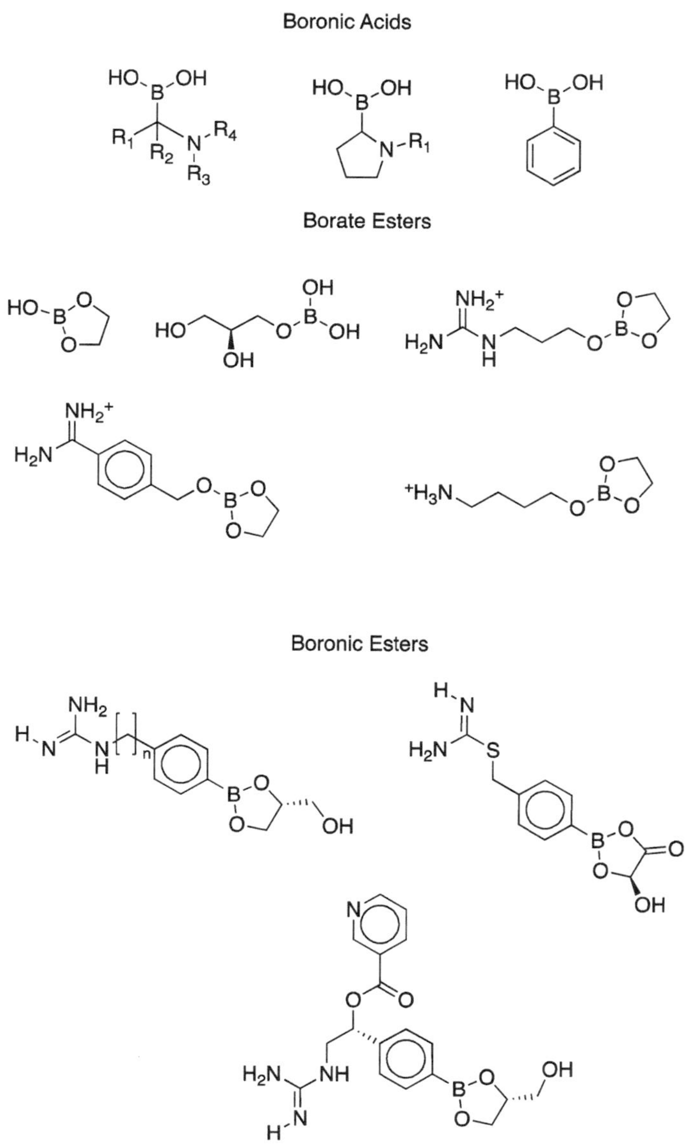

Fig. 12 Ligand types that make interactions with enzymes in the class presenting serine with a flanking histidine.

2.2.2 Flanking lysine/tyrosine pair

The next largest group of 45 complexes (24%) formed with serine feature a pair of flanking residues: lysine and tyrosine. These are exclusively β-lactamase enzymes. The role of these two residues in the catalysis caused by β-lactamases has been a subject of debate, informed by a range of studies including mutagenesis and structural studies.[46] The most likely role of these residues is that neutral lysine and neutral tyrosine between them are able to function as the acid and base catalysts necessary for the serine to be activated as an anion and then to protonate the amine as the amide cleaves (Fig. 13). Subsequently, they are able

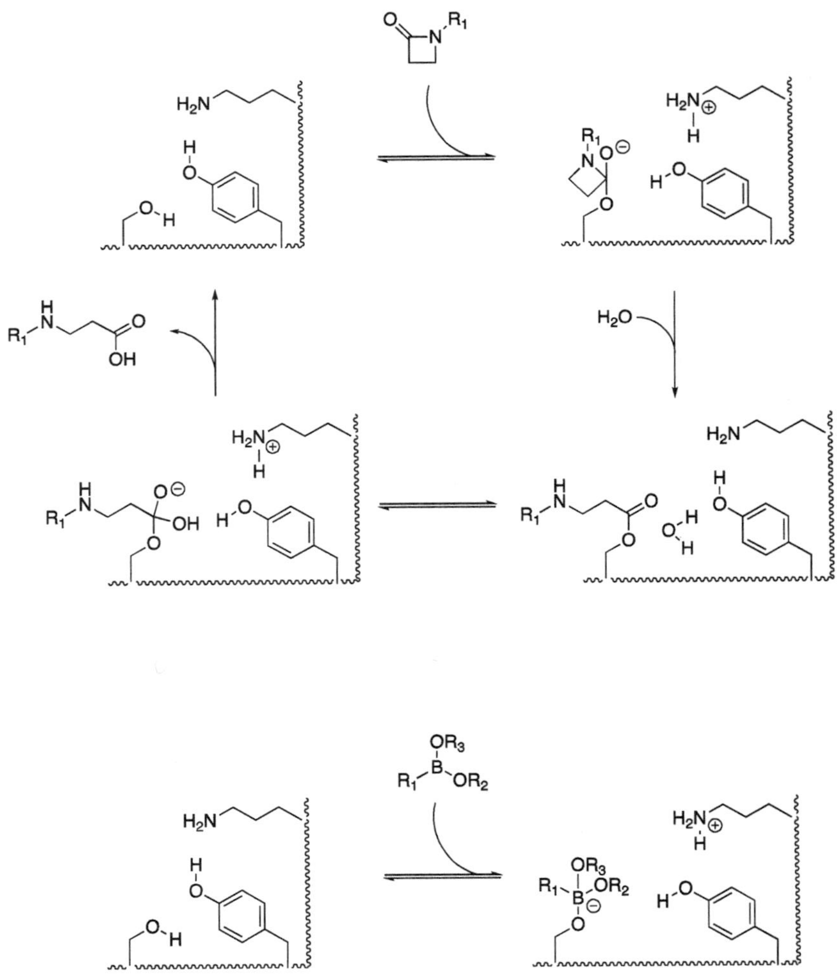

Fig. 13 (A) The catalytic reaction likely to operate for the serine with flanking Lysine/Tyrosine class of enzymes. (B) The reaction with boron-containing compounds.

to activate a water molecule that hydrolyses the serine–ester complex that forms. Understanding the reaction mechanism of the enzyme is important when designing boron-based inhibitors which must harness this reactivity in order to form the covalent B—O bond. The mechanism reveals that in this case, there is likely to be a concerted deprotonation of the serine by the adjacent tyrosine in the presence of an electrophile like the β-lactam or an appropriate boron-based inhibitor. As with the flanking histidine examples in Section 2.2.1, the boron compounds benefit from the stabilizing effect of the enzyme, which must stabilize transition states like those for the addition of either β-lactam or water, in both cases leading to tetrahedral anionic intermediates.

In this system, the other recognition elements between the enzyme and the boron-based inhibitor are almost universally an interaction with both a backbone carbonyl and backbone NH. This is illustrated in Fig. 14 with an example of a benzoxaborole in which (following addition of the serine alkoxide) the boronate acts as a hydrogen bond acceptor interacting with the tyrosine OH donor. The exocyclic OH interacts with the backbone NH and carbonyl of the serine residue making the two hydrogen bonds shown. This may reflect that whereas in the enzymes with an adjacent histidine, stabilizing the oxyanion is critical, for these enzymes, promoting the addition of water is the key role of the enzyme. The enhanced intrinsic reactivity of β-lactams as compared to amides in proteins may mean that formation of the acylated serine progresses more easily such that hydrolysis is rate limiting.

The type of boron containing compounds reported as inhibitors of this group of enzymes is also dominated by the amino acid analogues of alkyl boronic acids with an alpha-amino group (Fig. 15). There are also aromatic boronic acids including benzothiophenes. A number of structures featuring the cyclic boronic esters with either five- or six-membered rings are also present for this class possibly reflecting the nature of the binding site that must be occupied by the cyclic β-lactam. In a quest to discover β-lactamase

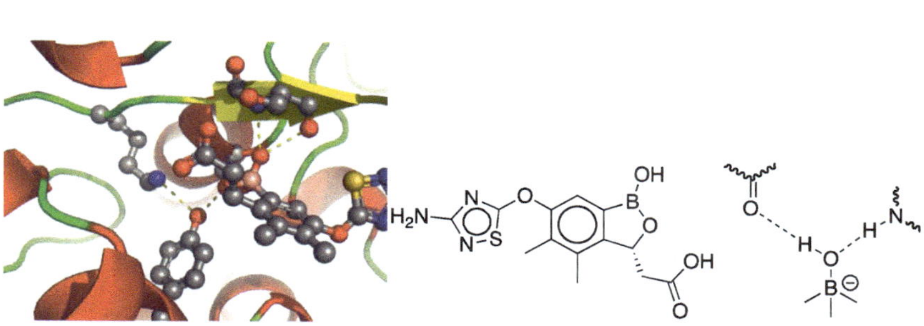

Fig. 14 The boronate complex of a benzoboroxole with the β-lactamase AmpC (PDB 4wz4). The likely hydrogen-bonding network is shown.

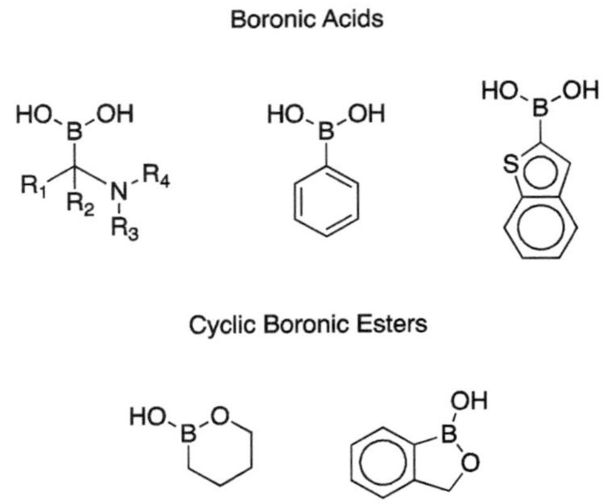

Fig. 15 Ligand types that make interactions with enzymes in the class presenting serine with a flanking Lysine/Tyrosine.

inhibitors with broad-spectrum activity, the ligand in Fig. 14 was found to bind to enzymes in classes A, C and D (spanning the serine β-lactamases).[47] It binds both in the mode described here and in the mode below in which an adjacent lysine has trapped carbon dioxide (Section 2.2.5) and this flexibility contributes to the wide activity.

2.2.3 Flanking lysine/serine pair

This is the third largest group of complexes with covalent bonds to serine and accounts for 35 (19%) of the complexes with serine. Enzymes in this category include penicillin-binding protein 1B (of the class of peptidoglycan glycosyltransferases that are key parts of cell wall synthesis) and β-lactamases. Studies of the mechanism of action of the enzymes in this class suggest a role for the flanking serine that differs to that of the tyrosine in the section above (Fig. 16).[48] Rather than acting as the residue that accepts the proton from the nucleophilic serine, the flanking serine plays an auxiliary role stabilizing the lysine in its deprotonated form (a neighboring glutamate accepts the proton). The flanking serine likely protonates the ring nitrogen as the C—N bond breaks and in turn the nascent serine alkoxide deprotonates the lysine.

The mechanism has different implications for the design of inhibitors as compared to the systems above. In particular, the association between the boronate and protonated lysine is more intimate and one of the groups attached to boron should be able to replicate the interactions that the ring nitrogen makes with the flanking serine in the reaction mechanism. This is reflected in the structures, with the majority featuring networks like those shown in Fig. 17 for a very high resolution (0.85 Å) structure of the complex of a boronic acid with TEM-1 β-lactamase.[49] In line with the proposed mechanism,

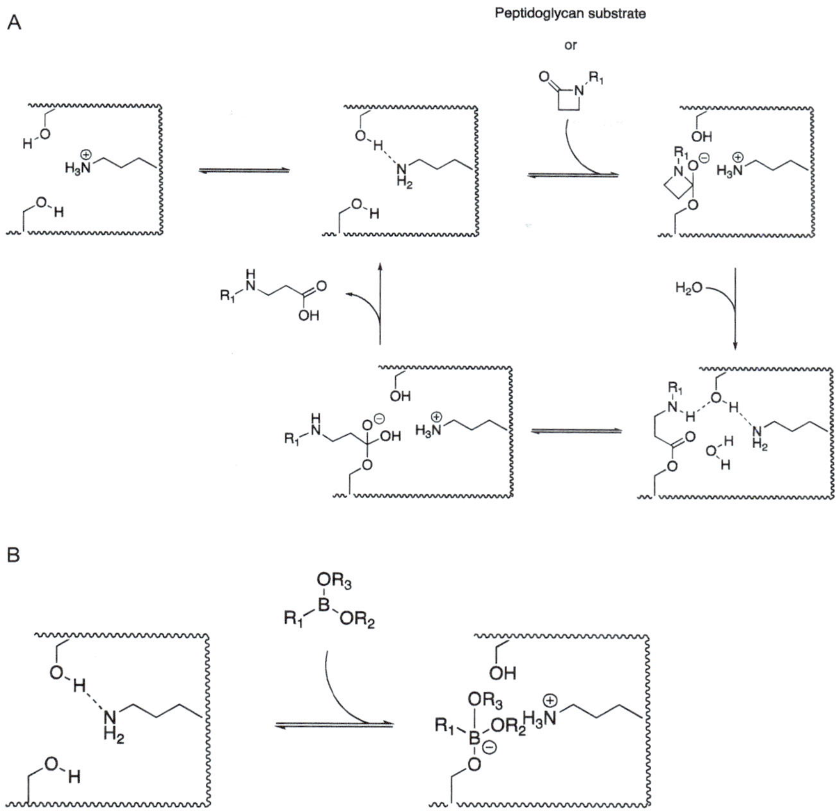

Fig. 16 (A) The catalytic reaction likely to operate for the serine with flanking Lysine/Serine class of enzymes. (B) The reaction with boron-containing compounds.

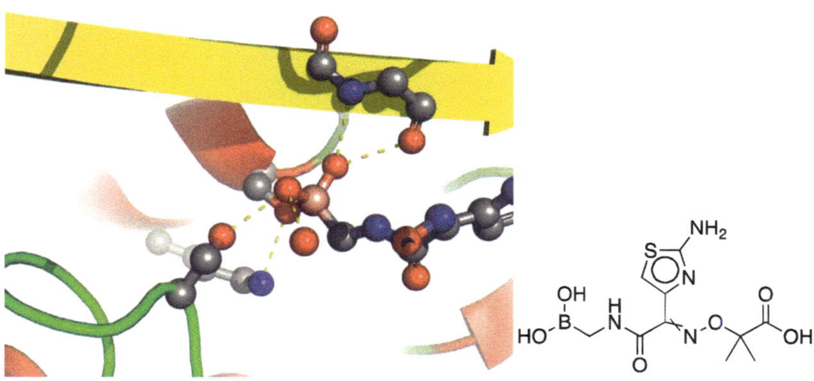

Fig. 17 Structure of a boronate complex of a boronic acid with TEM-1 β-lactamase (PDB 1m40).

the boronic acid has one B-OH group that makes the same set of interactions observed for the inhibitor in Fig. 14 (which features a flanking tyrosine) but the second B-OH (which can be thought of as taking the place of the ring nitrogen in a β-lactam) makes a hydrogen bond with the flanking serine and with a nearby water molecule. The flanking lysine associates with the boronate through the serine oxygen, having presumably accepted the proton that was lost during formation of the boronate. To date, the number of types of boron-containing compounds that have been reported in covalent structures with this group of enzymes is fairly limited: alkyl and aromatic boronic acids and a six-membered cyclic boronic ester (Fig. 18).

In some of the structures in this class, the boronate makes a set of interactions that go beyond those described so far. By rotating into the active site, the boronate can insert a B-OH into the hydrogen-bonding network holding a glutamate and asparagine sidechain together. In the structure shown in Fig. 19, this network is shown as placing the B-OH between two oxygen atoms. Ordinarily, this would involve two hydrogen bond acceptors being oriented toward the OH and make it challenging to avoid a repulsive interaction. However, given that the glutamate is likely to be the base that deprotonates the lysine in the mechanism shown in Fig. 16A, two alternative networks shown are possible and it would be challenging to distinguish between these three structures using X-ray crystallography. Again, the interactions reflect the reaction mechanism, given that the loss of a proton from lysine is required to initiate the catalytic cycle.

2.2.4 Flanking lysine

There is also a group of structures that have a lysine alone in contact with the boronate in their covalent complexes; some of these are a subset of the two

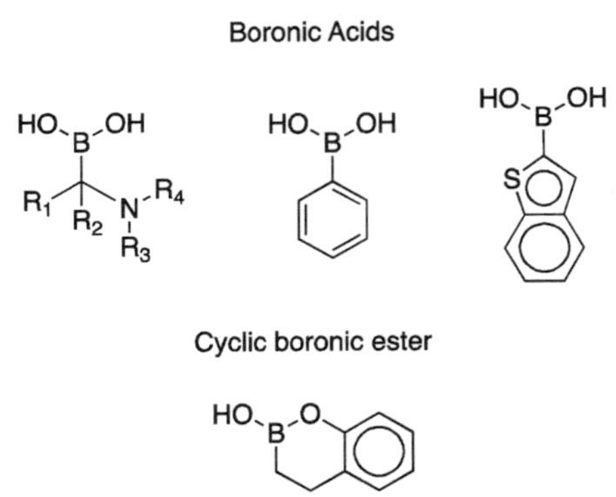

Fig. 18 Ligand types that make interactions with enzymes in the class presenting serine with a flanking Lysine/Serine.

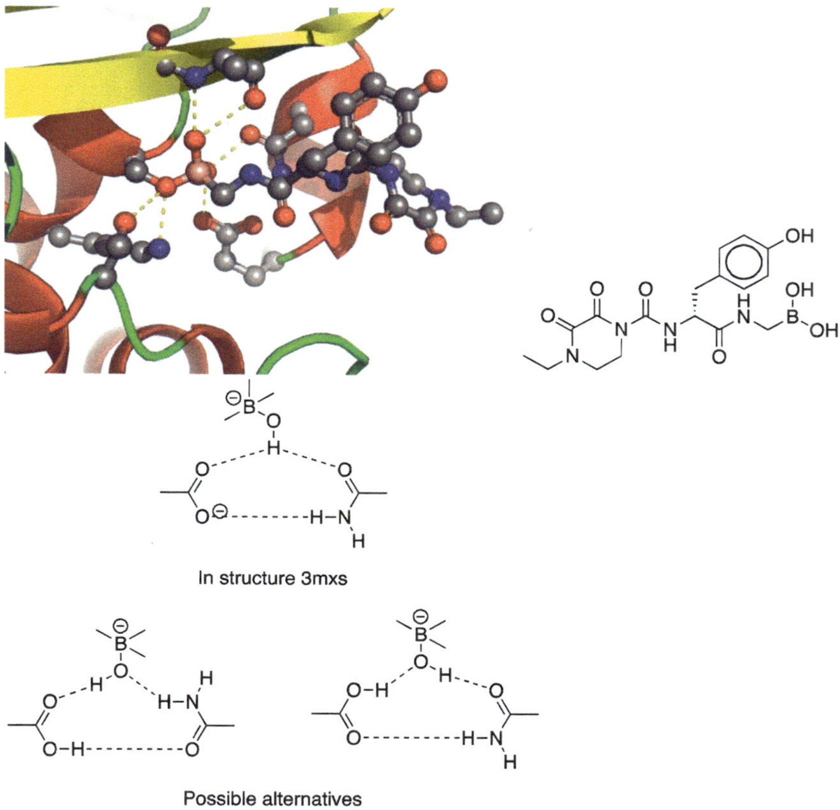

Fig. 19 Structure of a boronate complex of a boronic acid with SHV-1 β-lactamase (PDB 3mxs). The structure proposed requires the suboptimal set of interactions shown but an alternative interpretation of the electron density would support the alternatives shown.

groups above but where the complex does not involve intimate interaction with the tyrosine or serine. This group represents 20 (11%) of the structures that have a covalent bond to serine. Enzymes in this group include further β-lactamases and penicillin-binding proteins, this time of the serine-type D-Ala-D-Ala carboxypeptidase class, another enzyme type responsible for cell wall biosynthesis in bacteria. There are also 3 structures in this group in the EC class 3.4.21.53 of Endopeptidase La enzymes. These are ATP-dependent serine proteases and are LonA and Lon-like proteases and ATP may be involved in driving the process of moving unfolded proteins into the active site.[50] LonC does not have ATP-ase activity and is present at elevated levels in certain cancer types and the proteasome inhibitor bortezomib has been found to inhibit its protease activity. It has been speculated that the lysine in this case is deprotonated in the resting state because of the dipoles of two of the proximal helices. It is likely that the β-lactamases and carboxypeptidases follow a similar mechanism (Fig. 20) and also have a lysine that is either fully or partially deprotonated in the resting state, requiring a very particular environment for that lysine (such as a number of adjacent positively charged residues and/or a proximal aspartate or glutamate that can accept the lysine's proton).

In this group, the other interactions that the boronate forms depend on the enzyme type. For some of the β-lactamases, interactions (Fig. 21) feature the insertion into an asparagine/glutamic acid hydrogen bond, similar to that

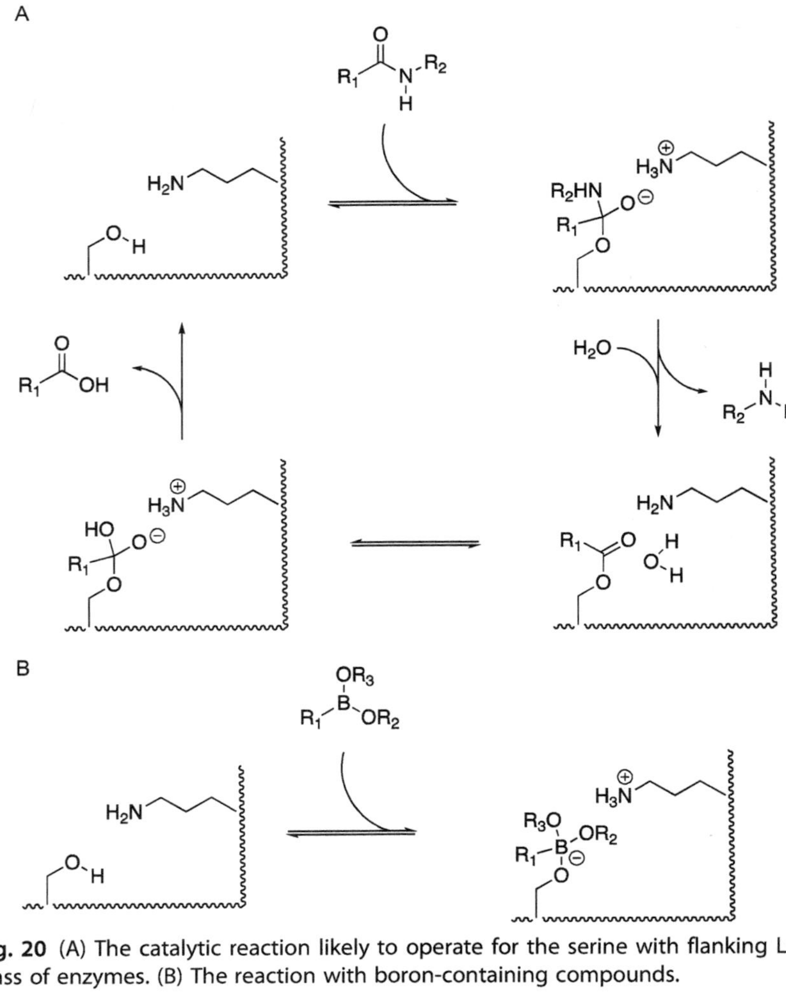

Fig. 20 (A) The catalytic reaction likely to operate for the serine with flanking Lysine class of enzymes. (B) The reaction with boron-containing compounds.

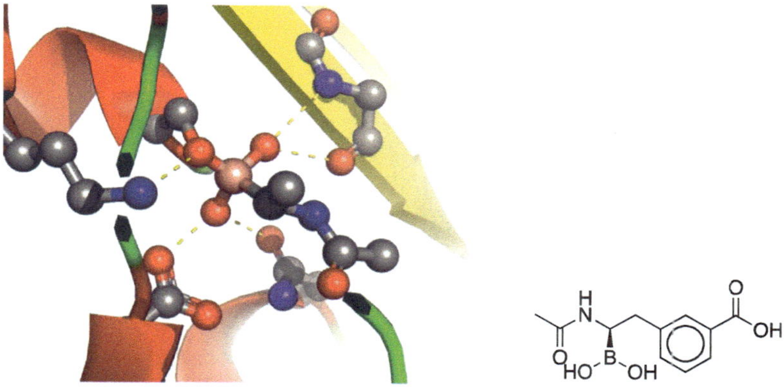

Fig. 21 Boronate complex of a boronic acid with TEM-1 β-lactamase (PDB 1erm). The ligand's aromatic ring is deleted for clarity.

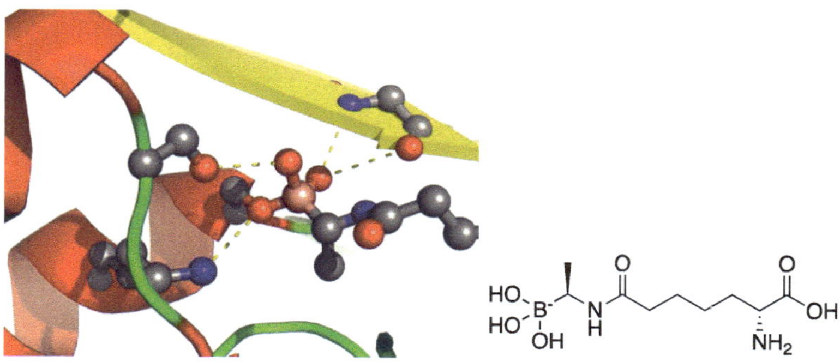

Fig. 22 Boronate complex of a boronic acid with Actinomadura R39 DD peptidase (PDB 2xdm).

noted in the previous section. It is likely that the glutamic acid is protonated and the details of the interaction carry some uncertainty, as noted in the previous section. The lysine in this case associates with the serine oxygen, supporting its role as the base for the serine OH. Other β-lactamases in this group lack the interaction with the asparagine/glutamic acid. The carboxypeptidases feature interaction with a serine in the active site. Unlike the system above, the serine is not in contact with the lysine although it is not implausible that it might play a similar role to that described in the flanking lysine/serine section above. The complex of a DD-peptidase (D-Ala-D-Ala carboxypeptidase) with a boronic acid inhibitor is shown in Fig. 22 and reveals a similar set of interactions. Finally, the Endopepticase La complex features interactions with an active site lysine (see for example Fig. 23 with a

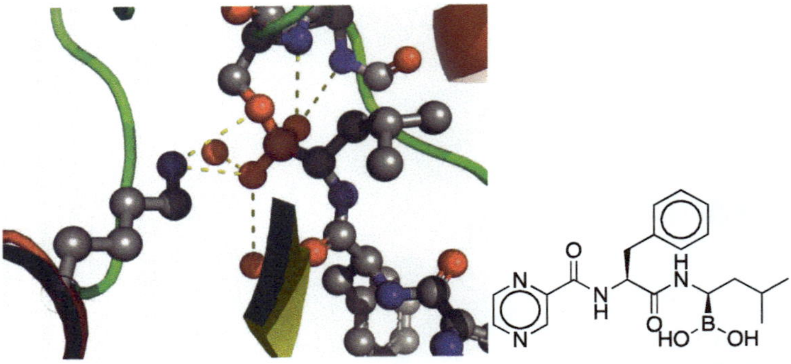

Fig. 23 Boronate complex of a boronic acid with Lon-like protease MtaLonC (PDB 4fwd).

structure of a boronic acid in complex with an ATP-independent Lon-like protease). In these enzymes, the lysine is closer to the B-OH groups of the boronate than to the serine suggesting that although it still presumably plays the same role of deprotonating the serine OH, it is possibly less efficient at this. One of the B-OH groups protrudes toward the backbone NHs of two adjacent amino acids and require the protein to adopt a specific and unusual conformation. This is doubtless facilitated by the presence of two glycine residues in close proximity, i.e., the sequence GDSAG in which the two residues that act as hydrogen bond donors are underlined, one of which is also the serine involved in the B—O bond.

2.2.5 Other flanking residues

An unusual mode of interaction is observed in seven structures that feature a lysine that has trapped ambient carbon dioxide to generate an oxamate (Fig. 24).[47,51,52] These are all β-lactamases (OXA-10 and OXA-24) and no explanation is given for why these residues should be present in this form. Likely, it reflects the nucleophilicity of a lysine that is in a predominantly unprotonated form. It is not clear from the structures whether the lysine is carbamoylated before or after addition of the boronic compound but given that the serine must be nucleophilic to form a B—O bond, it is most likely after. The presence of a second lysine residue close to the carbamoylated lysine changes the degree of protonation in the resting state. The type of inhibitor with reported structures that feature an oxamate include phenyl boronic acids and cyclic boronic esters featuring alkyl or aryl boronic acids in five- or six-membered rings (Fig. 25). One of the broad-spectrum β-lactamase inhibitors that binds to the systems with flanking lysine and tyrosine also binds in one of these binding modes.[47]

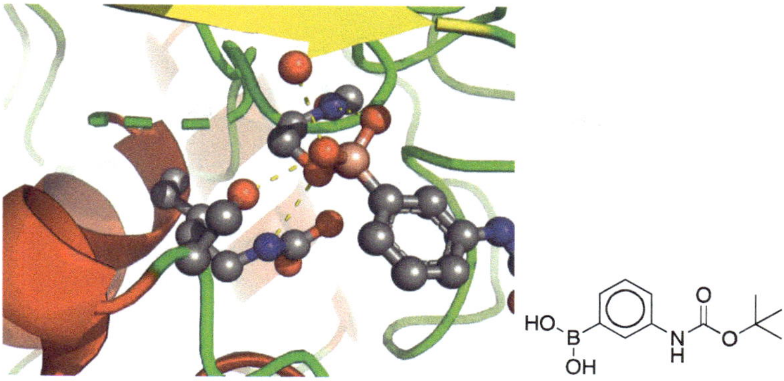

Fig. 24 The boronate complex of a boronic acid with β-lactamase OXA-24 (PDB 5tg4).

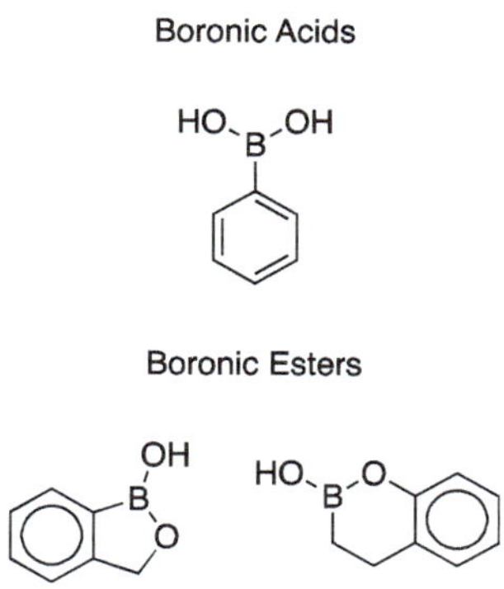

Fig. 25 Ligand types that make interactions with enzymes in the class presenting serine with a flanking carbamoylated lysine.

Two unusual binding modes are also present for acyl-homoserine-lactone acylase enzymes. In both, the residue that deprotonates the serine is not in contact with the serine. In 3 of these structures,[53,54] the serine is interacting with backbone NHs (Fig. 26) while in a further 2 structures,[53] interactions with an asparagine are also present. In these structures, there is a histidine in the correct placement to deprotonate the serine but its sidechain has swung out of the binding site in the reported structures. This suggests that the reaction mechanism for this set of structures likely mirrors that reported for flanking histidines in Section 2.2.1. One of these structures (with asparagine in contact with the serine) has a degree of ambiguity about whether it forms a boronic ester with serine or a boronate complex. The structure (Fig. 27) is given as the boronic ester but with the boron atom non-planar (one of the substituents is twisted 27° out of plane). All of the examples in this category have straight chain aliphatic boronic acids bound (Fig. 26).

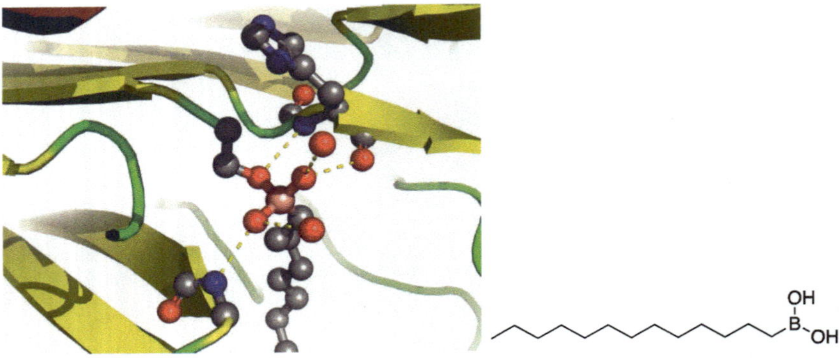

Fig. 26 Structure of the boronate complex of a boronic acid with Pseudomonas aeruginosa PvdQ (PDB 4m1j).

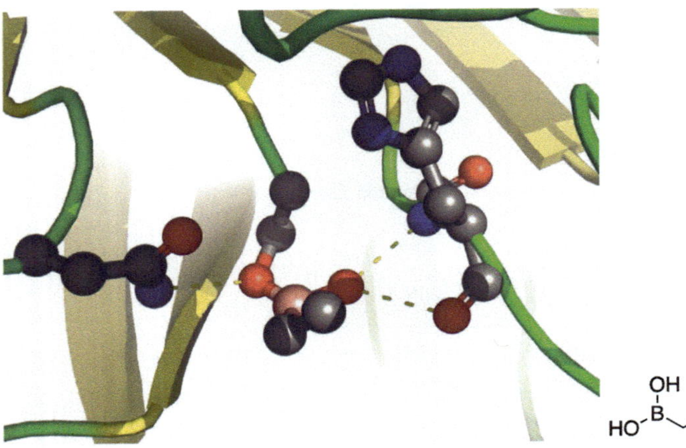

Fig. 27 Structure of the boronic ester of a boronic acid with Pseudomonas aeruginosa PvdQ (PDB 4wks).

A further class of proteins to which boronic acids form covalent bonds include transthyretin, which is one of the many proteins that aggregates forming amyloid (linked to a number of degenerative diseases).[55] Starting from resveratrol a series of boronic acid derivatives were investigated, most of which were found to be non-covalent. In order to overcome this, a bis-boronic acid approach was taken. In this system, a possible direct interaction between a non-protonated lysine and the boron atom was investigated and no evidence found for the formation of such an interaction. These boronic acids interact with a non-catalytically active serine that is solvent exposed (on the surface of the protein). Rather than form boronates, the structures

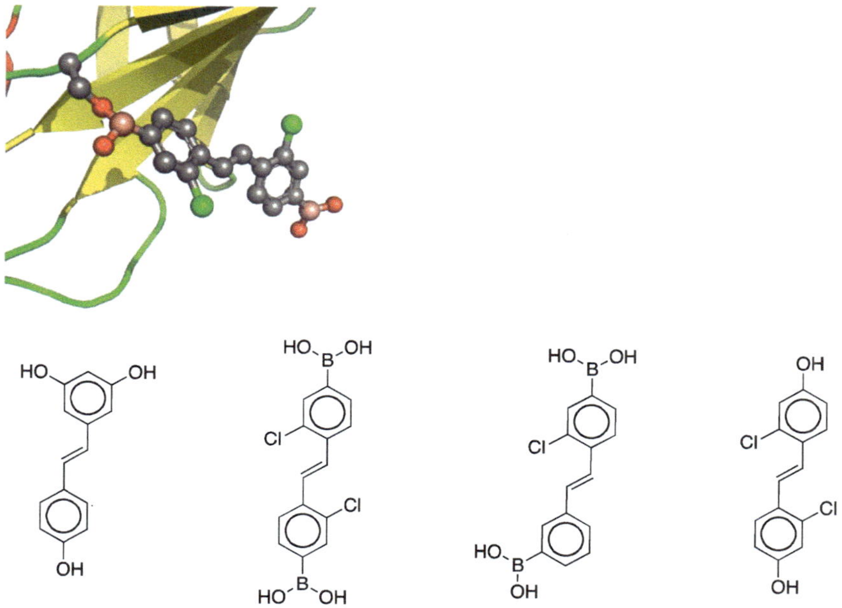

Resveratrol 1 2 3

Fig. 28 Structure of the boronic ester complex of bis-boronic acid **1** with transthyretin and some compounds studied in this system.

reveal that they form boronic esters (Fig. 28). This pair of structures reveals the importance of considering the mechanism of activation of the serine. It is likely that this system, which does not involve an activated serine, is unable to form boronate because no base that can deprotonate the serine is available and thus the boronic acid acts as its own base (one of its OH groups), leading to loss of water and formation of a boronic ester. Alternatively, the boron in this complex is highly pyramidalized, with one oxygen being 57° out of the O-B-C plane formed by the other oxygen. This suggests that the structure might be of a boronate with unobserved density for the final oxygen. No short-range interactions between the boron atom or any of those attached to it are observed.

It is interesting to note that for this series, the K_d values for the boronic acids (**1** and **2**) are indistinguishable from those for the phenol **3** (Table 2) suggesting that in the absence of an activated alcohol, covalent interactions of the boron are energetically indistinguishable from non-covalent interactions. The boron–oxygen bond can make inconsequential contributions to binding.

Table 2 Dissociation constants for binding to transthyretin.[55]

Compound	K_d (μM)
1	0.47
2	0.79
3	0.82

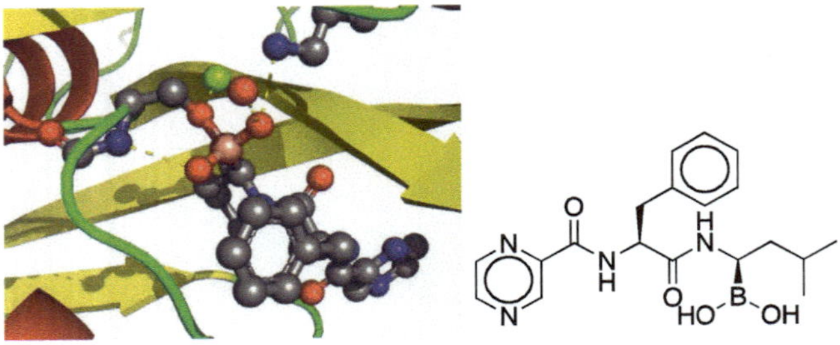

Fig. 29 Structure of the boronate complex of bortezomib with LonA protease (PDB 4ypm).

The final precedented environment around covalently modified serine is one that involves metal. Specifically, the LonA protease is Mg^{2+} dependent and the structure in complex with bortezomib has been solved (Fig. 29). It is possible that Mg^{2+} can activate the protease activity even in the absence of the ATP usually required. In a study that combined cryo-electron microscopy, X-ray diffraction and molecular dynamics simulations, it was found that Mg^{2+} activates the system with an adjacent lysine in a deprotonated form.[56] It is proposed that Mg^{2+} plays two roles in the activity of the system: it drives the protein into the active conformation and it activates the lysine into the deprotonated form required for catalytic activity. In that sense, the mechanism of the protease is likely to mimic that shown in Fig. 16 with Mg^{2+} being responsible for the pre-catalytic cycle deprotonation.

2.3 Binding to threonine

2.3.1 Proteasomes

This class of structures is almost exclusively proteasome endopeptidases; the target for the first approved boron-containing drug, bortezomib.[57] The proteasomes are a family of enzymes involved in protein degradation, with intricate regulatory factors but broad substrate specificity.[58,59] The active site threonine is the N-terminal residue and it is flanked by a lysine.

It has been proposed that the presence of this positively charged residue modifies the pK_a of the threonine so that it readily loses a proton.[58,59] Furthermore, the N-terminal location of the threonine means that this amino acid's own amino group is available to act as a base. A recent computational study has tried to disentangle these possibilities.[60] They predicted the pK_a of the N-terminal amine to be 3.0 and that of the lysine as 5.4 and thus both would be expected to be unprotonated and therefore available to act as the base that deprotonates the threonine. Their empirical valence bond calculations provide estimates for the key barrier heights for either residue acting as the base and find the N-terminal amine gives the lowest barrier. They posit a role for the flanking lysine as a catalyst for the proton transfer from threonine OH to the terminal amine (Fig. 30). The key barriers are those leading to the two tetrahedral intermediates and the second, for the deacylation of threonine by water, is computed to have the higher barrier and this is in

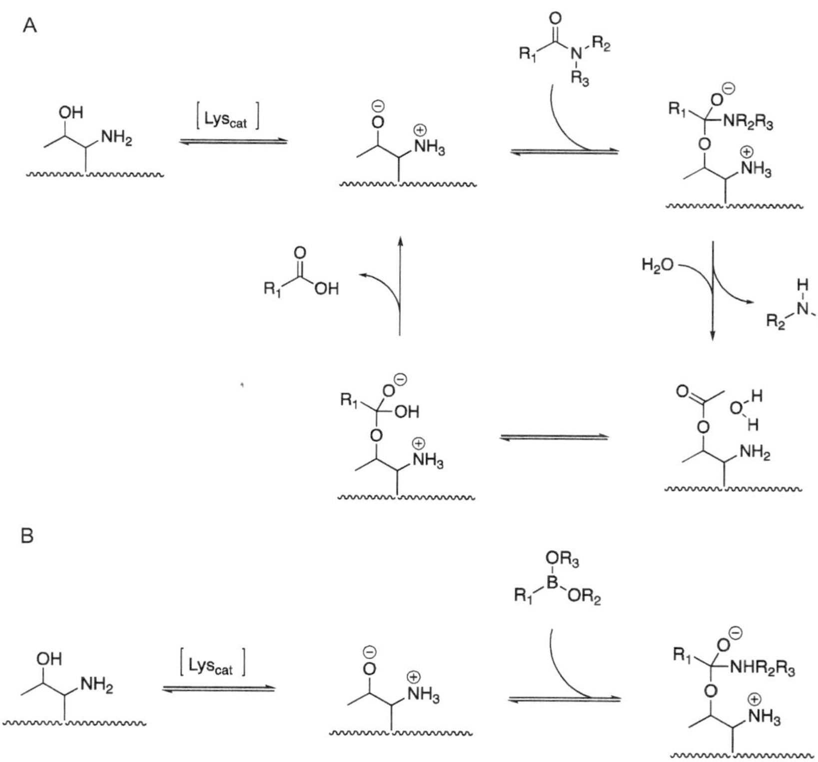

Fig. 30 (A) The catalytic reaction likely to operate for the proteasome endopeptidase. (B) The reaction with boron-containing compounds.

line with observed values of k_{cat}. This process is slightly different to that in other enzymes described and likely leads to specific requirements for covalent boron-based inhibitors. Most of the structures in this class (Figs. 31 and 32) contain a peptide mimic type of boron-containing compounds (a group that includes bortezomib).

2.3.2 Autotaxin

The enzyme autotaxin hydrolyses phosphate esters in lysophosphatidylcholines, as shown in Scheme 1. The enzyme active site contains two Zn^{2+} ions that are key to the reaction mechanism (Fig. 33). Many of the details of the reaction mechanism remain to be deduced but it seems highly likely that the boronate complex in this system mimics a phosphorylated threonine

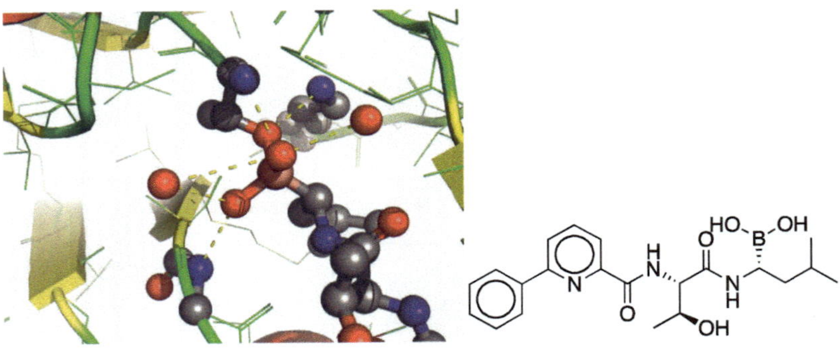

Fig. 31 The structure of the boronate complex of a boronic acid with the proteasome endopeptidase (PDB 5lf4).

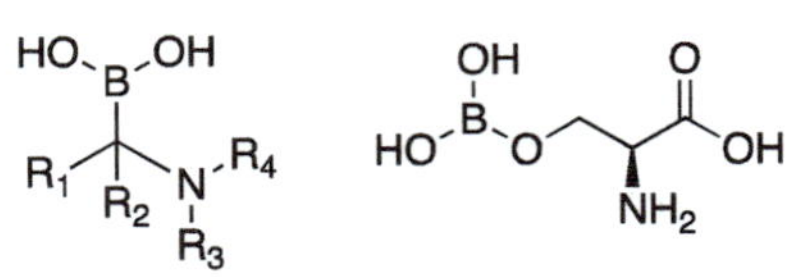

Fig. 32 Ligand types that make interactions with enzymes presenting an activated threonine.

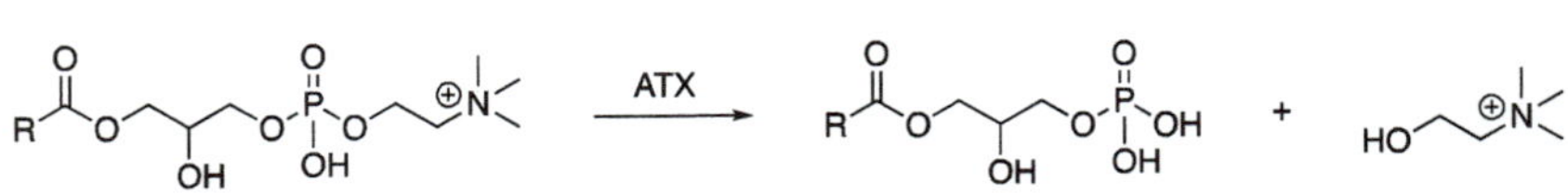

Scheme 1 The reaction catalyzed by Autotaxin.

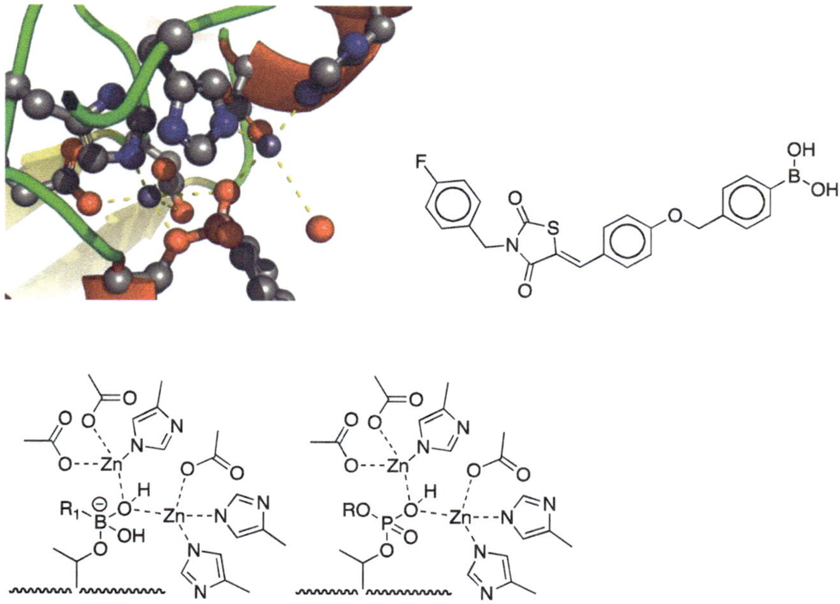

Fig. 33 The boronate complex of a boronic acid with autotaxin featuring two zinc ions. The analogy between boronate and phosphate is shown below.

intermediate in the reaction mechanism.[61,62] This is a contrasting role to that the boronate plays in every other structure and suggests an approach that would be amenable to broader application.

2.4 Binding to multiple amino acids (serine combined with serine, histidine or lysine)

Three reports describe situations in which boron makes a covalent bond with more than one amino acid sidechain.[63–65] In all of these cases, serine is one partner, reflecting its dominance in covalent interactions with boron-containing compounds. The first involves bridging between two serines in a serine-rich region of trypsin.[64] This is a weak interaction (despite involving two covalent bonds, the K_D is only 97 mM) and is observed for borate itself (see Fig. 34) and for phenyl boronic acid, for which the complex involves partial occupancy. There is an interesting comparison with the complexes formed on the periphery of the transthyretin protein that also involves inactivated serines (Fig. 28); the complex is weak but in transthyretin, the boronic ester forms whereas in the trypsin structure, a boronate forms

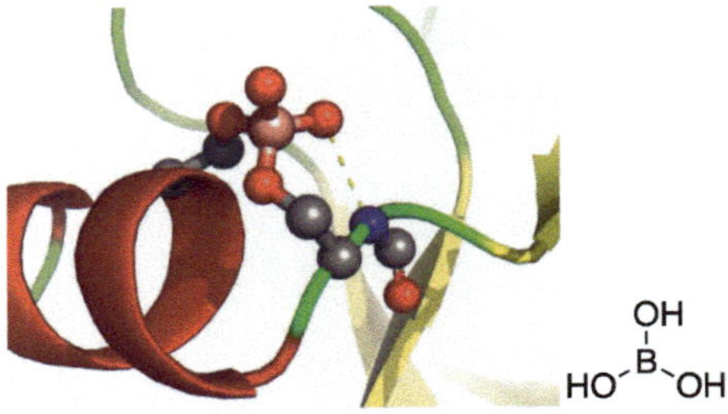

Fig. 34 Structure of the complex of boric acid with trypsin (PDB 2a31).

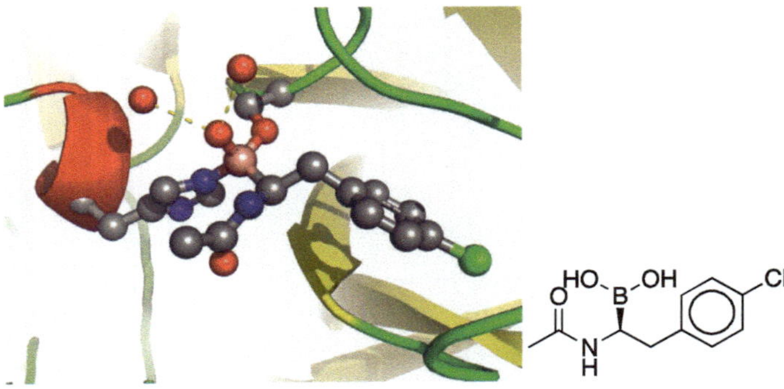

Fig. 35 Dicovalent interaction between one enantiomer of a boronic acid and γ-chymotrypsin (PDB 2vgc).

with two serine residues. Evidently the selection between tetrahedral and trigonal boron is relatively finely balanced and can be altered by the subtle details of the environment.

The system that forms interactions with serine and histidine is γ-chymotrypsin, where a remarkable change in complexation geometry occurs upon changing the stereochemistry of the boronic acid (Fig. 35).[63] Whereas the L-stereoisomers form complexes like those described above in Section 2.2.1 (flanking histidine), when the stereochemistry is inverted, a complex in which the histidine now acts as a Lewis base and forms a complex with a boronic ester of the serine. It is proposed that this is because the enzyme has evolved to selectively recognize one of the enantiomers of its substrates and the boronic acid that resembles the other enantiomer interacts with the catalytic machinery in a rather different fashion.

The final multiple amino acid complex involves the remarkable cross-linking of three amino acid sidechains in a penicillin binding protein (D-Ala-D-Ala-carboxypeptidase). In this case, the enzyme is one of the

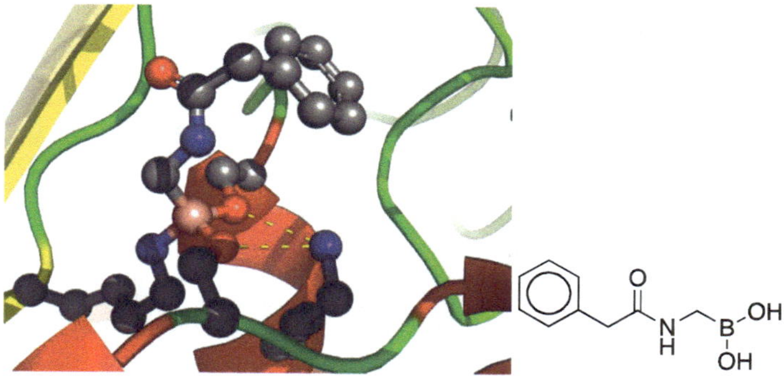

Fig. 36 Tricovalent interaction between a boronic acid and D-Ala-D-Ala-carboxypeptidase (PDB 3zvw).

group described above in the flanking lysine/serine pair section but remarkably, in the case of this enzyme from Actinomadura sp. R39, all three of the residues involved in the catalytic machinery are found to form a covalent bond to boron (Fig. 36).[65] Although this is an interesting case, it is hard to know if such adducts are likely to be of general relevance to the design of covalent boron-based inhibitors due to the improbability of having appropriately positioned amino acid sidechains. Such interactions would present a very attractive opportunity for selective covalent targeting. Given the reaction mechanism outlined above for this class of enzymes (Fig. 16), it is likely that small changes in the environment around this catalytic machinery may be sufficient to divert the system into forming the tricovalent structures that are observed. The two boronic ester oxygens make a bidentate interaction with a nearby lysine residue and this electrostatic stabilization may be important in forming this unusual interaction.

2.5 Binding to co-factors

In this class are complexes with two enzymes: enoyl acyl carrier protein reductase and leucine-tRNA ligase. In these two systems, boron forms covalent bonds to oxygen in molecules that are bound (non-covalently) to the enzyme, and in both cases through the alcohols on a ribose-type ring.

The synthesis of fatty acids is undertaken by quite different systems in mammals as compared to many lower organisms such as bacteria. This makes it an attractive target for selective anti-bacterials and is the target of several medicines, including isoniazid that is used for treatment of tuberculosis. In bacteria, each step in the synthesis is catalyzed by a distinct enzyme (unlike in mammals where a single multi-site system is active) and several of the steps have been targeted in drug discovery. The enoyl-acyl carrier protein reductase catalyzes the removal of a double bond, a key requirement for the generation of saturated fatty acids and isoniazid acts on this stage of the process. There are two key residues (lysine and tyrosine) and NAD(H) co-factor present in the enzyme active site and likely roles are indicated in Fig. 37.[66] It is believed that the active site causes the substrate to bind such that the π system stacks appropriately with the NADH and this is reflected in the complex of boronates in the same active site, shown in Fig. 38.

The binding of AN2690 (**4**) to Leucyl-tRNA synthetase was investigated with a range of techniques. It was found to be a slow-tight-binding inhibitor and ribonucleotides that include adenosine are required for binding.[67] A crystal structure (Fig. 39) revealed that the boron forms a diester with the two alcohols on the ribose ring of an adenosine monophosphate. Both of the ribose alcohols are required for activity and a range of structural variations revealed that the boron (specifically the five-membered oxaborole ring) is required for good potency.

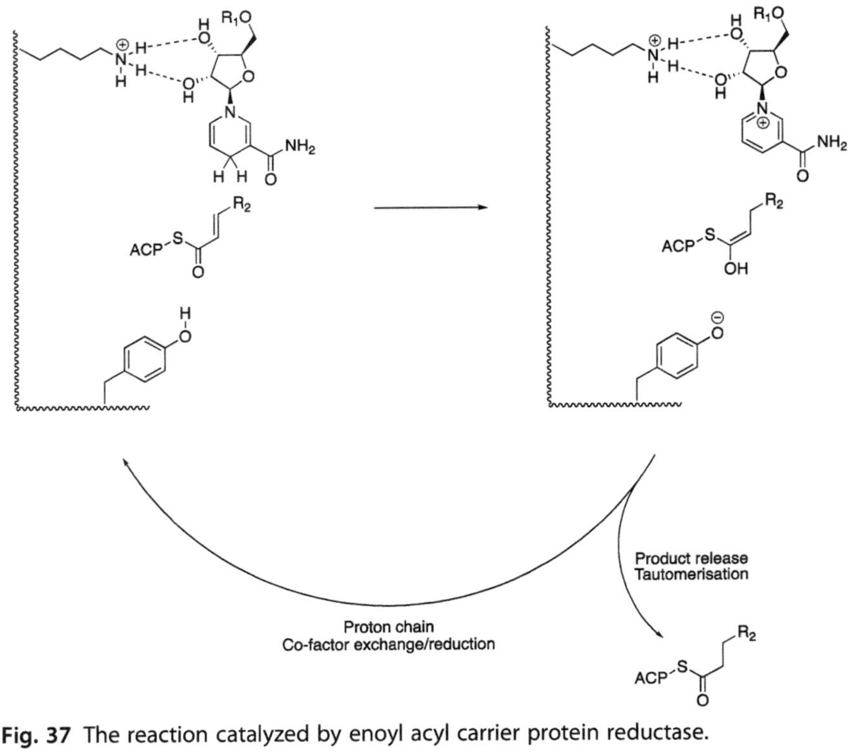

Fig. 37 The reaction catalyzed by enoyl acyl carrier protein reductase.

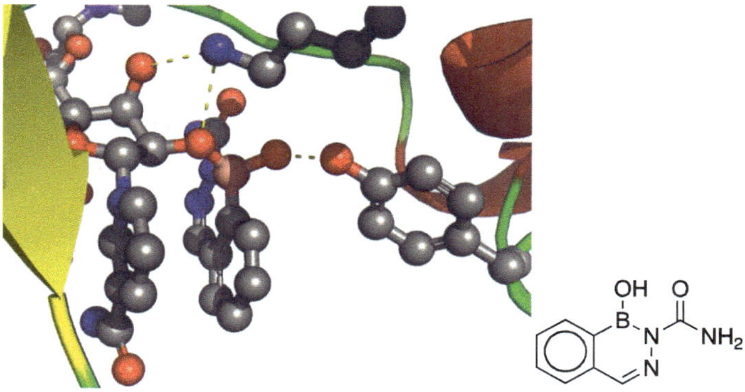

Fig. 38 Structure of the boronate complex of a cyclic boron-containing compound with enoyl acyl carrier protein reductase (PDB 5cg1).

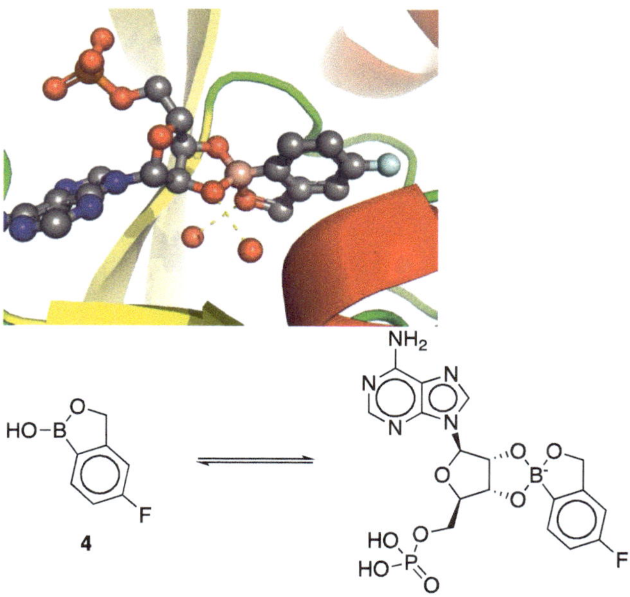

Fig. 39 The boronate complex of **4** with the diol of adenosine monophosphate in complex with Leucyl-tRNA synthetase and the likely pre-equilibrium process.

2.6 Binding to metals

In interactions with the arginase enzymes, boron interacts with the protein through metal-mediated covalent interactions. The enzyme is involved in the transformation of arginine to ornithine and urea (Scheme 2) and has at its core a binuclear manganese cluster.[68] It is likely that this cluster presents a metal bound hydroxide that attacks the guanidine; this primes the enzyme

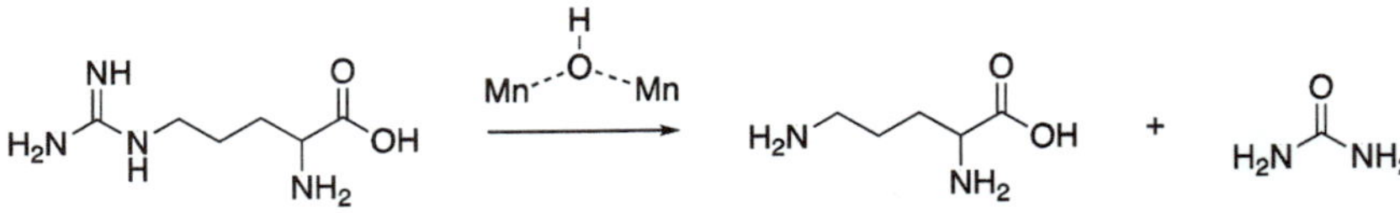

Scheme 2 The reaction catalyzed by arginase.

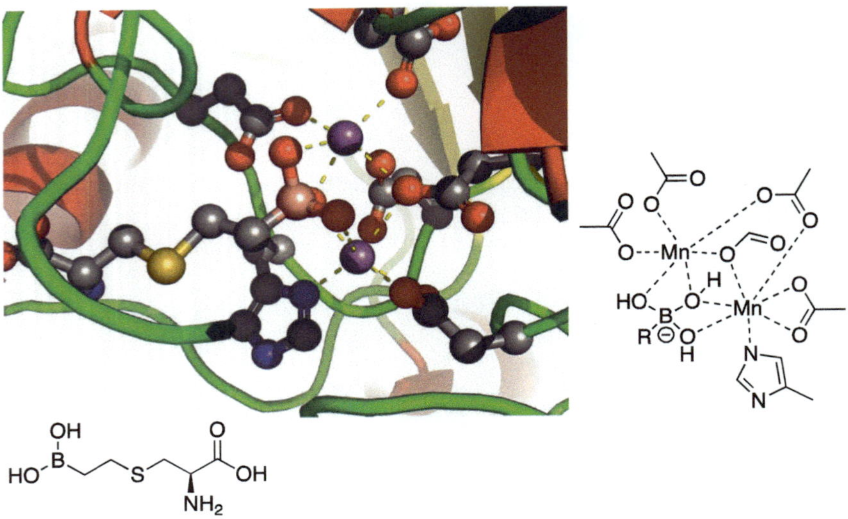

Fig. 40 The complex of a boronate species with arginase. The metal coordination state is illustrated (PDB 1p8r).

to add hydroxide to an appropriately selected boronic acid to generate a boronate that is highly complementary to the active site. The bound state sees two highly distorted, approximately octahedral coordination geometries of each of the manganese ions (Fig. 40).

The boroxole **5** is active against a resistant strain of *Plasmodium falciparum* and therefore it is a potential anti-malarial.[69] The challenge of docking (as described in Section 2.8) is noted but docking into PfCPSF3, a homologue of the mammalian Cleavage and Polyadenylation Specificity Factor subunit 3 was undertaken. The boronate in this case was placed so as to interact with the two zinc ions in the binding site and therefore it acts as phosphate mimic in a way noted previously (Fig. 33).

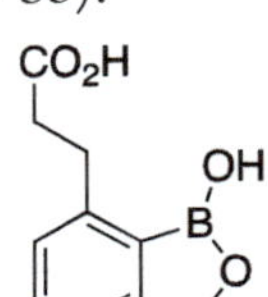

5

2.7 Other targets and phenotypic screening

A series of complexes of boronic compounds with 2-carboxylpyridines was investigated by Anacor as treatments for atopic dermatitis (Fig. 41). The compounds were found to be stable and to operate as anti-bacterials. An advantageous anti-inflammatory effect was also found.

A series of boroxoles that inhibit phosphodiesterase 4 were investigated as possible topical medicines with the strategy adopted being to include a group that would rapidly be transformed to an inactive form upon systemic administration in order to minimize any consequences arising from systemic exposure (Table 3).[70] This was reduced to practice with a range of groups that could be transformed into a carboxylic acid. As can be seen in Table 3, the acid has markedly less activity than the corresponding nitrile or ester and after 1 h, the esters are significantly transformed into the acid. A series of boroxoles with pendant carboxylic acids were also explored for their anti-malarial activity.[71] The best of the compounds was found to be soluble and non-cytotoxic.

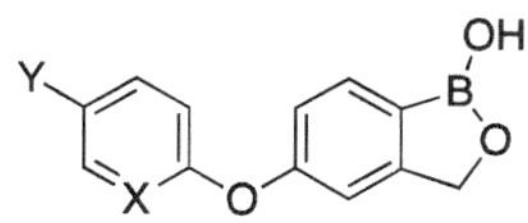

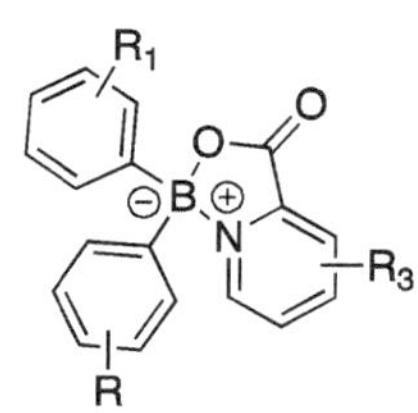

Fig. 41 Compound class tested against atopic dermatitis.

Table 3 IC_{50} values for PDE4 inhibitors compared with the amount of compound that is converted into the corresponding carboxylic acid in 1 h in plasma.

	IC_{50} (μM) [% acid formed after 1 h in plasma]	
Y	X = CH	X = N
CO_2H	6.42	[5% inhibition @ 10 μM]
CN	0.49	0.18
CO_2Me	0.29 [29%]	0.084 [32%]
CO_2Et	0.29 [54%]	0.047 [ND]
CO_2n-Pr	0.76 [59%]	0.13 [80%]

A series of boroxoles were optimized for anti-inflammatory activity leading to compound **6**.[72] The presence of ortho-substituents around the aromatic amide was essential for potency supporting a bioactive conformation with this ring twisted. Having optimized in vitro potency, pharmacokinetics were measured (Table 4) and found satisfactory. It was noted that the SAR for this series suggests that activity is not due to kinase or cyclooxygenase inhibition and the SAR is divergent from that against trypanosoma. Nonetheless, in vivo efficacy was demonstrated in a disease model.

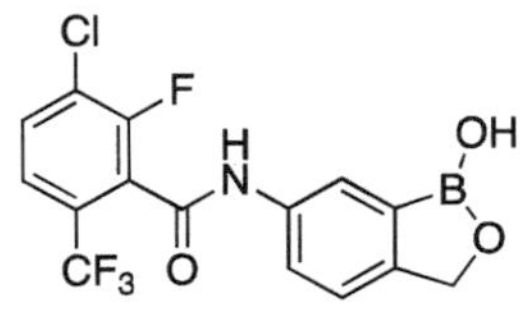

6

The borane complexes of some cyclic amines have been investigated for their anti-tumor activity.[77] These (Fig. 42) include complexes of piperidine, piperazine, morpholine and imidazole (and its *N*- or 2-methylated versions) with X being H, CN, CO_2H, CO_2Me, CONHEt or NMe_3. Activity was measured in a range of tumor cell lines suggesting potential utility although comparators with non-cancerous cell lines were not available so a general cytotoxicity cannot be ruled out. The activity was ascribed to reactivity with tissue nucleophiles and two examples were found to inhibit purine synthesis.

In the quest to identify inhibitors of hormone-sensitive lipase, the SAR of boronic acids deriving from the overlay of known inhibitors **7** and **8** was explored.[76] Replacement of the boronic acid in **9** with Bpin retained activity; moving the Bpin around the ring reduced activity. Replacing the boronic acid with a trifluoromethyl ketone or nitrile also reduced activity suggesting that boron is uniquely well placed to interact with the enzyme. The boronic acid, **9**, was found to be a mixed-type inhibitor that is able to bind to either the enzyme or the enzyme-substrate complex. Selectivity over the related acetyl- and butyryl-cholinesterases was found.

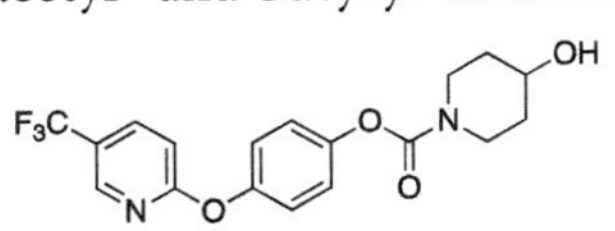

7

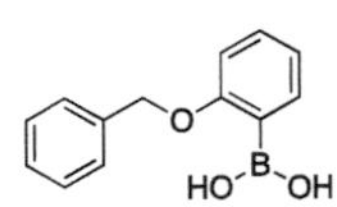

8

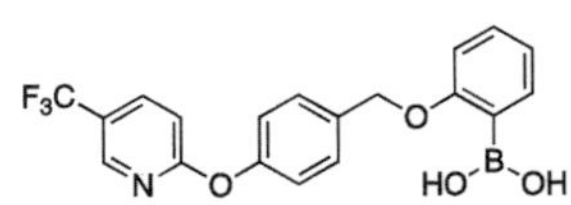

9

In the hunt for treatments for a range of infectious diseases, researchers screened the Anacor collection of benzoxaboroles against whole organisms for treatments of trypanosomiasis, malaria and Chagas' disease.[78] This phenotypic screening approach has yielded several boron-containing compounds that are active but without a known mode of action. Even where the target

Table 4 Pharmacokinetic parameters for boron-containing compounds with measured in vivo pharmacokinetics (in mouse, rat or monkey).

		i.v.					p.o.					
Compound	References	Dose (mg/kg)	CL (mL/h/kg)	V_{SS} (mL/kg)	AUC (h μg/mL)	$t_{1/2}$ (h)	Dose (mg/kg)	C_{max} (μg/mL)	T_{max} (h)	AUC (h μg/mL)	$t_{1/2}$ (h)	% F
Mouse												
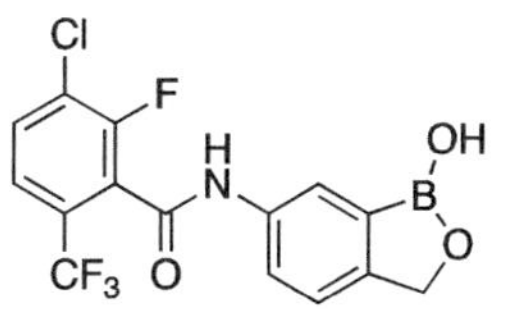 **13**	72	2	41	410	43.4	7.9	4	5.65	4	70	13.3	71
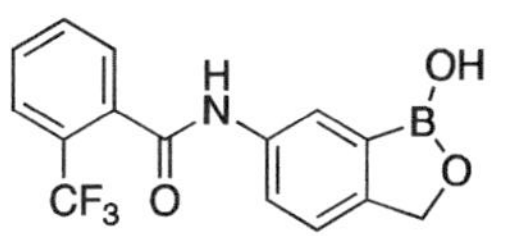 **10**	73	2	112	475	18.1	2.9	8.2	4.35	0.17	28.7	3.67	39
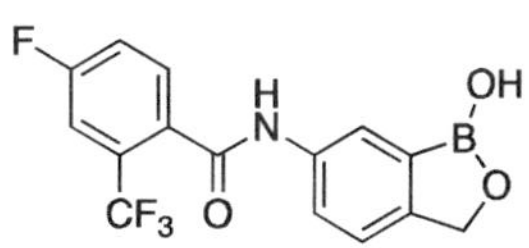 **11**	73	3.3	59	970	56.4	11.4	9	5.3	0.5	96.3	20.2	69

Continued

Table 4 Pharmacokinetic parameters for boron-containing compounds with measured in vivo pharmacokinetics (in mouse, rat or monkey).—cont'd

Compound	References	i.v. Dose (mg/kg)	i.v. CL (mL/h/kg)	i.v. V_{SS} (mL/kg)	i.v. AUC (h μg/mL)	i.v. $t_{1/2}$ (h)	p.o. Dose (mg/kg)	p.o. C_{max} (μg/mL)	p.o. T_{max} (h)	p.o. AUC (h μg/mL)	p.o. $t_{1/2}$ (h)	% F
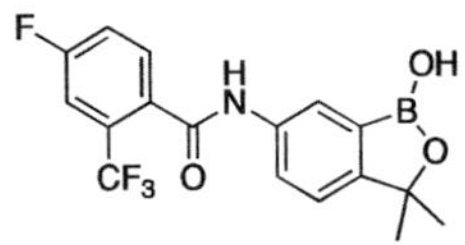 **14**	74	4.3	89	1690	48	26.6	13.4	6.96	6	82	–	55
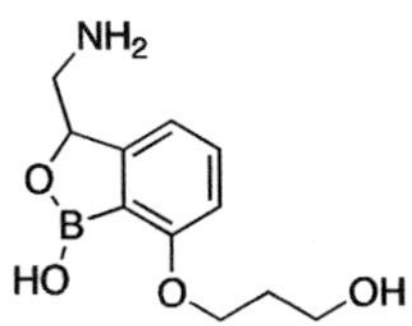 **15**	75	30	2914	3895	10.3	0.12	30	0.90	0.5	2.2	2.7	21
Rat												
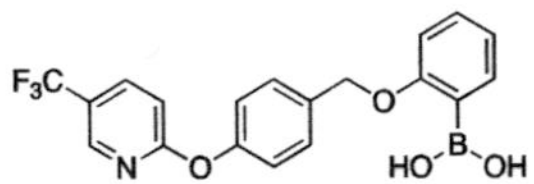 **9**	76	–	–	–	–	–	3	0.42		1.75	2.95	

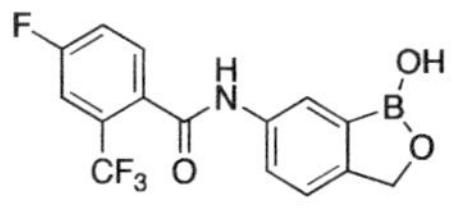 11	73	13.4	371	4610	26.4	13.8	13.4	6.28	2	–	–	–
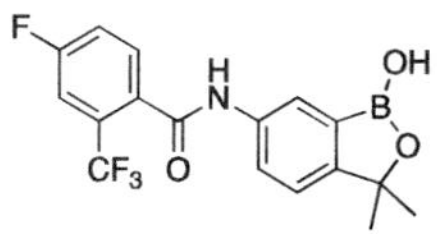 14	74	–	–	–	–	–	25	18.2	8	291	–	–
Monkey												
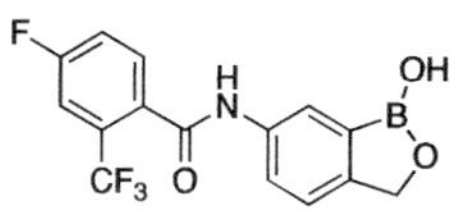 11	73	2	104	741	20.3	8.8	10	12.3	1.3	85.6	9	86
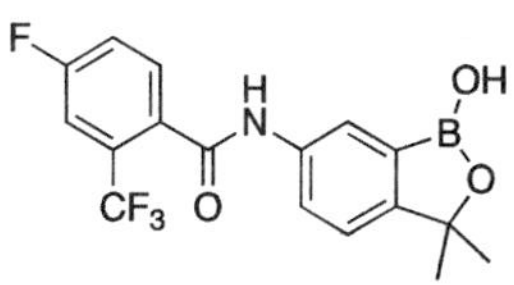 14	74	2	22	656	94.4			11	9.5	460		89

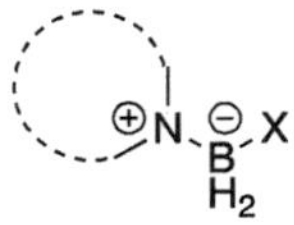

Fig. 42 The structure of borane complexes of cyclic amines tested for anti-tumor activity.

is not known, comparing compounds like **10** and **11** to **12** when searching for anti-trypanosomal treatments is illuminating; the absence of activity for **12** reveals that the boron functionality is likely key, although it does not clarify whether the mode of action involves covalent bond formation or not.[73]

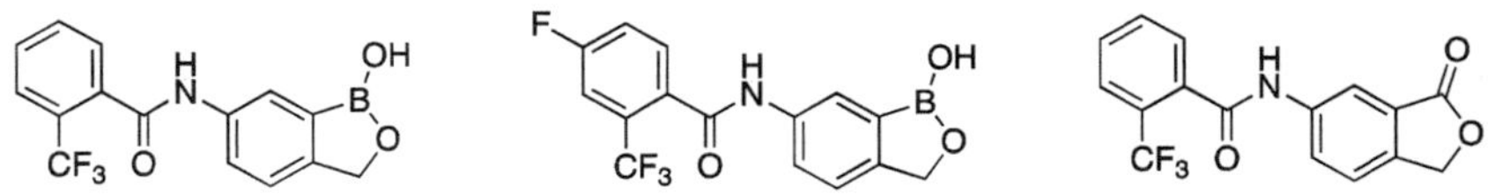

10 **11** **12**

2.8 Challenges for computational approaches

Unfortunately, many of the computational approaches to studying protein–ligand interactions were developed before including boron in drugs was considered a serious possibility. Therefore, parameters relevant to boron are often not included.[69] Indeed, the ability of boron to adopt two completely different atomic arrangements (either trigonal or tetrahedral) makes predicting the geometry of complexes challenging. The detailed energetics of the two arrangements and the interactions they can make can be extremely difficult to predict and in principle require a quantum mechanical approach.

In a 3D CoMFA/CoMSIA study of ubiquitin proteasome system inhibitors, 59 dipeptide boronates from the work of one group were studied.[79] In order to work around the problem that boron is not supported by the software, the boron atoms were all converted into carbon atoms and the bioactive geometry was derived from a crystal structure, specifically that of bortezomib in complex with the yeast 20S proteasome. A training set of 46 examples was used to build models with the two methods and these were tested on 13 examples that had been reserved for that purpose. The SYBYL software suite was used for the modelling. In the CoMFA analysis, both steric and electrostatic effects were found to contribute about equally whereas CoMSIA provides more details suggesting that hydrogen bonding groups and hydrophobic regions determine the activity levels for this group of compounds.

In a project aiming to obtain inhibitors of the β-lactamase AmpC, permeability was considered a potential problem and a strategy of varying pK_a and logD considered as a way of identifying compounds with enhanced efficacy by virtue of more getting into cells.[80] As part of these studies, a few values of these properties were measured and compared to those calculated; models for logP, logD and pK_a were all found to lack accurate agreement with experiment, likely reflecting the absence of significant numbers of boron-containing compounds in the training sets upon which most of the common models for these properties are built.

Docking methods that are able to process boron acting as a covalent recognition motif have been developed and some success has been derived using the GOLD covalent mode that was able to recapitulate the docking mode of bortezomib and to provide a view of the likely pharmacophore for binding.[81] The GOLDSCORE scoring function was proposed as being effective. As has been highlighted, the binding of boron-based covalent inhibitors is intimately linked to the network of water molecules around the binding site and this is a challenge that many docking programs struggle to treat appropriately. In a docking experiment profiling Autodock 4, Autodock Vina, Covalent dock and GOLD5.2, the variability in solvation gave GOLD the edge.[82] A training set of 23 boronic acids each of which has a crystal structure of its binding mode with the β-lactamase AmpC was used and the ChemPLP scoring function proved most effective for this challenge. The water toggling function (in which GOLD is able to include or exclude water molecules according to whether they improve the scoring function values or not) was found to be detrimental to success; selecting the active site waters to include was the preferred method but requires insight in advance of performing the study which makes the method challenging to apply in a prospective fashion.

2.9 Key considerations in designing for potency

Boron is only precedented as making covalent interactions with alkyl alcohols. This is not to say that interactions with other elements or with phenol (or other) types of oxygen is impossible but the means of achieving this have not been unraveled structurally yet. The interactions with boron are generally only of value to increasing interaction strength if catalytic machinery exists that can deprotonate the incoming alcohol in order to generate the boronate. This usually requires the presence of either a histidine or lysine in the active site. In the case of lysine, adjacent residues that perturb its pK_a are likely important and nearby serine or tyrosine can facilitate proton shuffling. Boronates are likely to bind well in environments that have evolved to stabilize alkoxide groups or to bind to phosphates. This comparison has been highlighted previously and similar opportunities to mimic sulfate are also likely.[3] Metalloenzymes that function by stabilizing hydroxide (or alkoxide) ion may also be amenable to the formation of boronates that complex the metals. Tuning the part of the molecule that is not involved in the B—O bond making can also be important and can impact on the type of covalent bond that is formed and its efficiency (for example changing from a single to several covalent bonds by changing the stereochemistry).

3. Designing for pharmacokinetic and chemical stability

Some of the previous reviews of the use of boron in medicinal chemistry have highlighted stability problems, particularly for boronic acids. However, in general, boron-containing compounds are not unusual and do not pose pharmacokinetic liabilities that are different to any other molecules with similar physical properties.[2] As shown in Table 4, a range of boron-containing compounds are found to have good oral pharmacokinetics. Benzoxaborole rings are generally more hydrolytically stable than boronic acids and often have good solublility.[2,71] Good stability to metabolism by microsomes or the S9 liver fraction from a range of species has been demonstrated for boroxoles **10** and **11**.[73] Furthermore, they can be permeable and not substrates for the efflux pump PGp.

In addition to measuring systemic exposure for **11**, levels in the CNS and CSF were also measured in the rat study.[73] At the 13.4 mg/kg dose, C_{max} in the brain was 0.99 μg/mL and in the CSF was 0.36 μg/mL. Similarly, for **14** in mice, the i.v. dose of 4.3 mg/kg gave a brain AUC of 5.57 h μg/mL and after an oral dose of 13.4 mg/kg C_{max} in the brain was 4.62 μg/mL. In rats, C_{max} in the brain was 8.5 μg/mL and in the CSF was 1.08 μg/mL with corresponding values of AUC of 129 and 14.6 h μg/mL. These suggest brain to plasma concentration ratios of 40%–60%. CNS exposure is important for efficacy in humans where trypanosomiasis in its most serious stages affects the CNS.

One of the interesting approaches in the area of promoting new boron-containing compounds for drug discovery is to devise new synthetic methods and then probe the metabolic stability and permeability of the resulting compounds. This has been applied to a range of spirocyclic B—O heterocycles based around **16**.[83] Compounds with logP in the range 3–3.7 and good stability in the presence of human liver microsomes (80%–94% remaining after 30 min) were generated and with at least one example showing good permeability also.

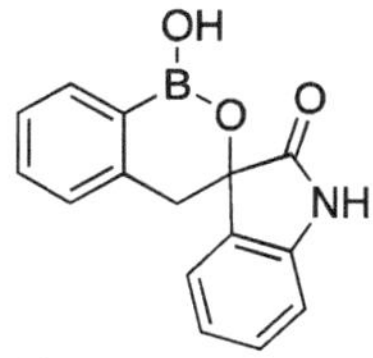

16

Given that benzoxaboroles have been a new group introduced to medicinal chemistry, some have introduced them as trial substituents in series that do not require covalent bonding. For instance, they have been introduced into some amino pyrimidine kinase inhibitors. Two examples (**17** and **18**) were subjected to human liver microsomes and both were found to have $t_{1/2}$ in the range of 55–60 min, and therefore have better stability than the closely related compound CC-292.

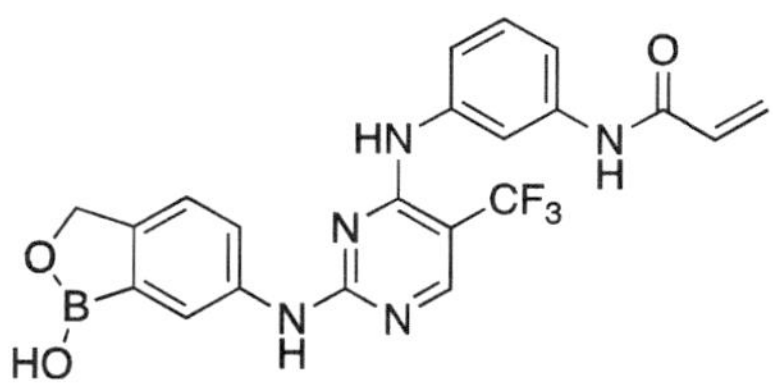

17

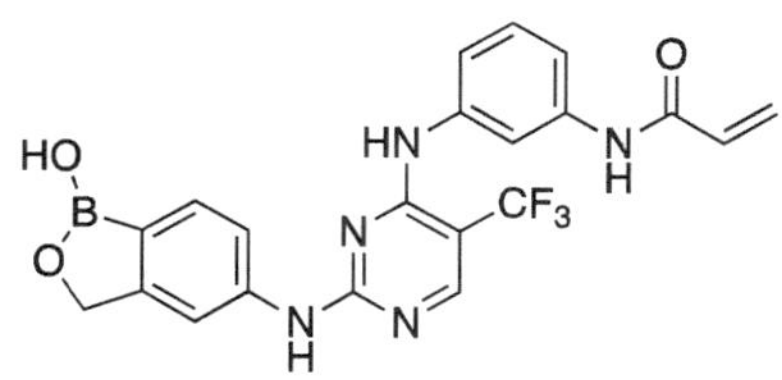

18

Two aryl boronic acids (**19a** and **19b**, Fig. 43) were studied for their stability in the presence of mouse and human liver microsomes as well as their permeability in Caco-2 monolayers.[84] Both have pharmacokinetic properties that would be useful for potential drugs.

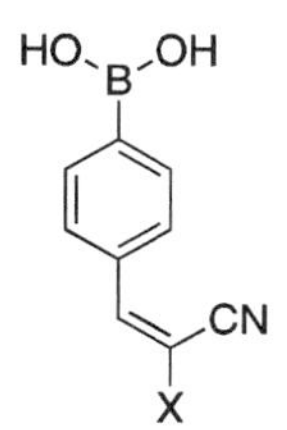

19a: X = CN
19b: X = CO_2Et

	19a	19b
CLint RLM (μl/min/mg protein)	20.2	36.3
CLint HLM (μl/min/mg protein)	20.0	35.8
Caco2 Papp (10^6 cm/s)	77	60

Fig. 43 Stability and permeability of two boronic acids.

3.1 Oxidative deboronation

In studies that identified the formation of reactive imines as a toxicological liability for bortezomib and ixazomib (see Section 4.2), it was found that these two drugs undergo oxidative deboronation mediated by the cytochrome P450 enzymes.[85] In particular, the isoforms 3A4 and 2C19 perform this reaction on these drugs and are the dominant oxidants although 1A2, 2C9, 2D6 and 2E1 also contribute.

In vivo pharmacokinetics studies of **20** revealed that whereas in rat the bioavailability was good, in dog and monkey, oxidative boronation leading to **21**, which has a long half-life, was problematic (Fig. 44).[86] Exploration next to the boron as a means to modulate this problem met with some success. The best compound found was **22** which has good oral bioavailability in rat (40%–50%) but this declines at higher doses suggesting a contribution from active uptake that can be saturated.

Samples taken from an in vivo toxicological study of the PDE4 inhibitor **23** revealed that in among an array of 10 metabolites, most represent the sort of reactions expected for metabolism of drug-like molecules (oxidation, sulfation, loss of alkyl groups) two pathways specific to boron make dominant contributions to the type of metabolites observed: protodeboronation and oxidative deboronation (Scheme 3).[87] The oxidative deboronation product is then itself transformed into a range of other metabolite suggesting that it is produced rapidly.

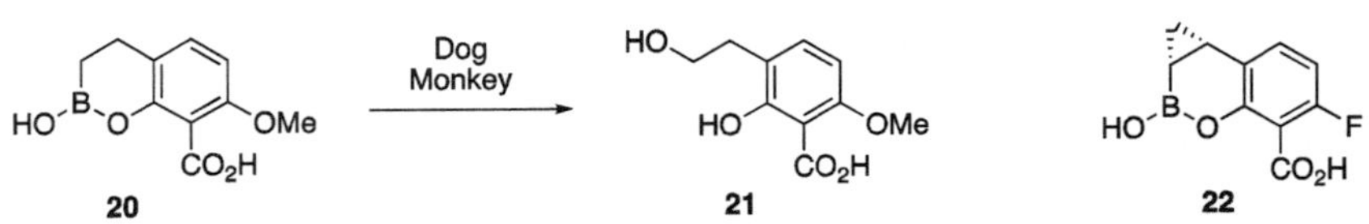

Fig. 44 In vivo metabolic transformation and an alternative that avoided the problem.

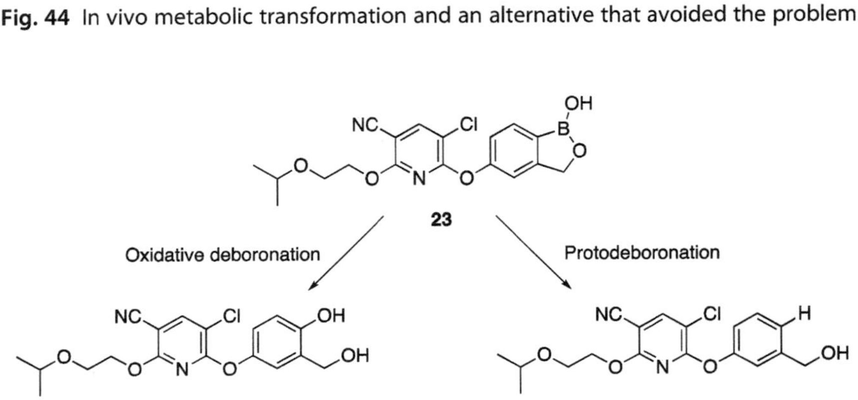

Scheme 3 An in vivo oxidative deboronation and protodeboronation.

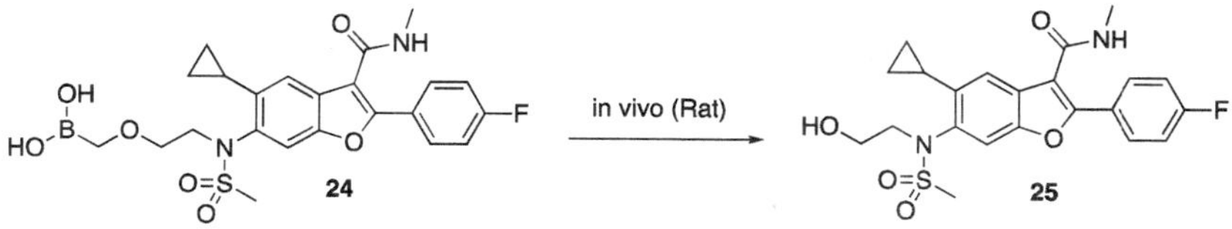

Scheme 4 An in vivo oxidative deboronation process.

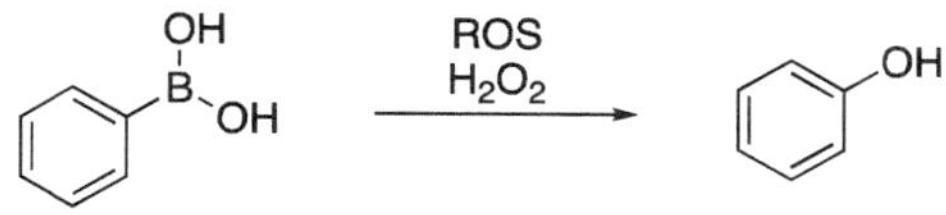

Scheme 5 Boronic acids as pro-drugs for phenols under ROS conditions.

The potential HCV-RNA dependent RNA polymerase inhibitor **24** suffers from high clearance in vivo in rats (Scheme 4), prominently due to oxidative deboronation that yields the alcohol **25**.[88]

Oxidative deboronation can be a desirable property and has been targeted as a mechanism to achieve selective activation of pro-drugs in cancer cells that feature increased levels of reactive oxygen species (Scheme 5).[89] There are a range of such reactive oxygen species and hydrogen peroxide has been found to be particularly effective for oxidative deboronation.

Similarly, **26a** is a selective estrogen receptor degrader (SERD), but the phenol arising from oxidative deboronation (**26b**) is also a known SERD. **26a** was dosed in a disease model in mice and concentrations of **26a** and **26b** were followed. The C_{max} for **26a** was 740 ng/g and 340 ng/g for **26b** when 5 mg/kg was dosed orally. At the termination of the in vivo experiments, **26a** and **26b** were both at about ~200 ng/g after dosing at 5 mg/kg whereas when dosed at 20 mg/kg levels of ~500 ng/g and ~300 ng/g of **26a** and **26b** respectively were obtained. This likely supports a saturable mechanism for the oxidation of **26a** to **26b**. When dosed in rats at 10 mg/kg, **26a** achieved a C_{max} of 3.51 μg/mL and resulted in concentrations of **26b** that had a C_{max} of 0.56 (approximately 15% that of **26a**), higher than the exposure achieved by dosing **26b** itself.[90]

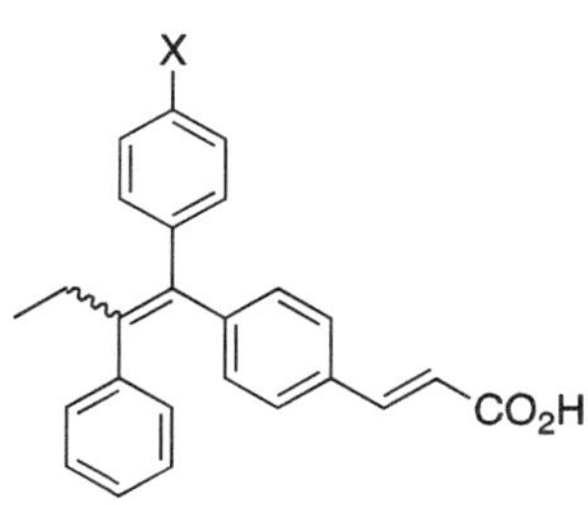

26a: X = $B(OH)_2$
26b: X = OH

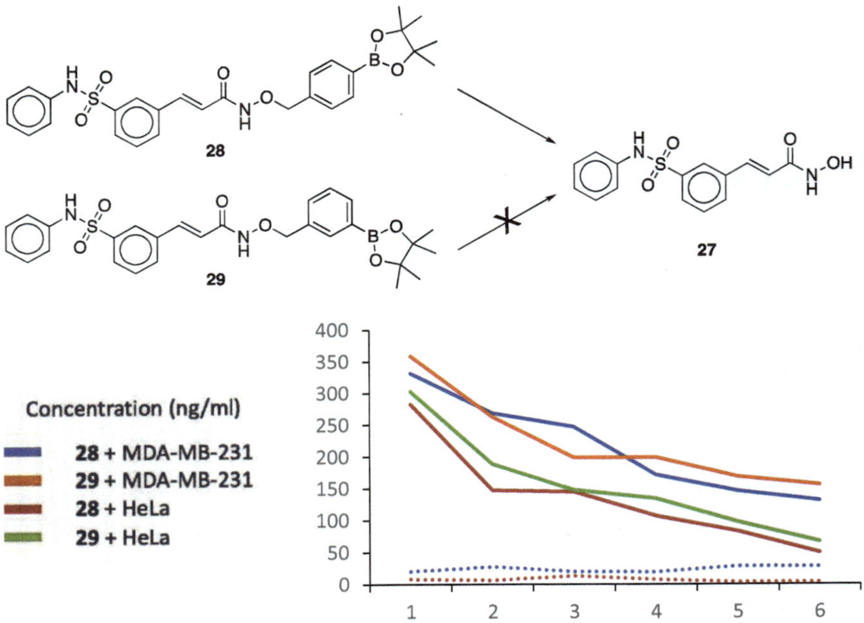

Fig. 45 Two potential pro-drugs for Belinostat (**27**). The time-course for the concentration of pro-drugs **28** and **29** in the presence of two tumor cell lines is plotted as well as the concentrations of **27** formed. Solid lines indicate the concentration of boronic acid whereas the dashed lines indicate belinostat concentrations.

The histone deacetylase inhibitor Belinostat (**27**) has been transformed into boron-containing pro-drugs **28** and **29** (Fig. 45). These have been incubated in the presence of MDA-MB-231 and HeLa cancer cell lines.[91] The concentration of the boronic acid arising from hydrolysis of the boronic ester and of Belinostat itself were measured over the course of 6 days. Whereas the para-substituted **28** leads to boronic acid and belinostat itself, the meta-isomer **29** does not. This is proposed to be because the oxidative deboronation process for **28** leads to a phenolic intermediate that rapidly fragments because the phenol is able to promote fragmentation of the benzylic C—O bond. The dosing of **28** in an in vivo model actually leads to higher concentrations of belinostat in the tumor than by dosing belinostat itself. A separate pharmacokinetic study found that dosing **28** i.p. (10 mg/kg) led to a concentration of belinostat of 223.1 ng/g after 4 h. At the same time, the boronic acid is at 2706.1 ng/g and the phenol arising from oxidative deboronation of **28** (or the boronic acid) is at 166.2 ng/g.[92] Belinostat represents about 10%–20% of the total drug/metabolite concentration.

In the dosing of **14** to mice, the oxidative deboronation product was found to be at approximately 1% of the concentration of the administered boroxole at all time points.[74] In rats, this was about 3%–5%. This same compound was found to have a long half-life in microsomes and in the presence of S9 fraction ($t_{1/2}$ >350 min). logD was measured as 3.51, solubility 25 μM, permeability (P_{app} A–B) in MDCK cells 776 nm/s with no effect of PGp measured.

The process of oxidative deboronation can even take place in the solid state. Boronic acid **30a** (and its trifluoroborate analog) was found to transform into phenol **30b** over the course of several days when stored as a solid (Fig. 46).[93] By contrast, **30c** in aqueous PBS solution degraded less than 5% in 3.5 h and less than 50% in a week. The difference in physical properties between the boronic ester **30c** and the corresponding phenol **30b** were measured and showed the boronic ester to have an increased logD and an increased solubility with almost identical permeability.

3.2 Protodeboronation

Detailed studies of the stability of boronic acids in buffered partially aqueous solutions across a pH range revealed an incredible span for the rates of protodeboronation with a number of mechanisms able to effect this transformation.[40] Whereas some examples of hetero-aryl boronic acids undergo an acid-catalyzed and a base-catalyzed protodeboronation at pHs below their pK_a (Group I, Fig. 47), others see the reaction inhibited by acid and base (Group II). Still others are more stable under acidic conditions but their stability is otherwise quite level across the pH range (Group III) whereas the final group (IV) undergoes a complex set of reactions catalyzed by both acid and base.

In further studies involving significant technological innovation to measure very fast rates of protodeboronation, it was found that ortho-fluorination of phenyl boronic acids can greatly destabilize the boronic acid.[41] Some boronic acids are known to degrade upon storage and this has been ascribed to protodeboronation in the solid state (the solution state protects the boronic acid).[94,95] The crystal structure of some boronic acids and boroxoles of interest as potential anti-fungals have been obtained and reveal some of the interaction patterns that these functionalities make but do not include any significant

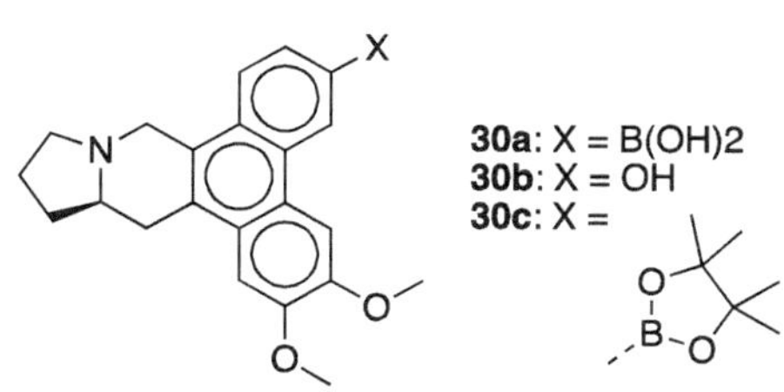

	30c	30b
Solubility in PBS (μM)	62	12
logD	4.48	3.41
Permeability	5.3	5.4

Fig. 46 Physical studies of a related set of compounds.

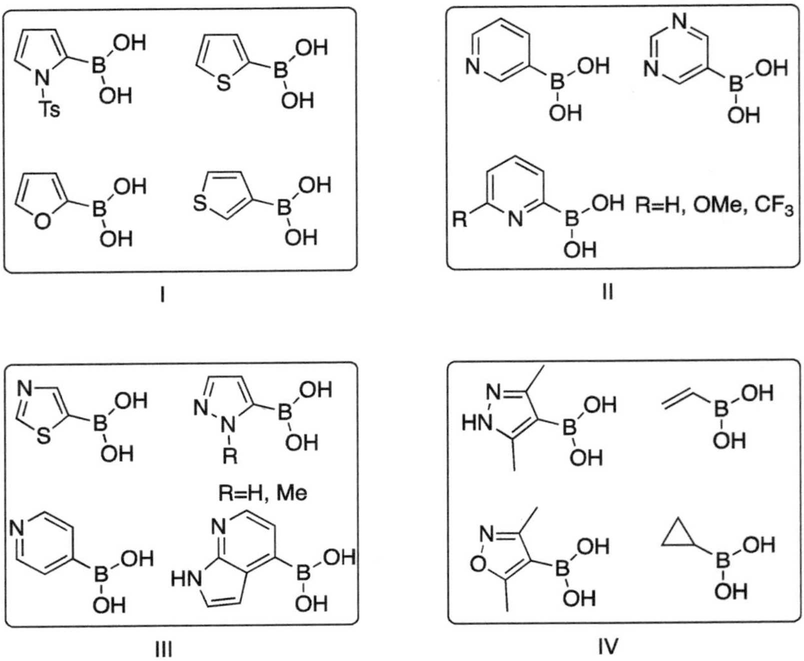

Fig. 47 Boronic acids organized according to their approximate stability type.

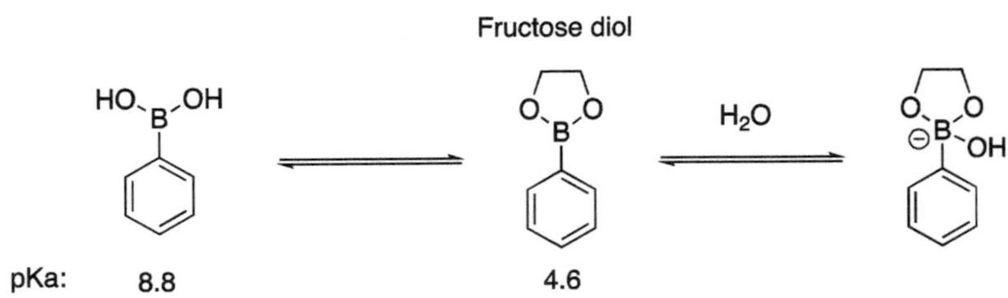

Fig. 48 Cyclic boronic esters and pK_a.

pyramidalization of the boron atoms suggesting that these examples would not be primed for protodeboronation.[96]

It was noted in the in vivo metabolite ID study of **23** (Scheme 3) that one of the metabolites arises from protodeboronation.[87]

3.3 Stability of boronic esters

Boronic esters are often used as a means to detect sugars wherein the diols present in some sugars permit the formation of stable esters.[38] There is a significant change in pK_a upon formation of the esters, with the ester being a better Lewis acid than the boronic acid (Fig. 48). This ability to form a complex can significantly stabilize the form with the diol bound. In cases where this form is required, the addition of an amine nitrogen that is able to form an intramolecular interaction can be beneficial. This is the basis of the MIDA boronates, that are used as slow release reservoirs for the equivalent boronic acid and which undergo slow hydrolysis in water requiring the cleavage of a weak B—N bond.[97]

3.4 Oligomerization and intramolecular complexation with boron

Boronic acids are known to trimerize to give boroxines (Scheme 6) although the details of the effect of aqueous conditions on this process are not clear.[95] An alternative mode of oligomerization was observed for some 6-membered boronic esters featuring an ortho-hydroxybenzoic acid substructure, that permitted them to form dimers and trimers with the structural features shown putatively assigned (Scheme 7).[86] Substituting ortho to the acid group was found to modulate and ultimately eliminate the oligomerization and presumably reflects the geometric requirements present in the dimer/trimer structure shown in which the carbonyl oxygen protrudes directly into the space that an ortho substituent would fill. All of the atoms in the newly formed ring are likely be sp^2 hybridized such that the ring itself will be flat.

In the quest for broad-spectrum activity against β-lactamases, it has been found that molecules like **31** are able to bind in one class in its bicyclic form

Scheme 6 The formation of boroxines.

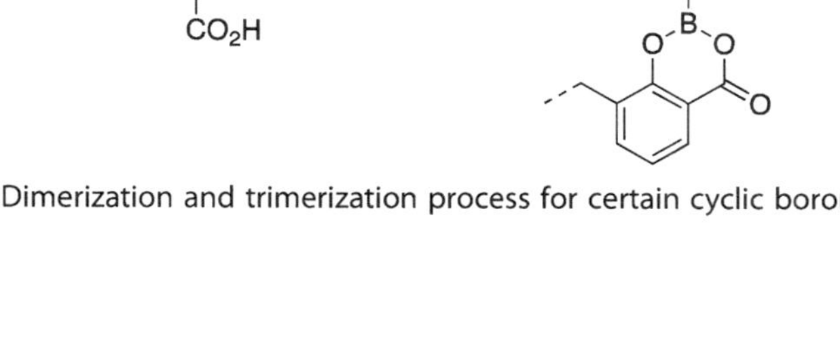

Scheme 7 Dimerization and trimerization process for certain cyclic boronic esters.

Fig. 49 Two alternative forms for compound **31**.

32 **33**

Fig. 50 Cyclisation of **32** to an inactive form, **33**.

(left hand side in Fig. 49) whereas in another, it binds in a mixture of the tricyclic (right hand side) and bicylic forms.[51] Whether this represents an equilibrium in free solution or a specific response to the presence of the enzymes requires further work. This chameleonic nature might offer other opportunities for enhanced or diverse activity. By contrast, the failure of the candidate drug talabostat, **32**, may have been due to cyclisation to an inactive form, **33** (Fig. 50).[3,98] The ability of intramolecular nitrogen groups to complex with boron has been probed computationally using the model compound **34**.[99] Calculations were benchmarked against the highly accurate CCSD(T) level of theory. A range of density functional theory approaches were used to compute the energy difference between open and closed forms. This revealed that achieving accuracy for such calculations is challenging and that many of the methods are not up to the task.

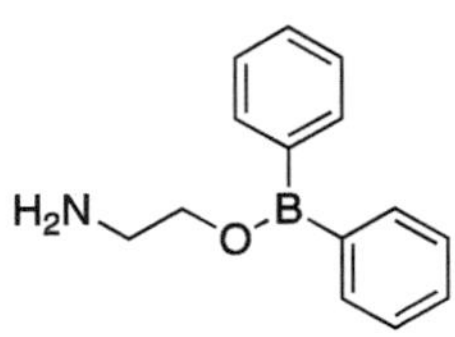

34

3.5 Key considerations in designing for stability

In general, boron-containing compounds have chemical and pharmacokinetic stability that is unremarkable. There are a number of processes that are specific to boron and which should be monitored. Oxidation deboronation can be biochemically or chemically facilitated while protodeboronation is likely promoted by the chemical environment and not promoted by enzymes. Some boron-containing compounds are chameleonic and can adopt a range of forms involving the forming and breaking of bonds and this can be a positive or negative feature.

4. Designing for safety

4.1 Boric acid

Given that oxidative deboronation is known to be a specific metabolic liability for compounds featuring a carbon boron bond, it is likely that this ensures that the administration of almost any boron-containing medicine will lead to exposure to boric acid or borates. Given that humans are exposed to borates in their diet, and plasma levels might be about 0.03 μg/mL,[100] a number of studies have investigated the safety aspects of this exposure. It was found that mice dosed at 1000 mg/kg of boric acid i.p. suffered from skeletal abnormalities in more than two thirds of animals.[101] The hypothesis that this might arise from the block of histone deacetylases leading to hyper-acetylation of histones was proposed although the evidence for this was weak. The LD_{50} for boric acid in mice was measured as being 2150 mg/kg from acute oral dosing.[102] In studies investigating its potential use to prevent histomoniasis in chickens, boric acid was found to be slightly detrimental to normal weight gain but to not influence histomoniasis infection.[103] The link between circulating levels of boron (in unspecified form but presumably predominantly as boric acid or borates) and thyroid hormones has been explored, as it has been proposed that boron may be responsible for the formation of goiters.[104] The levels in hair compared to the volume of the thyroid and between urinary boron levels and serum levels of the T4-thyroid hormone support a link but many details are lacking, particularly concerning the effect of the structures of the molecules containing boron. Boric acid was found to be non-mutagenic.[105]

Humans are estimated to consume approximately 1 mg of boron per day (predominantly as borates), mostly from fruit and vegetables although drinking wine increases the level of exposure.[100] Nonetheless, boron is a developmental toxin with a No-Observed Adverse Effect Level (NOAEL) of 9.6 mg/kg although those living near to a borate mine in Turkey were found to suffer from no effects on fertility. Understanding exposure during pregnancy is likely to be important to understand the risk posed by boron. A matched study of pregnant and non-pregnant women was undertaken and found lower levels of plasma boron in pregnant women.[100] Excretion and clearance rates are about the same in pregnant and non-pregnant women (about 1.3 mg/day) and the differences might be due to decreased tubular reuptake in pregnant women.

4.2 Boronic acids

The LD_{50} for a range of boronic acids after acute oral doses in mice was measured and shown in Table 5.[102] In several cases, damage to the CNS was noted post-mortem and behavioral effects were generally seen above 100 mg/kg. Raman spectroscopy was used to measure the levels of exposure in the CNS but no correspondence between exposure levels and CNS effects was found.

One of the dominant classes of boronic acid, noted in Section 2, features an α-amino group. Given that oxidative deboronation is a key metabolic pathway for many boronic acids, this group has the possibility of forming reactive imines (Fig. 51).[85] These imines can then undergo reaction with glutathione and such conjugates were detected with bortezomib and ixazomib (as its citrate complex (**35**), the form that is usually taken by patients) but not with **36**, the free form of ixazomib. Despite this clear toxicological liability, liver toxicity with bortezomib is extremely rare and this is attributed to the very low dose that is required. The key enzymes involved in the oxidative deboronation in these cases are the cytochrome P450s 3A4 and 2C19 although 1A2, 2C9, 2D6 and 2E1 all contribute.

A number of boronic acids (Fig. 52) were found to be mutagenic in the Ames test.[105] These results were not dependent upon activation with the S9 liver fraction used to mimic mammalian metabolism and the strain that detected the mutagenicity was not always the same. Given these findings and that boron remains a relatively new entrant into drug discovery with applications originally in oncology, the problems raised about exposure to boron-containing compounds has been a serious concern for many. It was particularly problematic for the portfolio of boron-containing compounds being developed by Anacor and this has prompted a series of studies detailed in the following section that has found these problems to not apply to boroxoles.[106] Given the potential problem with boronic acids, they are a concern not just in drug discovery but also for workers who must handle them when used as reaction intermediates and therefore sensitive methods for their detection are required and are being developed.[107] For instance, the detection of cyclopropyl boronic acid down to 0.8 ppm has been achieved.

Table 5 LD_{50} and specific organ effects of a range of boronic acids.

Boronic acid	**LD_{50} (mg/kg)**	**Specific damage notes**
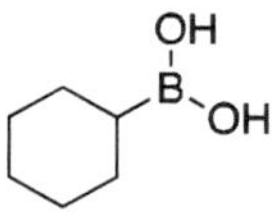	1250	–
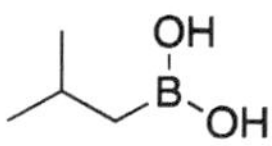	775	CNS
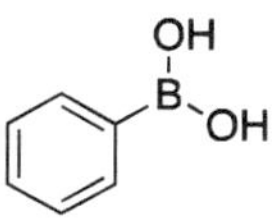	900	CNS
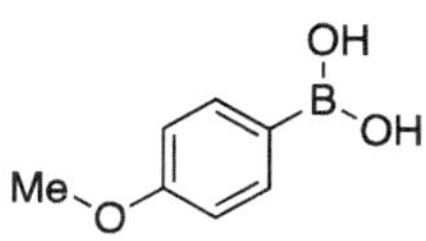	1250	CNS
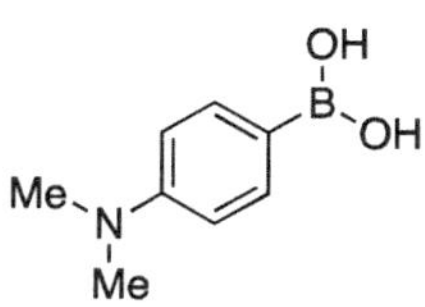	775	–
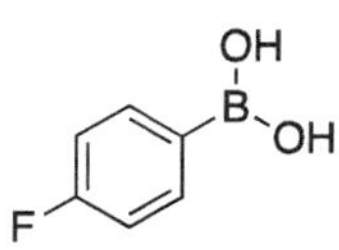	900	CNS
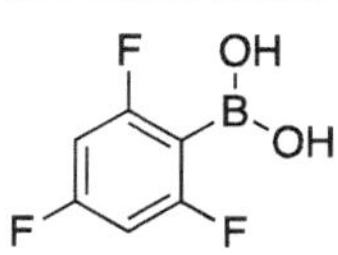	900	–
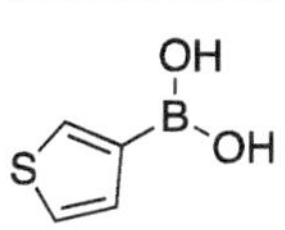	460	CNS, lungs, eyes, stomach, intestines

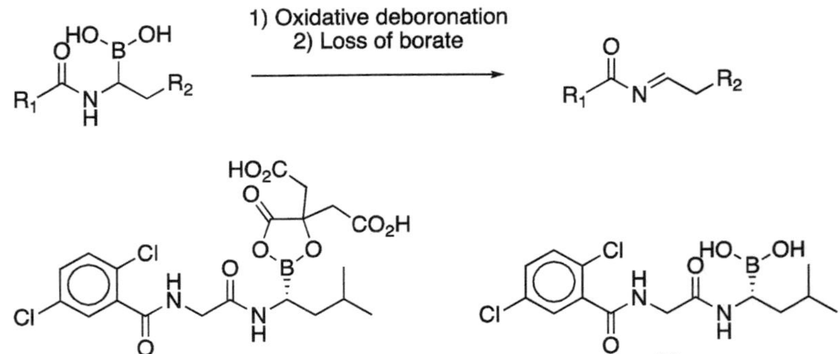

35 36

Fig. 51 A mechanism of potential toxicity concern: oxidative deboronation leading to imines that are reactive with biological nucleophiles such as glutathione.

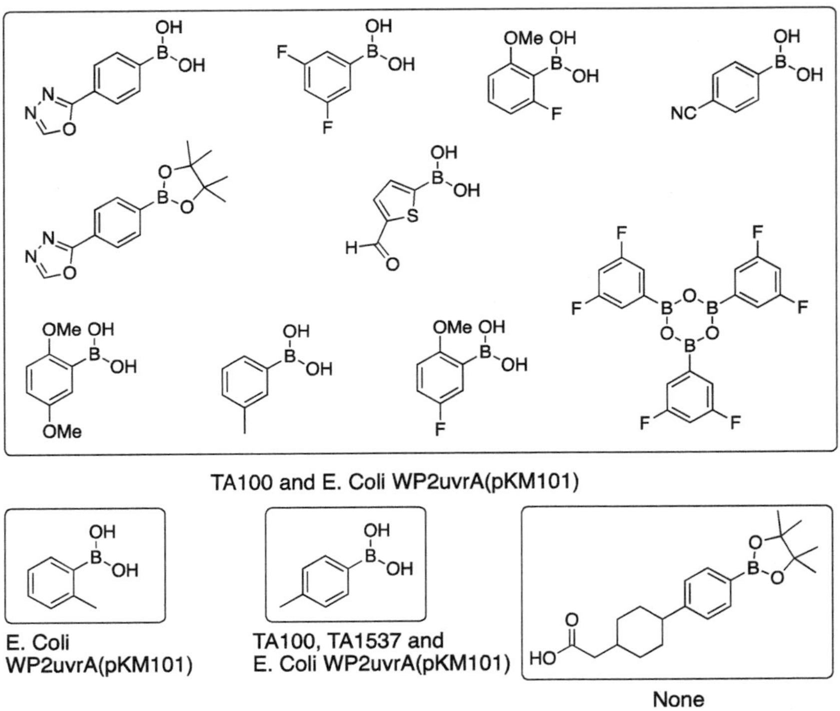

Fig. 52 Mutagenic activity of boron-containing compounds in the Ames test grouped according to the strains that were sensitive.

4.3 Boroxoles

Tavaborole is a boroxole-containing approved drug. It has been studied in great detail for a range of toxicological possibilities. It causes no significant inhibition of the cytochrome P450 enzymes that might raise concern for drug–drug interactions.[8]

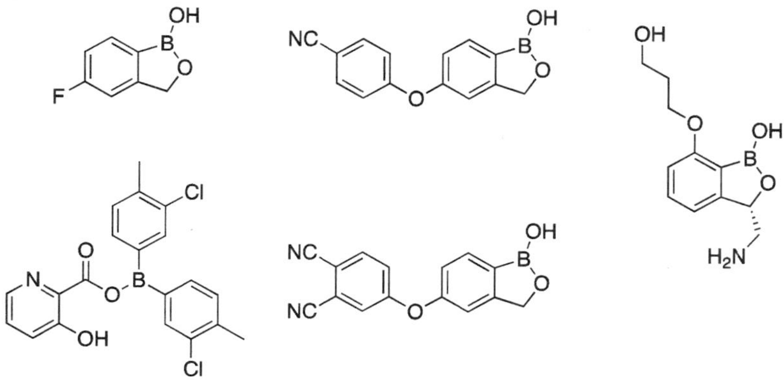

Fig. 53 Boron-containing compounds that were studied and found to be safe in a range of mutagenicity-related tests.

A series explored for antimalarial activity were tested and found to be non-cytotoxic.[71]

Following the mutagenicity results for boronic acids, significant scrutiny of other boron-containing compounds has been necessary and may be part of the reason that approvals outside of oncology so far have favored topical administration. By way of contrast, a very detailed study of several candidate drugs from Anacor (Fig. 53) saw investigations in a broader panel of tests.[106] In the Ames tests, there was a degree to which solubility limited some doses but no mutagenicity was detected in any of the strains. In the chromosome aberration study, the percentage of cells featuring structural aberrations were not higher than those with solvent control for any of the compounds with or without S9 activation. Furthermore, no change in polychromatic erythrocyte counts was found for any of the compounds in the in vivo micronucleus study. This encouraging finding suggests that boron can be consistent with drugs that will be safe for long term oral dosing but that this liability will require detailed study to alleviate concerns.

In separate studies, the safety of **37**, which has low clearance and therefore longer half-life was investigated and it was found to be negative in the Ames and micronucleus tests for mutagenicity, supporting it as a safe medicine for long term use.[108] Its NOAEL was assessed as being above 80 mg/kg/day.

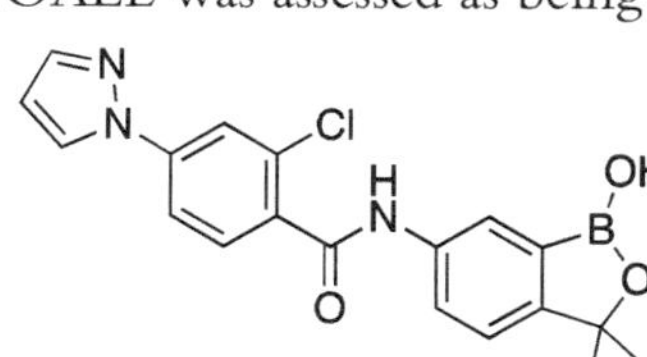

37

The related compound **14** was also found to have no Ames test positives up to 5000 μg/plate, hERG block was not observed up to concentrations of 100 μM, CYP inhibition by the compound and its oxidative deboronation metabolite was found to involve IC_{50} values no worse than 20 μM and the compound was well tolerated up to 100 mg/kg in mice.

Taking such studies even further and given that people will use tavaborole for very extended periods (even though it is a topical medicine), prolonged dosing effects were studied in life-long tests in mice to establish whether there is any indication of carcinogenicity.[109] Both dermal and oral dosing were performed and no effects were observed that were related to tavaborole for dermal dosing. Hyperplasia of the non-glandular stomach were the only observations possibly related to tavaborole in the oral-dosing arm. Therefore, tavaborole is deemed non-carcinogenic even upon prolonged exposure.

The reproductive toxicology of tavaborole has also been investigated.[110,111] At very high (300 mg/kg/day) oral doses in long term studies in rats, one of the rats suffered from damage related to non-glandular stomach effects and these are attributed to the drug. Rabbits dosed at 600 mg/kg/day suffered similar effects including breathing problems, also likely drug related. The NOAEL was established as being 100 mg/kg/d. No reproductive toxicity is seen until doses more than 100 times higher than the recommended human dose and no teratogenic effects are observed until 29 times the recommended human dose. This makes tavaborole safer than other medicines in the class. In a study of the impact of the drug on future generations, long-term studies in rats found no effect of tavaborole upon development and birth of offspring even if mothers were treated with oral tavaborole. Furthermore, there were no effects on their offspring's fertility or on their grandchildren.

4.4 Other boron-containing compounds

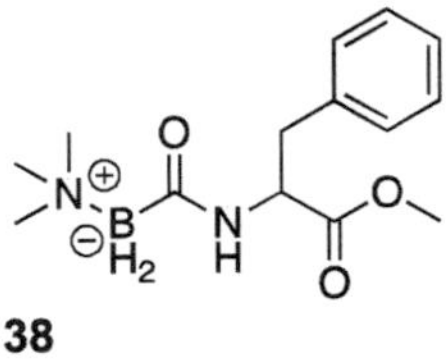

38

Compound **38** was investigated as a potential treatment for lymphoid leukemia and found to be well-tolerated at efficacious doses in an in vivo model.[112]

4.5 Key considerations in designing for safety

The safety of boron in the many forms in which humans are exposed to it is not clear cut. There are some concerns relating to borates that might be relevant (via metabolism) to any organic boron-containing compound. Despite the approval of bortezomib and ixazomib, the safety status of boronic acids with regards to mutagenicity remains a problem that any future compounds in this class, particularly for indications outside of oncology, must consider carefully. The status of boroxoles seems much more straightforward.

5. Opportunities and inspiration from nature

It has long been known that boron is required in plants and its role as an essential element for human health is being clarified.[3] Humans have boron at measurable levels in systemic circulation, presumably due to their diet.[100] A number of roles of boron-containing compounds in the natural world are beginning to be understood. For instance, covalent complexation with boron in molecules like **39** is essential in quorum sensing, a step in the production of biofilms that are a growing problem in the resistance to treatment of a range of microorganisms.[113] The preparation of inhibitors to interfere with the synthesis of **39** can disrupt biofilm formation, likely by disrupting quorum sensing. Nature is using the formation of covalent bonds with boron to radically alter the physical and molecular recognition elements of a simple sugar molecule. Two antibiotics that contain boron have been isolated from Streptomyces: boromycin and aplasmomycin both consist of complex macrolactones modified with the addition of boric acid to yield a borate complex.[3] Boromycin exerts its effect as a treatment against gram positive bacteria by acting as a potassium ionophore.[114] The use of natural boron-containing compounds (calcium fructoborate or sodium tetraborate) as adjuvants to enhance the efficacy of the antibody medicine etanercept has been tested although the evidence for such an effect is weak.[115]

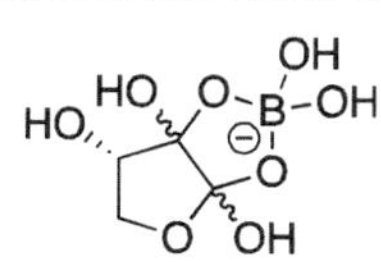

39

6. Concluding comments

We are currently in an exciting period for the application of boron-containing compounds as potential new therapeutics. The ability to form covalent bonds with oxygen (especially serine, threonine and RNA) present opportunities for potency that exceeds what can be achieved without covalency. Less well-explored possibilities for use as phosphate mimics and in complexes with metal centers merit attention. The pharmacokinetics and chemical stability of boron-containing compounds are generally unremarkable but a few specific liabilities should be monitored (oxidative- and proto-deboronation as well as inter- and intra-molecular complexation and bond exchange). The safety of boron-containing compounds, especially boronic acids, is a concern that requires further investigation although boroxoles appear to be safer compounds.

References

1. Paramore, A.; Frantz, S. Bortezomib. *Nat. Rev. Drug Discov.* **2003**, *2* (8), 611–612.
2. Fernandes, G. F. S.; Denny, W. A.; Dos Santos, J. L. Boron in Drug Design: Recent Advances in the Development of New Therapeutic Agents. *Eur. J. Med. Chem.* **2019**, *179*, 791–804.
3. Plescia, J.; Moitessier, N. Design and Discovery of Boronic Acid Drugs. *Eur. J. Med. Chem.* **2020**, *195*, 112270.
4. Shirley, M. Ixazomib: First Global Approval. *Drugs* **2016**, *76* (3), 405–411.
5. Markham, A. Tavaborole: First Global Approval. *Drugs* **2014**, *74* (13), 1555–1558.
6. Akama, T.; Baker, S. J.; Zhang, Y. K.; Hernandez, V.; Zhou, H.; Sanders, V.; Freund, Y.; Kimura, R.; Maples, K. R.; Plattner, J. J. Discovery and Structure-Activity Study of a Novel Benzoxaborole Anti-Inflammatory Agent (AN2728) for the Potential Topical Treatment of Psoriasis and Atopic Dermatitis. *Bioorg. Med. Chem. Lett.* **2009**, *19* (8), 2129–2132.
7. Mullard, A. 2017 FDA Drug Approvals. *Nat. Rev. Drug Discov.* **2018**, *17* (2), 81–85.
8. Jinna, S.; Finch, J. Spotlight on Tavaborole for the Treatment of Onychomycosis. *Drug Des. Dev. Ther.* **2015**, *9*, 6185–6190.
9. Diaz, D. B.; Yudin, A. K. The Versatility of Boron in Biological Target Engagement. *Nat. Chem.* **2017**, *9* (8), 731–742.
10. Hunter, P. Not Boring At All: Boron Is the New Carbon in the Quest for Novel Drug Candidates. *EMBO Rep.* **2009**, *10* (2), 125–128.
11. Hansch, C.; Leo, A.; Taft, R. A Survey of Hammett Substituent Constants and Resonance and Field Parameters. *Chem. Rev.* **1991**, *91* (2), 165–195.
12. Waring, M. J. Lipophilicity in Drug Discovery. *Expert Opin. Drug Discov.* **2010**, *5* (3), 235–248.
13. Davies, M.; Nowotka, M.; Papadatos, G.; Dedman, N.; Gaulton, A.; Atkinson, F.; Bellis, L.; Overington, J. P. ChEMBL Web Services: Streamlining Access to Drug Discovery Data and Utilities. *Nucleic Acids Res.* **2015**, *43* (W1), W612–W620.

14. Mendez, D.; Gaulton, A.; Bento, A. P.; Chambers, J.; De Veij, M.; Felix, E.; Magarinos, M. P.; Mosquera, J. F.; Mutowo, P.; Nowotka, M.; Gordillo-Maranon, M.; Hunter, F.; Junco, L.; Mugumbate, G.; Rodriguez-Lopez, M.; Atkinson, F.; Bosc, N.; Radoux, C. J.; Segura-Cabrera, A.; Hersey, A.; Leach, A. R. ChEMBL: Towards Direct Deposition of Bioassay Data. *Nucleic Acids Res.* **2019**, *47* (D1), D930–D940.
15. Kenny, P. W.; Sadowski, J. Structure Modification in Chemical Databases. In *Chemoinformatics in Drug Discovery*; Oprea, T. I., Ed.; Wiley, 2005; pp. 271–285.
16. Leach, A. G.; Jones, H. D.; Cosgrove, D. A.; Kenny, P. W.; Ruston, L.; MacFaul, P.; Wood, J. M.; Colclough, N.; Law, B. Matched Molecular Pairs as a Guide in the Optimization of Pharmaceutical Properties; a Study of Aqueous Solubility, Plasma Protein Binding and Oral Exposure. *J. Med. Chem.* **2006**, *49* (23), 6672–6682.
17. Różycka, D.; Leśnikowski, Z. J.; Olejniczak, A. B. Synthesis of Boron Cluster Analogs of Penicillin and Their Antibacterial Activity. *J. Organomet. Chem.* **2019**, *881*, 19–24.
18. Scholz, M. S.; Wingen, L. M.; Brunst, S.; Wittmann, S. K.; Cardoso, I. L.; Weizel, L.; Proschak, E. Soluble Epoxide Hydrolase Inhibitors with Carboranes as Non-natural 3-D Pharmacophores. *Eur. J. Med. Chem.* **2020**, *185*, 111766.
19. Ohta, K.; Ogawa, T.; Kaise, A.; Endo, Y. Novel Estrogen Receptor (ER) Modulators Containing Various Hydrophobic Bent-Core Structures. *Bioorg. Med. Chem.* **2014**, *22* (13), 3508–3514.
20. Asawa, Y.; Katsuragi, K.; Sato, A.; Yoshimori, A.; Tanuma, S. I.; Nakamura, H. Structure-Based Drug Design of Novel Carborane-Containing Nicotinamide Phosphoribosyltransferase Inhibitors. *Bioorg. Med. Chem.* **2019**, *27* (13), 2832–2844.
21. Olejniczak, A. B.; Adamska, A. M.; Paradowska, E.; Studzinska, M.; Suski, P.; Lesnikowski, Z. J. Modification of Selected Anti-HCMV Drugs with Lipophilic Boron Cluster Modulator. *Acta Pol. Pharm.* **2013**, *70* (3), 489–504.
22. Ohta, K.; Ogawa, T.; Oda, A.; Kaise, A.; Endo, Y. Design and Synthesis of Carborane-Containing Estrogen Receptor-Beta (ERbeta)-Selective Ligands. *Bioorg. Med. Chem. Lett.* **2015**, *25* (19), 4174–4178.
23. Ohta, K.; Ogawa, T.; Kaise, A.; Endo, Y. Enhanced Estrogen Receptor Beta (ERbeta) Selectivity of Fluorinated Carborane-Containing ER Modulators. *Bioorg. Med. Chem. Lett.* **2013**, *23* (24), 6555–6558.
24. Kojima, T.; Ogawa, T.; Kitao, S.; Sato, M.; Oda, A.; Ohta, K.; Endo, Y. Estrogenic Activity of Bis(4-Hydroxyphenyl)methanes With Cyclic Hydrophobic Structure. *Bioorg. Med. Chem.* **2015**, *23* (21), 6900–6911.
25. Mori, S.; Tsuemoto, N.; Kasagawa, T.; Nakano, E.; Fujii, S.; Kagechika, H. Development of Boron-Cluster-Based Progesterone Receptor Antagonists Bearing a Pentafluorosulfanyl (SF5) Group. *Chem. Pharm. Bull. (Tokyo)* **2019**, *67* (12), 1278–1283.
26. Scholz, M.; Blobaum, A. L.; Marnett, L. J.; Hey-Hawkins, E. Synthesis and Evaluation of Carbaborane Derivatives of Indomethacin as Cyclooxygenase Inhibitors. *Bioorg. Med. Chem.* **2011**, *19* (10), 3242–3248.
27. Fujii, S. Expanding the Chemical Space of Hydrophobic Pharmacophores: The Role of Hydrophobic Substructures in the Development of Novel Transcription Modulators. *MedChemComm* **2016**, 7 (6), 1082–1092.
28. Gabel, D. Boron Clusters in Medicinal Chemistry: Perspectives and Problems. *Pure Appl. Chem.* **2015**, *87* (2), 173–179.
29. Kahlert, J.; Austin, C. J.; Kassiou, M.; Rendina, L. M. The Fifth Element in Drug Design: Boron in Medicinal Chemistry. *Aust. J. Chem.* **2013**, *66* (10), 1118–1123.
30. Rak, J.; Dejlova, B.; Lampova, H.; Kaplanek, R.; Matejicek, P.; Cigler, P.; Kral, V. On the Solubility and Lipophilicity of Metallacarborane Pharmacophores. *Mol. Pharmaceutics* **2013**, *10* (5), 1751–1759.

31. Rokitskaya, T. I.; Khailova, L. S.; Makarenkov, A. V.; Shunaev, A. V.; Tatarskiy, V. V.; Shtil, A. A.; Ol'shevskaya, V. A.; Antonenko, Y. N. Carborane Derivatives of 1,2,3-Triazole Depolarize Mitochondria By Transferring Protons Through the Lipid Part of Membranes. *Biochim. Biophys. Acta* **2019**, *1861* (3), 573–583.
32. Vlasceanu, A.; Jessing, M.; Kilburn, J. P. BN/CC Isosterism in Borazaronaphthalenes Towards Phosphodiesterase 10A (PDE10A) Inhibitors. *Bioorg. Med. Chem.* **2015**, *23* (15), 4453–4461.
33. Boghi, M.; Hall, D. G. Valdecoxib vs. Borazavaldecoxib: Isoxazole BN/CC Isosterism as a Case Study in Designing and Stabilizing Boron Heterocycles. *Org. Biomol. Chem.* **2018**, *16* (26), 4849–4856.
34. Berman, H.; Henrick, K.; Nakamura, H. Announcing the Worldwide Protein Data Bank. *Nat. Struct. Biol.* **2003**, *10* (12), 980.
35. Krajnc, A.; Lang, P. A.; Panduwawala, T. D.; Brem, J.; Schofield, C. J. Will Morphing Boron-Based Inhibitors Beat the Beta-Lactamases? *Curr. Opin. Chem. Biol.* **2019**, *50*, 101–110.
36. Houk, K. N.; Leach, A. G.; Kim, S. P.; Zhang, X. Binding Affinities of Host-Guest, Protein-Ligand, and Protein-Transition-State Complexes. *Angew Chem. Int. Ed. Engl.* **2003**, *42* (40), 4872–4897.
37. Zhang, X.; Houk, K. N. Why Enzymes Are Proficient Catalysts: Beyond the Pauling Paradigm. *Acc. Chem. Res.* **2005**, *38* (5), 379–385.
38. McClary, C. A.; Taylor, M. S. Applications of Organoboron Compounds in Carbohydrate Chemistry and Glycobiology: Analysis, Separation, Protection, and Activation. *Carbohydr. Res.* **2013**, *381*, 112–122.
39. Baker, S. J.; Tomsho, J. W.; Benkovic, S. J. Boron-Containing Inhibitors of Synthetases. *Chem. Soc. Rev.* **2011**, *40* (8), 4279–4285.
40. Cox, P. A.; Leach, A. G.; Campbell, A. D.; Lloyd-Jones, G. C. Protodeboronation of Heteroaromatic, Vinyl, and Cyclopropyl Boronic Acids: pH-Rate Profiles, Autocatalysis, and Disproportionation. *J. Am. Chem. Soc.* **2016**, *138* (29), 9145–9157.
41. Cox, P. A.; Reid, M.; Leach, A. G.; Campbell, A. D.; King, E. J.; Lloyd-Jones, G. C. Base-Catalyzed Aryl-B(OH)2 Protodeboronation Revisited: From Concerted Proton Transfer to Liberation of a Transient Aryl Anion. *J. Am. Chem. Soc.* **2017**, *139* (37), 13156–13165.
42. Takahashi, L. H.; Radhakrishnan, R.; Rosenfield, R. E., Jr.; Meyer, E. F., Jr. Crystallographic Analysis of the Inhibition of Porcine Pancreatic Elastase by a Peptidyl Boronic Acid: Structure of a Reaction Intermediate. *Biochemistry* **1989**, *28* (19), 7610–7617.
43. Mace, J. E.; Agard, D. A. Kinetic and Structural Characterization of Mutations of Glycine 216 in Alpha-Lytic Protease: A New Target for Engineering Substrate Specificity. *J. Mol. Biol.* **1995**, *254* (4), 720–736.
44. Bone, R.; Frank, D.; Kettner, C. A.; Agard, D. A. Structural Analysis of Specificity: Alpha-Lytic Protease Complexes With Analogues of Reaction Intermediates. *Biochemistry* **1989**, *28* (19), 7600–7609.
45. Correy, G. J.; Zaidman, D.; Harmelin, A.; Carvalho, S.; Mabbitt, P. D.; Calaora, V.; James, P. J.; Kotze, A. C.; Jackson, C. J.; London, N. Overcoming Insecticide Resistance Through Computational Inhibitor Design. *Proc. Natl. Acad. Sci. U. S. A.* **2019**, *116* (42), 21012–21021.
46. Chen, Y.; McReynolds, A.; Shoichet, B. K. Re-Examining the Role of Lys67 in Class C Beta-Lactamase Catalysis. *Protein Sci.* **2009**, *18* (3), 662–669.
47. McKinney, D. C.; Zhou, F.; Eyermann, C. J.; Ferguson, A. D.; Prince, D. B.; Breen, J.; Giacobbe, R. A.; Lahiri, S.; Verheijen, J. C. 4,5-Disubstituted 6-Aryloxy-1,3-Dihydrobenzo[c][1,2]Oxaboroles Are Broad-Spectrum Serine Beta-Lactamase Inhibitors. *ACS Infect. Dis.* **2015**, *1* (7), 310–316.

48. Lewandowski, E. M.; Lethbridge, K. G.; Sanishvili, R.; Skiba, J.; Kowalski, K.; Chen, Y. Mechanisms of Proton Relay and Product Release by Class A Beta-Lactamase at Ultrahigh Resolution. *FEBS J.* **2018**, *285* (1), 87–100.
49. Minasov, G.; Wang, X.; Shoichet, B. K. An Ultrahigh Resolution Structure of TEM-1 Beta-Lactamase Suggests a Role for Glu166 as the General Base in Acylation. *J. Am. Chem. Soc.* **2002**, *124* (19), 5333–5340.
50. Liao, J. H.; Ihara, K.; Kuo, C. I.; Huang, K. F.; Wakatsuki, S.; Wu, S. H.; Chang, C. I. Structures of an ATP-Independent Lon-Like Protease and Its Complexes With Covalent Inhibitors. *Acta Crystallogr. D Biol. Crystallogr.* **2013**, *69* (Pt. 8), 1395–1402.
51. Krajnc, A.; Brem, J.; Hinchliffe, P.; Calvopina, K.; Panduwawala, T. D.; Lang, P. A.; Kamps, J.; Tyrrell, J. M.; Widlake, E.; Saward, B. G.; Walsh, T. R.; Spencer, J.; Schofield, C. J. Bicyclic Boronate VNRX-5133 Inhibits Metallo- and Serine-Beta-Lactamases. *J. Med. Chem.* **2019**, *62* (18), 8544–8556.
52. Werner, J. P.; Mitchell, J. M.; Taracila, M. A.; Bonomo, R. A.; Powers, R. A. Exploring the Potential of Boronic Acids as Inhibitors of OXA-24/40 Beta-Lactamase. *Protein Sci.* **2017**, *26* (3), 515–526.
53. Clevenger, K. D.; Wu, R.; Liu, D.; Fast, W. N-Alkylboronic Acid Inhibitors Reveal Determinants of Ligand Specificity in the Quorum-Quenching and Siderophore Biosynthetic Enzyme PvdQ. *Biochemistry* **2014**, *53* (42), 6679–6686.
54. Clevenger, K. D.; Wu, R.; Er, J. A.; Liu, D.; Fast, W. Rational Design of a Transition State Analogue With Picomolar Affinity for Pseudomonas Aeruginosa PvdQ, a Siderophore Biosynthetic Enzyme. *ACS Chem. Biol.* **2013**, *8* (10), 2192–2200.
55. Smith, T. P.; Windsor, I. W.; Forest, K. T.; Raines, R. T. Stilbene Boronic Acids Form a Covalent Bond With Human Transthyretin and Inhibit Its Aggregation. *J. Med. Chem.* **2017**, *60* (18), 7820–7834.
56. Su, S. C.; Lin, C. C.; Tai, H. C.; Chang, M. Y.; Ho, M. R.; Babu, C. S.; Liao, J. H.; Wu, S. H.; Chang, Y. C.; Lim, C.; Chang, C. I. Structural Basis for the Magnesium-Dependent Activation and Hexamerization of the Lon AAA+ Protease. *Structure* **2016**, *24* (5), 676–686.
57. Hideshima, T.; Richardson, P.; Chauhan, D.; Palombella, V. J.; Elliott, P. J.; Adams, J.; Anderson, K. C. The Proteasome Inhibitor PS-341 Inhibits Growth, Induces Apoptosis, and Overcomes Drug Resistance in Human Multiple Myeloma Cells. *Cancer Res.* **2001**, *61* (7), 3071–3076.
58. Groll, M.; Ditzel, L.; Lowe, J.; Stock, D.; Bochtler, M.; Bartunik, H. D.; Huber, R. Structure of 20S Proteasome From Yeast at 2.4 A Resolution. *Nature* **1997**, *386* (6624), 463–471.
59. Marques, A. J.; Palanimurugan, R.; Matias, A. C.; Ramos, P. C.; Dohmen, R. J. Catalytic Mechanism and Assembly of the Proteasome. *Chem. Rev.* **2009**, *109* (4), 1509–1536.
60. Saha, A.; Oanca, G.; Mondal, D.; Warshel, A. Exploring the Proteolysis Mechanism of the Proteasomes. *J. Phys. Chem. B* **2020**, *124* (27), 5626–5635.
61. Hausmann, J.; Kamtekar, S.; Christodoulou, E.; Day, J. E.; Wu, T.; Fulkerson, Z.; Albers, H. M.; van Meeteren, L. A.; Houben, A. J.; van Zeijl, L.; Jansen, S.; Andries, M.; Hall, T.; Pegg, L. E.; Benson, T. E.; Kasiem, M.; Harlos, K.; Kooi, C. W.; Smyth, S. S.; Ovaa, H.; Bollen, M.; Morris, A. J.; Moolenaar, W. H.; Perrakis, A. Structural Basis of Substrate Discrimination and Integrin Binding by Autotaxin. *Nat. Struct. Mol. Biol.* **2011**, *18* (2), 198–204.
62. Hausmann, J.; Keune, W. J.; Hipgrave Ederveen, A. L.; van Zeijl, L.; Joosten, R. P.; Perrakis, A. Structural Snapshots of the Catalytic Cycle of the Phosphodiesterase Autotaxin. *J. Struct. Biol.* **2016**, *195* (2), 199–206.

63. Stoll, V. S.; Eger, B. T.; Hynes, R. C.; Martichonok, V.; Jones, J. B.; Pai, E. F. Differences in Binding Modes of Enantiomers of 1-Acetamido Boronic Acid Based Protease Inhibitors: Crystal Structures of Gamma-Chymotrypsin and Subtilisin Carlsberg Complexes. *Biochemistry* **1998**, *37* (2), 451–462.
64. Transue, T. R.; Gabel, S. A.; London, R. E. NMR and Crystallographic Characterization of Adventitious Borate Binding by Trypsin. *Bioconjugate Chem.* **2006**, *17* (2), 300–308.
65. Zervosen, A.; Herman, R.; Kerff, F.; Herman, A.; Bouillez, A.; Prati, F.; Pratt, R. F.; Frere, J. M.; Joris, B.; Luxen, A.; Charlier, P.; Sauvage, E. Unexpected Tricovalent Binding Mode of Boronic Acids Within the Active Site of a Penicillin-Binding Protein. *J. Am. Chem. Soc.* **2011**, *133* (28), 10839–10848.
66. White, S. W.; Zheng, J.; Zhang, Y. M.; Rock, C. O. The Structural Biology of Type II Fatty Acid Biosynthesis. In *Annual Review of Biochemistry*; vol. 74; Annual Reviews: Palo Alto, 2005; pp. 791–831.
67. Rock, F. L.; Mao, W.; Yaremchuk, A.; Tukalo, M.; Crepin, T.; Zhou, H.; Zhang, Y. K.; Hernandez, V.; Akama, T.; Baker, S. J.; Plattner, J. J.; Shapiro, L.; Martinis, S. A.; Benkovic, S. J.; Cusack, S.; Alley, M. R. An Antifungal Agent Inhibits an Aminoacyl-tRNA Synthetase by Trapping tRNA in the Editing Site. *Science* **2007**, *316* (5832), 1759–1761.
68. Cama, E.; Emig, F. A.; Ash, D. E.; Christianson, D. W. Structural and Functional Importance of First-Shell Metal Ligands in the Binuclear Manganese Cluster of Arginase I. *Biochemistry* **2003**, *42* (25), 7748–7758.
69. Sonoiki, E.; Ng, C. L.; Lee, M. C.; Guo, D.; Zhang, Y. K.; Zhou, Y.; Alley, M. R.; Ahyong, V.; Sanz, L. M.; Lafuente-Monasterio, M. J.; Dong, C.; Schupp, P. G.; Gut, J.; Legac, J.; Cooper, R. A.; Gamo, F. J.; DeRisi, J.; Freund, Y. R.; Fidock, D. A.; Rosenthal, P. J. A Potent Antimalarial Benzoxaborole Targets a Plasmodium Falciparum Cleavage and Polyadenylation Specificity Factor Homologue. *Nat. Commun.* **2017**, *8* (1), 14574.
70. Zhang, Y.-K.; Plattner, J. J.; Akama, T.; Baker, S. J.; Hernandez, V. S.; Sanders, V.; Freund, Y.; Kimura, R.; Bu, W.; Hold, K. M. Design and Synthesis of Boron-Containing PDE4 Inhibitors Using Soft-Drug Strategy for Potential Dermatologic Anti-Inflammatory Application. *Bioorg. Med. Chem. Lett.* **2010**, *20* (7), 2270–2274.
71. Zhang, Y.-K.; Plattner, J. J.; Freund, Y. R.; Easom, E. E.; Zhou, Y.; Gut, J.; Rosenthal, P. J.; Waterson, D.; Gamo, F.-J.; Angulo-Barturen, I. Synthesis and Structure–Activity Relationships of Novel Benzoxaboroles as a New Class of Antimalarial Agents. *Bioorg. Med. Chem. Lett.* **2011**, *21* (2), 644–651.
72. Akama, T.; Dong, C.; Virtucio, C.; Freund, Y. R.; Chen, D.; Orr, M. D.; Jacobs, R. T.; Zhang, Y.-K.; Hernandez, V.; Liu, Y. Discovery and Structure–Activity Relationships of 6-(Benzoylamino) Benzoxaboroles As Orally Active Anti-Inflammatory Agents. *Bioorg. Med. Chem. Lett.* **2013**, *23* (21), 5870–5873.
73. Nare, B.; Wring, S.; Bacchi, C.; Beaudet, B.; Bowling, T.; Brun, R.; Chen, D.; Ding, C.; Freund, Y.; Gaukel, E. Discovery of Novel Orally Bioavailable Oxaborole 6-Carboxamides That Demonstrate Cure in a Murine Model of Late-Stage Central Nervous System African Trypanosomiasis. *Antimicrob. Agents Chemother.* **2010**, *54* (10), 4379–4388.
74. Jacobs, R. T.; Nare, B.; Wring, S. A.; Orr, M. D.; Chen, D.; Sligar, J. M.; Jenks, M. X.; Noe, R. A.; Bowling, T. S.; Mercer, L. T.; Rewerts, C.; Gaukel, E.; Owens, J.; Parham, R.; Randolph, R.; Beaudet, B.; Bacchi, C. J.; Yarlett, N.; Plattner, J. J.; Freund, Y.; Ding, C.; Akama, T.; Zhang, Y. K.; Brun, R.; Kaiser, M.; Scandale, I.; Don, R. SCYX-7158, an Orally-Active Benzoxaborole for the Treatment of Stage 2 Human African Trypanosomiasis. *PLoS Negl. Trop. Dis.* **2011**, *5* (6), e1151.

75. Hernandez, V.; Crepin, T.; Palencia, A.; Cusack, S.; Akama, T.; Baker, S. J.; Bu, W.; Feng, L.; Freund, Y. R.; Liu, L.; Meewan, M.; Mohan, M.; Mao, W.; Rock, F. L.; Sexton, H.; Sheoran, A.; Zhang, Y.; Zhang, Y. K.; Zhou, Y.; Nieman, J. A.; Anugula, M. R.; Keramane El, M.; Savariraj, K.; Reddy, D. S.; Sharma, R.; Subedi, R.; Singh, R.; O'Leary, A.; Simon, N. L.; De Marsh, P. L.; Mushtaq, S.; Warner, M.; Livermore, D. M.; Alley, M. R.; Plattner, J. J. Discovery of a novel class of boron-based antibacterials with activity against gram-negative bacteria. *Antimicrob. Agents Chemother.* **2013**, *57* (3), 1394–1403.
76. Ogiyama, T.; Yamaguchi, M.; Kurikawa, N.; Honzumi, S.; Yamamoto, Y.; Sugiyama, D.; Inoue, S. Identification of a Novel Boronic Acid as a Potent, Selective, and Orally Active Hormone Sensitive Lipase Inhibitor. *Bioorg. Med. Chem.* **2016**, *24* (16), 3801–3807.
77. Sood, C. K.; Sood, A.; Spielvogel, B. F.; Yousef, J. A.; Burnham, B.; Hall, I. Synthesis and Antineoplastic Activity of some Cyano-, Carboxy-, Carbomethoxy-, and Carbamoylborane Adducts of Heterocyclic Amines. *J. Pharm. Sci.* **1991**, *80* (12), 1133–1140.
78. Jacobs, R. T.; Plattner, J. J.; Keenan, M. Boron-Based Drugs as Antiprotozoals. *Curr. Opin. Infect. Dis.* **2011**, *24* (6), 586–592.
79. Zhu, Y.-Q.; Lei, M.; Lu, A.-J.; Zhao, X.; Yin, X.-J.; Gao, Q.-Z. 3D-QSAR Studies of Boron-Containing Dipeptides as Proteasome Inhibitors With CoMFA and CoMSIA Methods. *Eur. J. Med. Chem.* **2009**, *44* (4), 1486–1499.
80. Venturelli, A.; Tondi, D.; Cancian, L.; Morandi, F.; Cannazza, G.; Segatore, B.; Prati, F.; Amicosante, G.; Shoichet, B. K.; Costi, M. P. Optimizing Cell Permeation of an Antibiotic Resistance Inhibitor for Improved Efficacy. *J. Med. Chem.* **2007**, *50* (23), 5644–5654.
81. Milani, P.; Demasi, M.; de Rezende, L.; Antonia, T.; Andrade, L. H. Synthesis of L-Cysteine-Based Boron Compounds and Their Evaluation as Proteasome Inhibitors. *New J. Chem.* **2014**, *38* (10), 4859–4871.
82. Sgrignani, J.; Novati, B.; Colombo, G.; Grazioso, G. Covalent Docking of Selected Boron-Based Serine Beta-Lactamase Inhibitors. *J. Comput. Aided Mol. Des.* **2015**, *29* (5), 441–450.
83. Natsutani, I.; Iwata, R.; Yamai, Y. S.; Ishida, K.; Nagaoka, Y.; Sumiyoshi, T. Design, Synthesis and Evaluations of Spiro-Fused Benzoxaborin Derivatives as Novel Boron-Containing Compounds. *Chem. Biol. Drug Des.* **2019**, *93* (5), 657–665.
84. Faria, R. X.; de Jesus Hiller, N.; Salles, J. P.; Resende, J.; Diogo, R. T.; von Ranke, N. L.; Bello, M. L.; Rodrigues, C. R.; Castro, H. C.; de Luna Martins, D. Arylboronic Acids Inhibit P2X7 Receptor Function and the Acute Inflammatory Response. *J. Bioenerg. Biomembr.* **2019**, *51* (4), 277–290.
85. Li, A. C.; Yu, E.; Ring, S. C.; Chovan, J. P. Boronic Acid-Containing Proteasome Inhibitors: Alert to Potential Pharmaceutical Bioactivation. *Chem. Res. Toxicol.* **2013**, *26* (4), 608–615.
86. Hecker, S. J.; Reddy, K. R.; Lomovskaya, O.; Griffith, D. C.; Rubio-Aparicio, D.; Nelson, K.; Tsivkovski, R.; Sun, D.; Sabet, M.; Tarazi, Z.; Parkinson, J.; Totrov, M.; Boyer, S. H.; Glinka, T. W.; Pemberton, O. A.; Chen, Y.; Dudley, M. N. Discovery of Cyclic Boronic Acid QPX7728, an Ultrabroad-Spectrum Inhibitor of Serine and Metallo-Beta-Lactamases. *J. Med. Chem.* **2020**, *63* (14), 7491–7507.
87. Bu, W.; Akama, T.; Chanda, S.; Sullivan, D.; Ciaravino, V.; Jarnagin, K.; Freund, Y.; Sanders, V.; Chen, C. W.; Fan, X.; Heyman, I.; Liu, L. Early Rapid Identification of In Vivo Rat Metabolites of AN6414, a Novel Boron-Containing PDE4 Inhibitor by QTRAP LC/MS/MS to Support Drug Discovery. *J. Pharm. Biomed. Anal.* **2012**, *70*, 344–353.

88. Maynard, A.; Crosby, R. M.; Ellis, B.; Hamatake, R.; Hong, Z.; Johns, B. A.; Kahler, K. M.; Koble, C.; Leivers, A.; Leivers, M. R.; Mathis, A.; Peat, A. J.; Pouliot, J. J.; Roberts, C. D.; Samano, V.; Schmidt, R. M.; Smith, G. K.; Spaltenstein, A.; Stewart, E. L.; Thommes, P.; Turner, E. M.; Voitenleitner, C.; Walker, J. T.; Waitt, G.; Weatherhead, J.; Weaver, K.; Williams, S.; Wright, L.; Xiong, Z. Z.; Haigh, D.; Shotwell, J. B. Discovery of a Potent Boronic Acid Derived Inhibitor of the HCV RNA-Dependent RNA Polymerase. *J. Med. Chem.* **2014**, *57* (5), 1902–1913.
89. Wang, L.; Xie, S.; Ma, L.; Chen, Y.; Lu, W. 10-Boronic Acid Substituted Camptothecin as Prodrug of SN-38. *Eur. J. Med. Chem.* **2016**, *116*, 84–89.
90. Liu, J.; Zheng, S.; Guo, S.; Zhang, C.; Zhong, Q.; Zhang, Q.; Ma, P.; Skripnikova, E. V.; Bratton, M. R.; Wiese, T. E. Rational Design of a Boron-Modified Triphenylethylene (GLL398) as an Oral Selective Estrogen Receptor Downregulator. *ACS Med. Chem. Lett.* **2017**, *8* (1), 102–106.
91. Zheng, S.; Guo, S.; Zhong, Q.; Zhang, C.; Liu, J.; Yang, L.; Zhang, Q.; Wang, G. Biocompatible Boron-Containing Prodrugs of Belinostat for the Potential Treatment of Solid Tumors. *ACS Med. Chem. Lett.* **2018**, *9* (2), 149–154.
92. Zhang, C.; Guo, S.; Zhong, Q.; Zhang, Q.; Hossain, A.; Zheng, S.; Wang, G. Metabolism and Pharmacokinetic Study of the Boron-Containing Prodrug of Belinostat (ZL277), a Pan HDAC Inhibitor with Enhanced Bioavailability. *Pharmaceuticals (Basel)* **2019**, *12* (4), 180.
93. Omran, Z.; Abdalla, A. N.; Ibrahim, M. M.; Hossain, M. A.; Alarja, M.; Chen, L.; Liu, Y.; Wang, Q. Boronic Analogues of (R)-6-O-Desmethylantofine as Anticancer Agents. *Chem. Pharm. Bull.* **2019**, *67*, c19-00692.
94. Noonan, G.; Leach, A. G. A Mechanistic Proposal for the Protodeboronation of Neat Boronic Acids: Boronic Acid Mediated Reaction in the Solid State. *Org. Biomol. Chem.* **2015**, *13* (9), 2555–2560.
95. Hall, D. G. Structure, Properties, and Preparation of Boronic Acid Derivatives. Overview of Their Reactions and Applications. In *Boronic Acids*; Wiley, 2006; pp. 1–99.
96. Borys, K. M.; Matuszewska, A.; Wieczorek, D.; Kopczyńska, K.; Lipok, J.; Madura, I. D.; Adamczyk-Woźniak, A. Synthesis and Structural Elucidation of Novel Antifungal N-(Fluorophenyl) Piperazinyl Benzoxaboroles and Their Analogues. *J. Mol. Struct.* **2019**, *1181*, 587–598.
97. Gonzalez, J. A.; Ogba, O. M.; Morehouse, G. F.; Rosson, N.; Houk, K. N.; Leach, A. G.; Cheong, P. H.; Burke, M. D.; Lloyd-Jones, G. C. MIDA Boronates Are Hydrolysed Fast and Slow by Two Different Mechanisms. *Nat. Chem.* **2016**, *8* (11), 1067–1075.
98. Poplawski, S. E.; Lai, J. H.; Sanford, D. G.; Sudmeier, J. L.; Wu, W.; Bachovchin, W. W. Pro-Soft Val-boroPro: A Strategy for Enhancing In Vivo Performance of Boronic Acid Inhibitors of Serine Proteases. *J. Med. Chem.* **2011**, *54* (7), 2022–2028.
99. Dong, H.; Li, W.; Sun, J.; Li, S.; Klein, M. L. Understanding the Boron-Nitrogen Interaction and Its Possible Implications in Drug Design. *J. Phys. Chem. B* **2015**, *119* (45), 14393–14401.
100. Pahl, M. V.; Culver, B. D.; Strong, P. L.; Murray, F. J.; Vaziri, N. D. The Effect of Pregnancy on Renal Clearance of Boron in Humans: A Study Based on Normal Dietary Intake of Boron. *Toxicol. Sci.* **2001**, *60* (2), 252–256.
101. Di Renzo, F.; Cappelletti, G.; Broccia, M. L.; Giavini, E.; Menegola, E. Boric Acid Inhibits Embryonic Histone Deacetylases: A Suggested Mechanism to Explain Boric Acid-Related Teratogenicity. *Toxicol. Appl. Pharmacol.* **2007**, *220* (2), 178–185.
102. Soriano-Ursua, M. A.; Farfan-Garcia, E. D.; Lopez-Cabrera, Y.; Querejeta, E.; Trujillo-Ferrara, J. G. Boron-Containing Acids: Preliminary Evaluation of Acute Toxicity and Access to the Brain Determined by Raman Scattering Spectroscopy. *Neurotoxicology* **2014**, *40*, 8–15.

103. Beer, L.; Vuong, C.; Barros, T.; Latorre, J.; Tellez, G.; Fuller, A.; Hargis, B. Research Note: Evaluation of Boric Acid as a Chemoprophylaxis Candidate to Prevent Histomoniasis. *Poult. Sci.* **2020**, *99*, 1978–1982.
104. Popova, E. V.; Tinkov, A. A.; Ajsuvakova, O. P.; Skalnaya, M. G.; Skalny, A. V. Boron–A Potential Goiterogen? *Med. Hypothes.* **2017**, *104*, 63–67.
105. O'Donovan, M. R.; Mee, C. D.; Fenner, S.; Teasdale, A.; Phillips, D. H. Boronic Acids-a Novel Class of Bacterial Mutagen. *Mutat. Res.* **2011**, *724* (1–2), 1–6.
106. Ciaravino, V.; Plattner, J.; Chanda, S. An Assessment of the Genetic Toxicology of Novel Boron-Containing Therapeutic Agents. *Environ. Mol. Mutagen.* **2013**, *54* (5), 338–346.
107. Patel, I.; Venkatramani, C.; Stumpf, A.; Wigman, L.; Yehl, P. Trace Analysis of Potentially Mutagenic Boronic Acids and Esters in Drug Substance by ICP-MS. *Org. Process Res. Dev.* **2017**, *21* (2), 182–186.
108. Lunde, C. S.; Stebbins, E. E.; Jumani, R. S.; Hasan, M. M.; Miller, P.; Barlow, J.; Freund, Y. R.; Berry, P.; Stefanakis, R.; Gut, J.; Rosenthal, P. J.; Love, M. S.; McNamara, C. W.; Easom, E.; Plattner, J. J.; Jacobs, R. T.; Huston, C. D. Identification of a Potent Benzoxaborole Drug Candidate for Treating Cryptosporidiosis. *Nat. Commun.* **2019**, *10* (1), 2816.
109. Ciaravino, V.; Coronado, D.; Lanphear, C.; Shaikh, I.; Ruddock, W.; Chanda, S. Tavaborole, a Novel Boron-Containing Small Molecule for the Topical Treatment of Onychomycosis, Is Noncarcinogenic in 2-Year Carcinogenicity Studies. *Int. J. Toxicol.* **2014**, *33* (5), 419–427.
110. Ciaravino, V.; Coronado, D.; Lanphear, C.; Hoberman, A.; Chanda, S. Tavaborole, a Novel Boron-Containing Small Molecule Pharmaceutical Agent for Topical Treatment of Onychomycosis: I. Reproductive and Developmental Toxicity Studies. *Int. J. Toxicol.* **2016**, *35* (5), 530–542.
111. Ciaravino, V.; Coronado, D.; Lanphear, C.; Hoberman, A.; Chanda, S. Tavaborole, a Novel Boron-Containing Small Molecule Pharmaceutical Agent for Topical Treatment of Onychomycosis: II. Prenatal and Postnatal Developmental Toxicity and Maternal Function Study. *Int. J. Toxicol.* **2016**, *35* (5), 543–557.
112. Hall, I.; Hall, E.; Miller, M.; Sood, A.; Spielvogel, B. The Effects of Boron Containing Peptides on L 1210 Lymphoid Leukemia Metabolism. *Amino Acids* **1993**, *4* (3), 287–302.
113. Kadirvel, M.; Fanimarvasti, F.; Forbes, S.; McBain, A.; Gardiner, J. M.; Brown, G. D.; Freeman, S. Inhibition of Quorum Sensing and Biofilm Formation in Vibrio Harveyi by 4-Fluoro-DPD; a Novel Potent Inhibitor of AI-2 Signalling. *Chem. Commun.* **2014**, *50* (39), 5000–5002.
114. Moreira, W.; Aziz, D. B.; Dick, T. Boromycin Kills Mycobacterial Persisters Without Detectable Resistance. *Front. Microbiol.* **2016**, *7*, 199.
115. Hussain, S. A.; Abood, S. J.; Gorial, F. I. The Adjuvant Use of Calcium Fructoborate and Borax with Etanercept in Patients With Rheumatoid Arthritis: Pilot Study. *J. Intercult. Ethnopharmacol.* **2017**, *6* (1), 58.

CHAPTER SIX

Measuring and predicting warhead and residue reactivity

Ernest Awoonor-Williams, Jacob Kennedy, and Christopher N. Rowley*
Department of Chemistry, Memorial University of Newfoundland, St. John's, NL, Canada
*Corresponding author: e-mail address: crowley@mun.ca

Contents

1. Introduction

Descriptors of the reactivity of covalent warheads and their targets are an important tool in the development of new covalent inhibitors. By predicting the rates and spontaneity of the covalent modification of an amino acid by an inhibitor, inhibitors can be synthesized with more effective reactive functional groups (warheads) and amino acids more susceptible to covalent modification can be targeted.

The choice of an optimal reactive functional group is one of the factors that determine the activity of a covalent drug.[1,2] Acrylamides,[3–5] cyanoacrylamides,[6] and vinyl sulfones[7] have all been employed as reactive functional groups in covalent modifier drugs. These warheads must satisfy a number of criteria. They must be sufficiently reactive toward its target

Annual Reports in Medicinal Chemistry, Volume 56
ISSN 0065-7743
https://doi.org/10.1016/bs.armc.2020.09.001

amino acid to form a covalent linkage spontaneously and at rates sufficient to inhibit the target. The reactivity of an amino acid inside a protein varies due to effects like inter-residue electrostatic interactions, solvation, and the degree of its accessibility to the bound inhibitor, so prediction of amino acid reactivity is also needed. Descriptors to predict these types of reactivity would aid in the development of new covalent inhibitors.

In this chapter, we will briefly summarize the physical parameters that describe covalent inhibition, discuss conventional descriptors for these properties and those determined by electronic structure calculations, and discuss descriptors that predict the reactivity of residues in their targets.

2. Physical parameters of covalent inhibition

2.1 Binding kinetics and thermodynamics

Descriptors of the activity of a covalent inhibitor are directly or indirectly intended to serve as methods to predict the underlying physical parameters of covalent inhibition. For both covalent and non-covalent inhibitors, the first step in binding is the formation of a protein-ligand complex where the ligand is bound to the protein through intermolecular interactions. In a kinetic model, this can be defined in terms of a rate of association (k_{on}) and a rate of dissociation (k_{off}). Recently, there have been efforts to determine these rates and design inhibitors to have slower dissociation rates,[8] although these rates are generally so fast that this component of protein-ligand binding can be described as a rapid equilibrium,

$$P + L \overset{K_d}{\rightleftharpoons} P \cdot L. \tag{1}$$

This equilibrium is commonly defined in terms of its dissociation constant (K_d), which is simply the ratio of the dissociation rate constant (k_{off}) to the association rate constant (k_{on}) of the complex. This equilibrium can also be defined in terms of its Gibbs energy, which is related to K_d by,

$$\Delta G_{binding} = RT \ln K_d. \tag{2}$$

While the binding equilibrium constants have been the primary focus of analysis, the kinetics of an inhibitor binding to a protein have also been examined.[9,10] The rate constants of association and dissociation (k_{on} and k_{off}, respectively) can be considered directly, but the average time required for dissociation (τ_d) is often used as a metric for the kinetics of drug binding. For reversibly binding drugs, τ_d can be calculated from k_{off},

$$\tau_D = \frac{1}{k_{off}}. \tag{3}$$

The kinetics of covalent inhibition is more complicated because the non-covalent complex ($P{\cdot}L$) will undergo a chemical reaction to form a covalently linked protein-ligand complex (P–L),

$$P + L \overset{K_d}{\rightleftharpoons} P{\cdot}L \underset{k_{inact}^{-1}}{\overset{k_{inact}}{\rightleftharpoons}} P - L. \tag{4}$$

The reaction to form P–L is unimolecular in the concentration of $P{\cdot}L$, with rate constant k_{inact}. If this reaction is strongly exergonic, the rate of the reverse reaction (P–$L \rightarrow P{\cdot}L$, ${k_{\text{inact}}}^{-1}$) can be considered negligible and the formation of the covalent-adduct is effectively irreversible. Some covalent modification reactions are reversible (e.g., thio-Michael additions), so the ligand can still dissociate, albeit at low rates.[11–15] As a consequence, there are some covalent-but-reversible inhibitors.[6,16–18]

The measured rates of activation for covalent modifiers are often much slower than the binding/unbinding rates. For example, Schwartz et al. measured the kinetics of covalent modifier drug neratinib inhibiting EGFR-L858R/T790M.[19] A k_{on} of $37{,}000 \pm 46{,}000\ \mu\text{m}^{-1}\ \text{s}^{-1}$ was measured, while k_{inact} was $0.0011 \pm 0.0002\ \text{s}^{-1}$, so the rate of covalent modification will typically be rate-limiting.

Assuming that the covalent-modification step is effectively irreversible, the non-covalent binding step can be described by the equilibrium inhibition constant (K_i) and the rate of covalent bond formation can be described by the inactivation rate constant (k_{inact}). The ratio of these constants (i.e., k_{inact}/K_i) corresponds to the second-order rate constant for the formation of the covalent adduct directly from the free protein and ligand (i.e., $P + L \rightarrow P - L$, k_{inact}/K_i).[20]

The more complex kinetics of covalent inhibition presents additional challenges for the experimental determination binding parameters. A range of spectroscopic, electrochemical, and calorimetric techniques have been applied for this purpose.[21] Isothermal titration calorimetry (ITC) measures the heat released or consumed by protein-ligand complexes during a biomolecular binding event. Other kinetic measures that rely on determining product/substrate concentrations are less effective than ITC which measures the enzyme catalysis rate directly. The data provided from ITC can be used to determine k_{on}, k_{off}, and K_i in a single experiment, where other methods require multiple separate enzyme assays.[22]

The IC_{50} is another way to quantify the potency of a drug. The IC_{50} is the concentration of an inhibitor that produces half-maximum inhibitory effect against a biological component, such as the enzyme target or entire cells. The definition of the IC_{50} is more complicated for a covalent inhibitor than it is for a non-covalent inhibitor because the extent of inhibition changes over time, as more of the target is converted to the covalent adduct.[23] For this reason, the IC_{50}'s of covalent inhibitors are often reported along with the concentration of the enzyme and length of the incubation period (the time between the time the inhibitor is introduced and the time when the percentage inhibition is measured). The IC_{50} of an enzyme inhibited by a covalent inhibitor measured at two incubation periods can be used to determine k_{inact} and K_i.[24]

2.2 Mechanisms of covalent inhibition

The rate of covalent bond formation is generally described as a single, unimolecular net reaction rate, defined by rate constant k_{inact}, although these covalent modification reactions generally occur through multistep mechanisms. For example, covalent modification of cysteine, serine, and tyrosine typically requires the side-chain to be deprotonated to form the reactive nucleophilic form (Scheme 1). The familiar reference pK_a's of amino acid side chains are measured for model amino acid in solution, where the charged state is stabilized only by interactions with water molecules. When these residues are located in the active site of an enzyme and participate in acid-catalysis, the anionic protonation state is often stabilized by salt-bridges and electrostatic interactions, lowering the pK_a so that the enzyme predominately adopts these reactive protonation states. Some covalent inhibitors target non-catalytic active site residues instead. In these instances, the anionic state will not necessarily be stabilized by inter-residue interactions, so the pK_a can be closer to its solution value or even elevated. Scheme 1 shows the deprotonation reaction of some common nucleophilic amino acids in proteins and their intrinsic pK_a reference in solution.[25]

The mechanism of the modification of an amino acid typically proceeds through an ionic mechanism, where the targeted amino acid side-chain of the protein is deprotonated. This deprotonation yields a nucleophile that will react with the electrophilic warhead functional group of the inhibitor. The covalent adduct is then protonated to yield a neutral covalent adduct. In some instances, there is a basic amino acid present in the active site that is capable of donating or accepting the protons required for the protonation

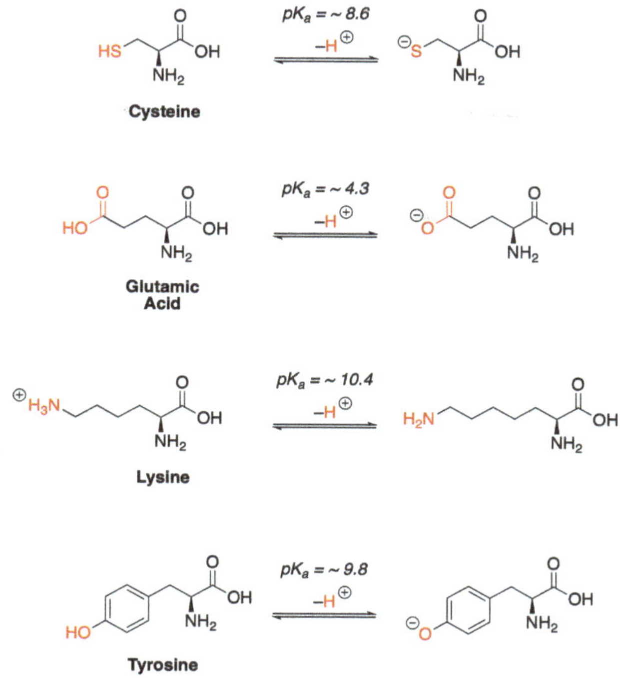

Scheme 1 Deprotonation reaction of selected nucleophilic amino acid residues in proteins. The nucleophilic side-chain of the amino acid residue is indicated in red.

and deprotonation steps, although there are also instances where no basic residue is nearby but the reaction proceeds anyway.

Despite the multistep mechanism of covalent inhibitors, their binding mechanism is typically considered to consist of at least two steps; a non-covalent and a covalent binding component. In theory, it is possible to define an equilibrium constant between the covalently bound state and the completely dissociated state, which would be an absolute binding energy that includes both covalent and non-covalent binding components. This type of approach will require the use of combined multiscale simulation methods such as hybrid quantum mechanics/molecular mechanics calculations due to the chemical bond formed. In addition, the non-covalent component of binding can be modeled by advanced molecular dynamics simulation techniques such as free energy perturbation methods. Routine and accurate first-principles models of covalent inhibition are currently impractical, especially for screening a large number of potential inhibitors against a number of targets.

2.3 Need for descriptors

A variety of functional groups are currently used as warheads in covalent inhibitors and many variants of these structures are possible. Likewise, there are several amino acid side-chains that can be covalently modified, but their intrinsic reactivity toward a given warhead and their relative reactivity with respect to each other can vary considerably. In many instances, experimental reactivity data is not currently available and can require elaborate experiments to obtain. In particular, both the inhibitor and target would have to be obtained in sizable, purified amounts to measure K_i and k_{inact} experimentally, which is not practical for screening a large number of potential inhibitors against multiple targets. The prediction of the reactivity of a new covalent inhibitor with a target based on the structures would greatly simplify the development of new covalent inhibitors. To a first approximation, we can attempt to predict k_{inact} for the reaction of a covalent inhibitor with a given amino acid based on chemical descriptors. The state of this field is described in Section 3 of this chapter.

3. Predicting warhead reactivity

3.1 Conventional physical organic parameters

Established physical organic descriptors have been used successfully to predict the reactivity of electrophilic covalent warheads.[26] Much of this work has focused on developing models to predict the reactivity of covalent warheads toward biologically relevant model nucleophiles such as glutathione (GSH) and *N*-α-acetyl-L-lysine.[26–28] Although the rate of reaction of a covalent inhibitor with glutathione is relevant to its off-target activity and toxicology, Schwartz et al. found no correlation of these rates with the k_{inact} for their inhibition of EGFR, suggesting that factors determining the rate of covalent modification of a specific protein are determined by different factors than those that determine their reaction rates with glutathione in solution.[19]

Cee et al. found a correlation between NMR chemical shifts (^{1}H and ^{13}C) and density functional theory computed activation parameters (E_a and $\Delta G^{\ddagger}$) with GSH reaction rates for a set of *N*-phenylacrylamides.[29] These trends could also be predicted using a simple linear Hammett correlation for substitution of the phenyl group, although this type of analysis is difficult to extend to more widely varied structural motifs.[29] In a similar vein, Ward et al. showed that the inductive Hammett parameter (σ_{ind} or σ_{meta}) predicted the log of the rates of reaction for a set of aryl acrylamides ($N=15$) reacting with glutathione with R^2 values near ~0.9, although limitations exist for more complex compounds.[30] In a recent study, Lonsdale et al. assessed the ability of several experimental and computational methods to predict the rates of reaction between cysteine-targeting covalent warheads and glutathione.[31] The half-life for the reaction of the covalent warheads with glutathione (i.e., GSH $t_{1/2}$) was used as an indicator for warhead reactivity for the different methods investigated. Among these methods, a pK_a approach that consists of calculating the pK_a of acrylamide warhead nitrogens in their corresponding amine matched molecular pairs was found to be a reliable predictor of GSH $t_{1/2}$. This predictor was shown to be very effective for acrylamide warheads that are connected to aromatic groups ($R^2=0.91$) but was less successful for acrylamide warheads connected to nonaromatic groups ($R^2=0.59$). Ultimately, a general and high-throughput method is needed that can provide consistent predictions across a range of warhead, including more complex motifs that have not yet been used as warheads.

3.2 Electronic structure based predictors

A number of metrics have been developed to describe the reactivity of an electrophile toward a nucleophile based on calculated electronic structure properties. These descriptors can be calculated routinely for arbitrary molecules using methods like density functional theory, allowing descriptors to be calculated for potential covalent inhibitors that have not yet been synthesized.

Smith and Rowley published an exhaustive search of DFT-calculated reaction energies and carbanion stability for the addition of a model thiol (methylthiol) to 325 alkenes with all permutations of -CN, -C(=O)OCH_3, -C(=O)$NHCH_3$, H, and CH_3 substituents.[32] Steric, atomic charges, and Natural Bond Orbital theory orbital energies were evaluated as descriptors to predict these reaction energies from the electronic structure of the alkene. Natural Bond Orbital (NBO) theory serves to transform the canonical Kohn-Sham molecular orbitals into a set of Natural Orbitals, so they are defined in terms of chemically intuitive Lewis-type orbitals, localized on the electrophile.

Smith and Rowley evaluated the NBO sigma, π and π* orbital energies, the natural population analysis charges of the α and β carbons of the olefin, the electrophilicity index, and the refractive indices of the substituents (a metric for steric bulk) as descriptors for the intermediate and product energies. The intermediate stability was taken as a measure of kinetic stability while the stability of the product was used as a measure of the spontaneity and reversibility of the formation of the thioether. The NBOs are shown for illustration in Fig. 1.

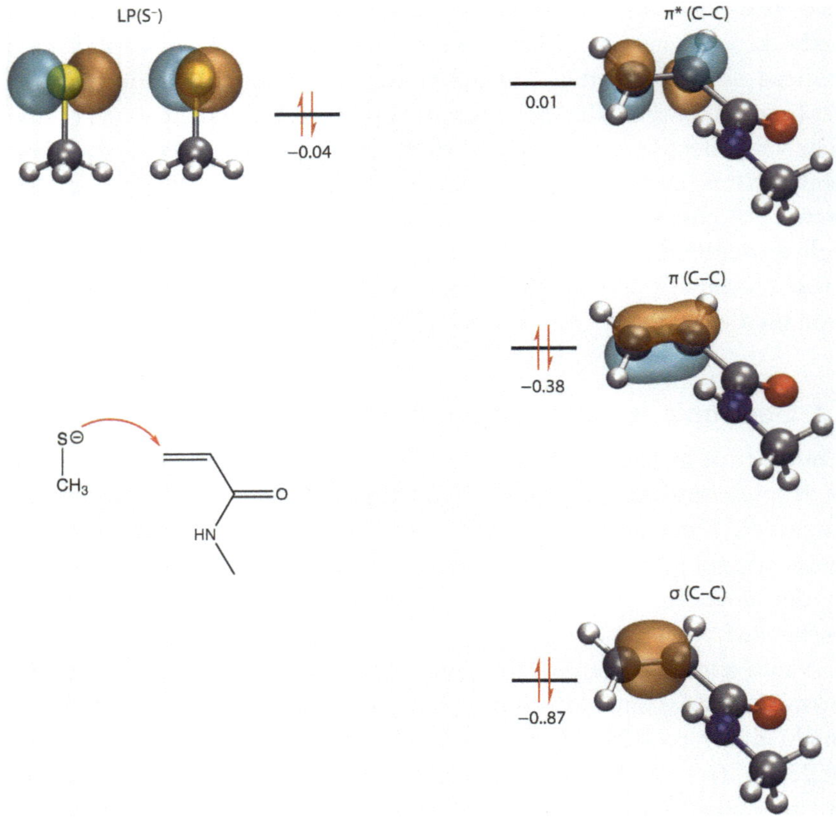

Fig. 1 Natural Bond Orbitals (NBO) used by Smith and Rowley to predict thiol addition carbanion and thioether stability. In the orbital description of the model thio-Michael addition of methylthiolate undergoing addition to acrylamide, the anionic charge centered in degenerate sulfur lone pairs is transferred to the pi* orbital of the electrophile. Calculations were performed using density functional theory, with the wB97X-D exchange-correlation functional and the TZVP basis set.

Analysis of electronic structure descriptors showed that the electrophilicity index was the best single descriptor for the stability of the carbanion intermediate. The electrophilicity index of a functional group was defined by Parr et al. and can be approximated by the orbital energies of the highest occupied molecular orbital (HOMO) and lowest unoccupied molecular orbital (LUMO) of the electrophile. This electrophilicity index can be approximated by the frontier orbitals of the electrophile,[33]

$$\omega = \frac{E_{\pi}^2 + 2E_{\pi}E_{\pi*} + E_{\pi*}^2}{4\left(E_{\pi*} - E_{\pi}\right)}. \tag{5}$$

The electrophilicity index predicts the stability of the carbanion intermediate with an R^2 value of 0.48, consistent with more electrophilic compounds yielding more stable carbanion intermediates. Multiple linear regression provided a model with an R^2 value of 0.83 that included the α-carbon charge and the steric bulk of the α-carbon substituents.

The stability of the products proved more difficult to predict. No single descriptor proved effective in predicting the ΔG_{prod}. Multiple linear regression found that a multiparameter model of the atomic charges of the α and β cartons and steric terms resulted in an R^2 value of 0.67. Linear regression identified that the magnitude of the atomic charges on the α and β carbons correlated to more positive Gibbs energies. Generally, this analysis suggested that the steric size of the substituents of the α-carbon of the electrophile must be limited and there should be at least one electrophilic group. Electrophiles with multiple electron-withdrawing groups are typically unstable compounds, which would be unsuitable for use in biological systems.

Palezzesi et al. also applied the electrophilicity index to estimate the reactivity of acrylamide warheads in covalent drug design. The HOMO and LUMO of the canonical Kohn-Sham molecular orbitals were used to calculate the electrophilicity index. This index was correlated to published experimental rates of reaction of these electrophiles with glutathione (a model for cysteine modification) and *N*-α-acetyl-L-lysine (a model for lysine modification). This method was found to be reliable for small molecules but was unreliable in larger molecules where the atomic orbitals contributing to the frontier orbitals were often unrelated to the electrophile. This illustrates the need for orbital localization, such as Boys,[34] or Pipek-Mezey[35] methods to describe the electrophilicity of a functional group within molecular orbital theory, or the use of a method like Natural Bond Orbital theory. Secondly, this descriptor was less effective at predicting the reactivity of nonterminal acrylamide warheads, indicating that electrophilicity alone is insufficient to predict the activity of all types of acrylamides.

3.2.1 Fukui function and the local electrophilicity index

Conceptual density functional theory provides additional methods for predicting the susceptibility of a site in a molecule to the type of nucleophilic attack that occurs in covalent modification of a cysteine residue. This analysis employs the Fukui function, which is defined as the rate at which the electron density at a point in space (**r**) changes as the number of electrons in the molecule is varied (Eq. 6).

$$f(\mathbf{r}) = \frac{\partial \rho(\mathbf{r})}{\partial N} \tag{6}$$

A nucleophilic attack on an electrophilic molecule can be thought of as the transfer of electron density. Locations on the molecular surface with more positive values of the Fukui function are more likely to be attacked by a nucleophile.

The standard practical method of calculating the Fukui function is to use an electronic structure method (e.g., density functional theory). This method is used to geometry-optimize a molecule. At the same molecular geometry, another calculation is performed where an electron is added to the molecule (e.g., the radical anion). This additional electron will be distributed across the molecule but concentrated near atoms with greater electrophilicity. This is defined as the positive Fukui function ($f+$),

$$f_{+}(\mathbf{r}) = \rho_{N+1}(\mathbf{r}) - \rho_{N}(\mathbf{r}). \tag{7}$$

By plotting this function at the molecular surface, the areas of the surface most susceptible to a nucleophilic attack are visually apparent. To a first approximation, these are the accessible regions of the molecular surface where the LUMO is concentrated, so the electron density from the HOMO of the nucleophile can be readily transferred into this empty orbital in these regions. For thio-Michael additions, this corresponds to transfer of electron density from the lone-pair orbital on the sulfur of the thiolate, where the anionic charge is localized, to the π^* orbital of the Michael acceptor (see, for example, the NBOs of Fig. 1). The nucleophilic Fukui function surface for three electrophilic warheads, phenyl vinyl sulfone, acrylamide, and cyanoacrylamide are plotted in Fig. 2, which shows the β-carbon as being a highly electrophilic center for all three electrophiles.

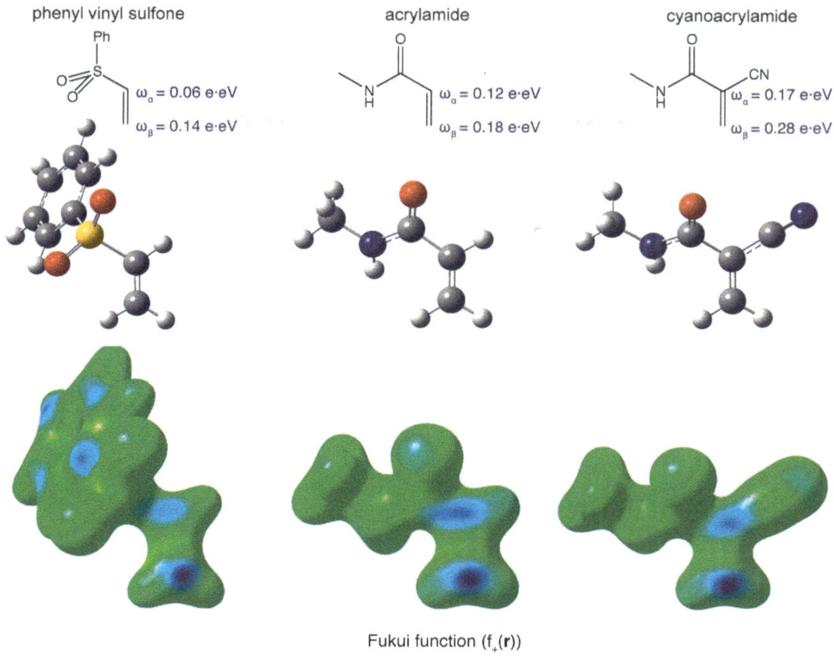

Fig. 2 The electrophilicity indices of phenyl vinyl sulfone (left), acrylamide (center), cyanoacrylamide (right) with the 3-dimensional structures and surface of the electrophilic Fukui function ($f+(r)$). The blue region indicates an area where a nucleophile is most likely to attack.

The same calculations can be used to calculate the atomic (a.k.a., condensed) Fukui function, which provides a quantitative measure of the propensity for an atom in a molecule to be the site of a nucleophilic attack. The condensed Fukui function is typically calculated by using a charge analysis method to assign partial atomic charges for the parent molecule and the anion. Commonly, Hirshfeld[36] or Natural Population Analysis[37] is used as charge analysis methods. Using the atomic charges for the neutral molecule and the radical anion ($q_{i,N}$ and $q_{i,N+1}$) respectively,

$$f_{i,+} = q_{i,N} - q_{i,N+1} \tag{8}$$

This form of the Fukui function is most relevant to covalent modification, which are typically nucleophilic protein side-chains reacting with an electrophilic functional group in the inhibitor, but the nucleophilic Fukui function can also be defined ($f-$) to indicate the areas of the molecule most susceptible to an electrophilic attack.

The condensed Fukui function can be combined with the molecular electrophilicity to provide the local electrophilicity index (ω_i), which indicates the propensity of an atom in a molecule to be the site of a nucleophilic attack (Eq. 9).

$$\omega_i = \omega \cdot f_{i,+} \tag{9}$$

The electrophilicity indices of C_β in phenyl vinyl sulfone, acrylamide, and cyanoacrylamide have large magnitudes and are correctly predicted to be more electrophilic than the C_α, although this index does not correctly predict that vinyl phenylsulfone is more reactive toward thiols than acrylamide.

These descriptors simplify the reaction between the anionic side chain nucleophile and the electrophile warhead to localize charge transfer reactions. This analysis is inherently approximate, as a full set of complex orbital interactions determines the activation energy of a nucleophilic attack and the electronic structure at the transition state differs from those of the reactants. These methods are most effective for gauging the relative electrophilicity for a common set of structurally related electrophiles and as a simple method to predict the site of nucleophilic attack.

3.2.2 Direct calculation of reaction energetics

Computational chemistry allows the stability of species on the reaction profile to be estimated by calculating the electronic structure of the species on the reaction profile. The stability of these species can serve as descriptors for analogous species of the covalent modification reaction inside the protein target. These calculations avoid the need for descriptors based purely on the electronic structure. One notable example of this was reported by Lonsdale et al., who found that the log of the half-life for the reaction between glutathione and acrylamide-based electrophiles was correlated to DFT calculations of both the activation energy and the stability of the thioether product.[31]

Thiol additions to some electrophiles are only modestly exergonic, so these additions are reversible in some cases. Although inhibitors featuring these warheads form a covalent bond with their target, they can dissociate from the target, albeit with exceptionally long residence times. Taunton and coworkers used the DFT-calculated proton affinity of the corresponding carbanion of acrylonitrile-based thiol adducts as a descriptor to predict the rate at which these covalent adducts undergo β-elimination to revert back to the cysteine thiol and dissociated ligand.[17] For a series of 10 model acrylonitriles substituted with aryl and heteroaryl electron-withdrawing groups reacting with a model thiol, β-mercaptoethanol, the calculated proton affinity predicted the thiol β-elimination rate for the adducts with an R^2 value of 0.96. These covalent adducts were found to undergo β-elimination at rates that correlate to the inverse of the proton affinity of their corresponding carbanions (Scheme 2).

Although the reaction energies were calculated using DFT, this type of modeling is very sensitive to the computational methods employed. The earliest ab initio models of covalent modification of thiols predicted a canonical ionic mechanism. Later models have described concerted or radical mechanisms. Smith et al. showed that several popular DFT methods (e.g., B3LYP, PBE, DFTB) predict that the enolate intermediate of a thiol-Michael addition is not an intermediate in this reaction mechanism, but high-level ab initio methods like CCSD(T) predict a stable enolate intermediate.[38] Analysis by Awoonor-Williams et al. showed that delocalization error in popular DFT methods is at fault, which spuriously favors states where the thiolate forms a non-covalent complex with the electrophile.[39] In many cases, this results in a significant underestimation of the stability of the carbanion intermediate and can even result in the lack of any potential energy minimum corresponding to this intermediate. Some DFT functionals (e.g., ωB97X-D, PBE0) describe these reactions accurately, so this type of modeling can be performed using cost-effective DFT methods provided an appropriate functional is used.

3.3 Predicting of side-chain reactivity

The mechanism of covalent modification is typically characterized by the reaction of a nucleophilic amino acid residue of a target protein with an

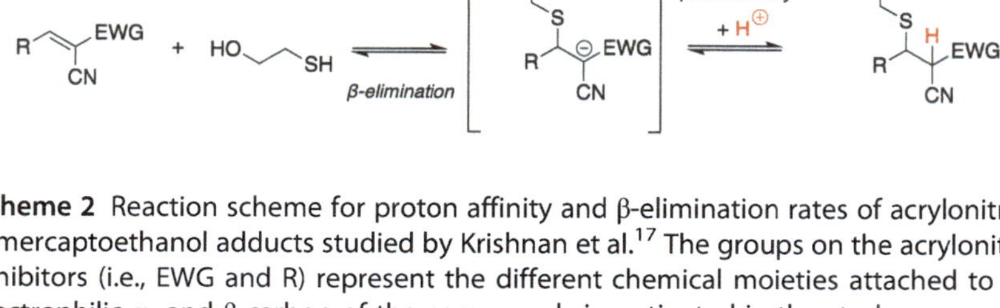

Scheme 2 Reaction scheme for proton affinity and β-elimination rates of acrylonitrile/β-mercaptoethanol adducts studied by Krishnan et al.[17] The groups on the acrylonitrile inhibitors (i.e., EWG and R) represent the different chemical moieties attached to the electrophilic α- and β-carbon of the compounds investigated in the study.

electrophilic reactive group, commonly referred to as "warhead," of an inhibitor or drug molecule. Typically, the first step in the covalent modification reaction process is the deprotonation of the nucleophilic amino acid side-chain in the target protein. Nucleophilic amino acids such as cysteine, serine, and lysine can be deprotonated to form a nucleophilic species ($-S^-$ for cysteine, $-O^-$ for serine, $-NH_2$ for lysine) that can react with the electrophilic functional group of an inhibitor. Currently, residues are targeted for covalent modification based on their location but prediction of the stability of a reactive, deprotonated form of these residues would help drug developers choose residues that are sufficiently reactive to be a viable target for a covalent inhibitor. The reactivity of the nucleophilic moiety towards an electrophilic warhead of an inhibitor or drug molecule depends on a complex range of factors, including steric repulsion, inductive effects, conjugative effects, and the protein environment. The nucleophilic amino acid side-chain to be deprotonated can be characterized by its pK_a. The pK_a of a titratable residue reflects the tendency of the ionizable residue to lose a proton and is a descriptor that can be used to assess the relative stabilities of both the protonated and deprotonated states of a titratable residue in a target protein. In principle, amino acids with low pK_a's are more prone to be reactive toward electrophilic warheads than amino acids with high pK_a's because the deprotonated form of the amino acid residue occurs with a high probability, allowing it to react readily with a covalent inhibitor. Beyond the pK_a, the orientation and accessibility of the deprotonated side-chain, and its proximity to a protein site appropriate for binding small molecular inhibitors, are also important factors. Scheme 3 shows the addition reaction of selected nucleophilic amino acid residues with different electrophilic warhead chemotypes.

Among the nucleophilic amino acids, cysteines are the most common targets for covalent-modifier drugs.[40] The modest pK_a of the cysteine thiol side-chain (p$K_a = 8.6$) allows for deprotonation of the neutral thiol form (-SH) to the more reactive nucleophilic thiolate state ($-S^-$) for covalent modification. Several studies have employed diverse approaches including bioinformatic studies to identify cysteines that are amenable to covalent modification within therapeutically relevant enzyme targets (e.g., kinases),[41–47] for selective covalent inhibition. More recently, studies are looking beyond cysteine,[48,49] as a means of targeting other nucleophilic amino acid residues such as lysine and tyrosine.

Although several experimental techniques are available for the determination the pK_a of an amino acid in a protein,[50] this can be challenging where

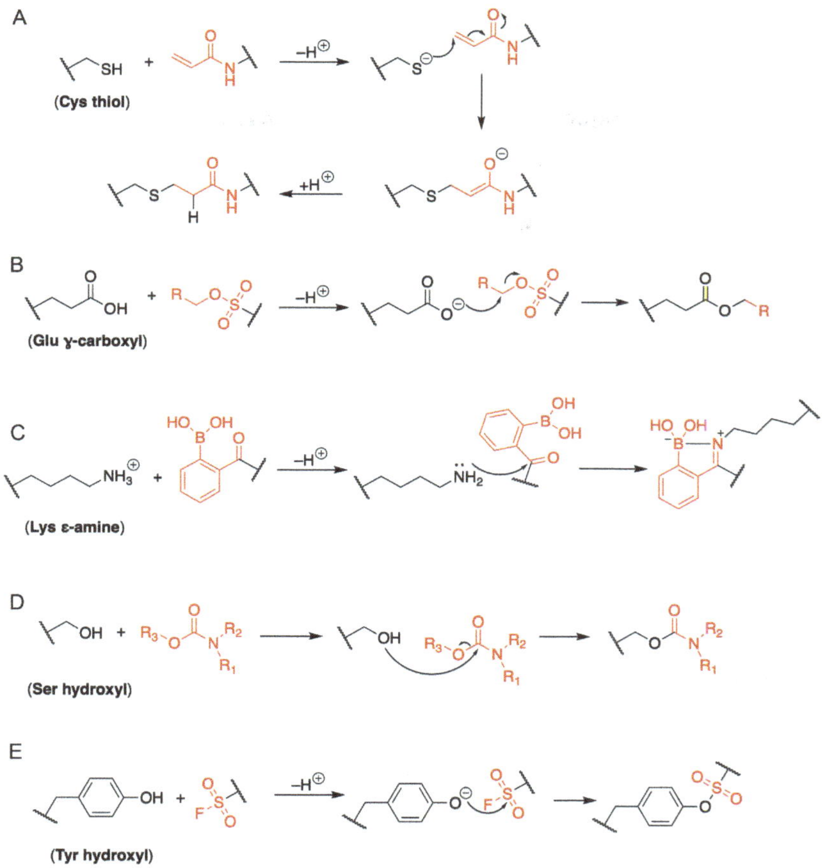

Scheme 3 Deprotonation of amino acid residues to form reactive nucleophile that reacts with different electrophilic warhead chemotypes (electrophilic warhead is shown in red). (A) Cysteine thiol group is covalently modified by acrylamide warhead. (B) Carboxyl group of glutamic acid is covalently modified by sulfonate ester. (C) Amine group of lysine covalently modified by boronic acid derivative (D) Hydroxyl group of serine covalently modified by carbamate. (E) Hydroxyl group of tyrosine covalently modified by sulfonyl fluoride.

individual titratable have strong electrostatic interactions with each other, which can stabilize or destabilize a given protonation state, shifting the pK_a's of the residues. Furthermore, experimental determination of pK_a requires synthesis and purification of significant quantities of the protein.

Computational pK_a calculation methods, which provide pK_a estimates based on the relative stabilities of protonation states, can offer a simpler route. Although these methods show limited accuracy in absolute pK_a predictions, particularly for cysteine pK_a's,[51] they can yield accurate relative pK_a's of titratable residues within enzyme targets.

3.3.1 Computational methods for calculating sidechain pK_a's

Empirical models and those using implicit solvent models can provide almost-immediate prediction of pK_a's using web-based or workstation-based methods. Popular examples of these methods include Delphipka,[52] PROPKA[53], H++,[54] and MCCE.[55] Awoonor and Rowley found that these methods were generally inefficient for predicting the pK_a's of cysteine residues from a test set where experimental pK_a's were available.[51] Instead, they found that techniques based on molecular simulation, where the pK_a of a residue is estimated by simulating the molecular dynamics of the charged and neutral states of an amino acid were more promising. Thermodynamic integration[56] and constant-pH molecular dynamics[57–60] are both molecular simulation based pK_a methods, where the pK_a of a residue is estimated based on its shift from its reference solution value due to its environment. The pK_a of a residue can then be estimated by its reference value and the calculated pK_a shift (ΔpK_a).

$$pK_a\ (\text{residue}) = pK_a\ (\text{reference}) + \Delta pK_a\ (\text{solvent} \rightarrow \text{protein}) \qquad (10)$$

where pK_a (reference) refers to the intrinsic pK_a reference of the titratable amino acid residue in solution, and ΔpK_a (solvent→protein) is a resultant term that comprises of all the intermolecular interactions experienced by the titratable residue when removed from the solvent and embedded into the protein.[61]

One drawback of simulation-based predictions of pK_a is that they require long simulations to achieve a converged pK_a. For example, a simulation cell of Bruton's tyrosine kinase (BTK) in an explicit aqueous solution contains approximately 40,000 atoms. A molecular dynamics simulation run on a contemporary compute node (a 32-core Intel Xeon E5-2557 @ 3.2 GHz) generates 8 ns per day. A constant-pH MD simulation of this simulation would require 10 replicas to sample a sufficient ranges of pH's and require upwards of 100 ns to achieve convergence, so the calculation of the pK_a for one residue would require a wall-clock time of 12 days, consuming 90,000 core-hours of computing time. Secondly, these simulations are sensitive to the parameters used to calculate the potential energy of a configuration (the molecular mechanical force field). Benchmarks showed that the CHARMM36 force field was more accurate than the AMBER ff99SB-ILDN.[51] This could be partially attributed to the lack of distinct non-bonded parameters for the cysteine thiolate state, which are included in the CHARMM force field.[62]

3.3.2 Applications of side-chain pK_a prediction

The pK_a of an ionizable residue in a target is a descriptor that can be used to estimate the relative tendency of the residue toward covalent modification by an electrophilic inhibitor. Using these acidity prediction methods, the calculated pK_a of an ionizable residue can be used as a descriptor to predict its susceptibility toward covalent modification. Relatively few studies have reported on the pK_a's of targetable ionizable residues in therapeutically relevant enzyme targets such as kinases.[47,63] In an effort to identify and predict the intrinsic reactivity of druggable kinase cysteines, Awoonor-Williams and Rowley employed replica-exchange thermodynamic integration and constant-pH molecular dynamics simulation methods to calculate the pK_a's of select druggable cysteines in clinically relevant protein kinase targets.[47] The druggable cysteines were classified based on their positions within the kinase domain and were confirmed to be covalently modified by experimental methods, such as X-ray crystallography or mass spectrometry. The calculated cysteine pK_a's were generally found to be elevated relative to the reference solution cysteine pK_a of ~8.5, suggesting that these cysteines rarely exist in the nucleophilic anionic cysteinate state, making them less reactive toward electrophiles. Paradoxically, the slow reactivity observed for the druggable kinase cysteines suggests that these residues may avoid S-oxidation or modification by non-specific electrophiles, so these elevated pK_a's may protect the target thiol for reaction with the targeted covalent inhibitor.

The key contributors for the perturbation in the selected kinase cysteine pK_a's include inter-residue electrostatic interactions and the degree to which the thiolate state of the protein can be stabilized by interactions with the solvent. Fig. 3 highlights some examples of druggable cysteines and their positions within the epidermal growth factor (EGFR) kinase domain color-coded according to their predicted pK_a's (data from Ref. 47). The red sphere indicates the P-Loop region of the kinase domain where a cysteine residue has a depressed pK_a (i.e., <9 pK units), while the blue sphere indicates the catalytic-loop region of the kinase domain where a selected cysteine residue (Cys788 of c-KIT) has a significantly elevated predicted pK_a (>13 pK units). The yellow spheres indicate positions for which

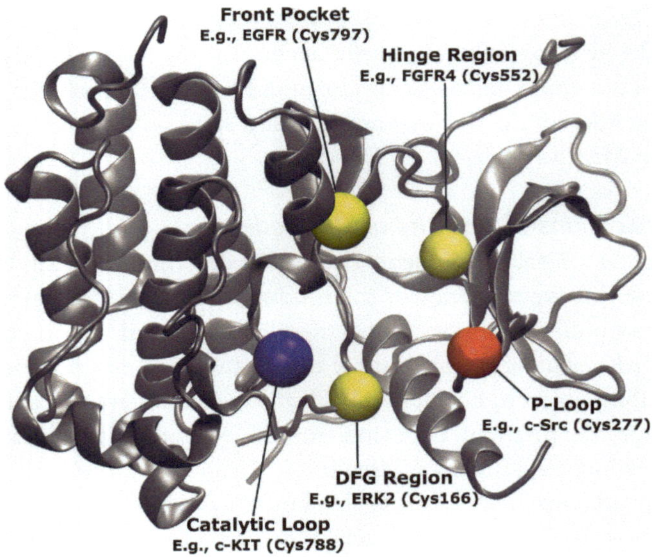

Fig. 3 Representation of positions of select druggable cysteines within EGFR kinase domain (PDB ID: 6VHN). The positions are color-coded based on the predicted pK_a's of the indicated cysteine residues for each region (data from Ref. 47). The red sphere represents the P-Loop region of the kinase domain and indicates a cysteine residue in that region with a predicted pK_a less than 9 pK units, for which Cys277 in c-Src kinase is an example. The yellow spheres represent the hinge, front, and DFG regions of the kinase domain. Cysteine residues indicated for this region were predicted to have moderate elevated pK_a's (i.e., pK_a from 9 to 13). The blue sphere represents the catalytic loop region of the kinase. Cysteine pK_a's within this region were predicted to have very elevated pK_a's (i.e., >13 pK units).

cysteine pK_a's were predicted to be moderately elevated (i.e., 9–13 pK units). In a similar vein, Lui et al. employed continuous constant-pH molecular dynamics simulations to predict the nucleophilicity of catalytic lysines in a group of kinases, based on their computed pK_a's.[63] They found that eight unique kinases when in a rare conformational state possess reactive catalytic lysines that can be targeted for covalent kinase inhibitor design.

4. Limitations of descriptor-based analysis

Accurate descriptors for the reactivity of functional groups and amino acid targets are relevant in covalent inhibitor design and discovery. The development of targeted covalent inhibitors (TCIs) is based on combining a scaffold with a high affinity and selectivity to its target based on non-covalent interactions. Schwartz et al.'s found that the inhibitory activity of a covalently binding ligand toward a kinase target resulted from both covalent and non-covalent interactions with the ligand.[19] Formation of a covalent linkage results in an even higher affinity and additional selectivity from the presence of a reactive side-chain in the target protein capable of undergoing reaction with the inhibitor.

Schwartz et al. showed that the k_{inact} alone was only a moderate predictor of the cellular IC_{50} of a covalent inhibitor ($R^2=0.60$), while the K_i was much more effective ($R^2=0.89$).[19] This suggests that the selectivity and affinity of a covalent inhibitor for its target in a biological setting is largely determined by the non-covalent interactions of the ligand. This phenomenon was also seen in a computational study by Luo and coworkers, who showed that the activity of a set of covalent ketoamide inhibitors of calpastatin could be predicted by the relative non-bonded interactions between the ligand and the target.[64,65] This indicates that descriptors for k_{inact} alone are not sufficient for predicting the activity of a covalent modifier drug; both non-covalent (i.e., K_i) and covalent interactions are important.

Protein turnover and off-target inhibition is yet another complicating factor in predicting the activity of a covalent inhibitor in a cellular environment. Backus et al. reported that a library of covalent inhibitor fragments showed a high degree of specificity toward their targets in situ, although these fragments were not designed to bind to a specific target.[66] One possible explanation for this is that the complex kinetics for both the covalent and non-covalent interactions between inhibitors and the cell contents gives rise to a specific pairing between inhibitors and proteins. Zaro et al. determined that an analogue of ibrutinib where a fumarate ester replaced the acrylamide warhead had similar activity to its target (Bruton's tyrosine kinase (BTK)), but had lower off-target activity in vivo.[67] They showed that the fumarate ester underwent metabolism at a greater rate than the acrylamide, therefore the free ligand in the intracellular solution was inactivated by metabolism rather than inhibiting off-target proteins. In this instance, the rate of metabolic degradation is the factor that determines the efficacy of the covalent inhibitor, so prediction of reactivity toward their target alone would not be effective for predicting the activity in a cellular environment.

5. Conclusions and future directions

Prediction of the reactivity of covalent warheads with model thiols or other biological reactive molecules have matured significantly over the last 10 years. The reactivity of a warhead to a model sidechain can be predicted using conventional physical organic parameters, electronic-structure-calculated descriptors, or direct calculation using ab initio methods. The calculated pK_a of an amino acid in a protein can also be used as a predictor of its activity. These methods are available immediately to researchers working to develop new warheads and identify new covalently druggable targets. These descriptors can help optimize aspects of covalent inhibition, but activity of these compounds is complicated by off-target activity and metabolism, so single descriptors are unlikely to be predictive of cellular inhibition.

The most immediate limitation to the development of a more complete and accurate set of descriptors is the limited amount of some types of quantitative experimental data to calibrate and validate models. The measurement of a broader set of rates of covalent adduct formation between various electrophilic fragments with models for reactive amino acids like cysteine and lysine would improve predictions of k_{inact}. The determination of inhibition parameters between covalent inhibitors and their targets (and off-targets) would also be a valuable source of data for the development of new modeling techniques.

The prediction of the susceptibility to covalent modification of an amino acid in a protein remains a major difficulty. Simple pK_a prediction methods are ineffective for the prediction of cysteine pK_a and even thermodynamic-integration simulations predict pK_a with an uncertainty of 1 pK_a unit. Additionally, there are relatively few experimental reports of the pK_a's of non-catalytic cysteine residues, so it is difficult to validate these computational methods. For this reason, the experimental determination of many of these pK_a's is an important task. The development of constant-pH MD methods, polarizable force fields,[68–70] QM/MM methods, and neural network potentials[71–73] are all promising routes for improving the accuracy of the predictions. Ultimately, prediction of the activity of a covalent inhibition will require rigorous, physics based models of the complete dynamics of binding and covalent bond formation.[74]

References

1. Jackson, P. A.; Widen, J. C.; Harki, D. A.; Brummond, K. M. Covalent Modifiers: A Chemical Perspective on the Reactivity of α,β-Unsaturated Carbonyls with Thiols via Hetero-Michael Addition Reactions. *J. Med. Chem.* **2017**, *60* (3), 839–885.
2. Gehringer, M.; Laufer, S. A. Emerging and Re-Emerging Warheads for Targeted Covalent Inhibitors: Applications in Medicinal Chemistry and Chemical Biology. *J. Med. Chem.* **2019**, *62* (12), 5673–5724.
3. Wakeling, A. E.; Guy, S. P.; Woodburn, J. R.; Ashton, S. E.; Curry, B. J.; Barker, A. J.; Gibson, K. H. ZD1839 (Iressa): An Orally Active Inhibitor of Epidermal Growth Factor Signaling with Potential for Cancer Therapy. *Cancer Res.* **2002**, *62* (20), 5749–5754.

4. Li, D.; Ambrogio, L.; Shimamura, T.; Kubo, S.; Takahashi, M.; Chirieac, L. R.; Padera, R. F.; Shapiro, G. I.; Baum, A.; Himmelsbach, F.; Rettig, W. J.; Meyerson, M.; Solca, F.; Greulich, H.; Wong, K.-K. BIBW2992, an Irreversible EGFR/HER2 Inhibitor Highly Effective in Preclinical Lung Cancer Models. *Oncogene* **2008**, *27* (34), 4702–4711.
5. Janne, P. A.; Ramalingam, S. S.; Yang, J. C.-H.; Ahn, M.-J.; Kim, D.-W.; Kim, S.-W.; Planchard, D.; Ohe, Y.; Felip, E.; Watkins, C.; Cantarini, M.; Ghiorghiu, S.; Ranson, M. Clinical Activity of the Mutant-Selective EGFR Inhibitor AZD9291 in Patients (pts) with EGFR Inhibitor-resistant Non-Small Cell Lung Cancer (NSCLC). *J. Clin. Oncol.* **2014**, *32*, 8009. https://doi.org/10.1200/jco.2014.32.15_suppl.8009.
6. Serafimova, I. M.; Pufall, M. A.; Krishnan, S.; Duda, K.; Cohen, M. S.; Maglathlin, R. L.; McFarland, J. M.; Miller, R. M.; Frödin, M.; Taunton, J. Reversible Targeting of Noncatalytic Cysteines with Chemically Tuned Electrophiles. *Nat. Chem. Biol.* **2012**, *8* (5), 471–476.
7. Kerr, I. D.; Lee, J. H.; Farady, C. J.; Marion, R.; Rickert, M.; Sajid, M.; Pandey, K. C.; Caffrey, C. R.; Legac, J.; Hansell, E.; McKerrow, J. H.; Craik, C. S.; Rosenthal, P. J.; Brinen, L. S. Vinyl Sulfones as Antiparasitic Agents and a Structural Basis for Drug Design. *J. Biol. Chem.* **2009**, *284*, 25697–25703. https://doi.org/10.1074/jbc.m109.014340.
8. Copeland, R. A. The Dynamics of Drug-Target Interactions: Drug-Target Residence Time and Its Impact on Efficacy and Safety. *Expert Opin. Drug Discovery* **2010**, *5*, 305–310. https://doi.org/10.1517/17460441003677725.
9. Nunes-Alves, A.; Kokh, D. B.; Wade, R. C. Recent Progress in Molecular Simulation Methods for Drug Binding Kinetics. *Curr. Opin. Struct. Biol.* **2020**, *64*, 126–133.
10. Copeland, R. A. Drug-Target Residence Time. *Thermodyn. Kinetics Drug Binding* **2015**, 155–167. https://doi.org/10.1002/9783527673025.ch8.
11. Gabizon, R.; Shraga, A.; Gehrtz, P.; Livnah, E.; Shorer, Y.; Gurwicz, N.; Avram, L.; Unger, T.; Aharoni, H.; Albeck, S.; Brandis, A.; Shulman, Z.; Katz, B.-Z.; Herishanu, Y.; London, N. Efficient Targeted Degradation Via Reversible and Irreversible Covalent PROTACs. *J. Am. Chem. Soc.* **2020**, *142* (27), 11734–11742.
12. Patel, H. M.; Pawara, R.; Surana, S. J. *Third Generation EGFR Inhibitors: Overcoming EGFR Resistance and Toxicity Problems*; Elsevier, 2018.
13. Tuley, A.; Fast, W. The Taxonomy of Covalent Inhibitors. *Biochemistry* **2018**, *57* (24), 3326–3337.
14. Dalby, K. N.; Jencks, W. P. General Acid Catalysis of the Reversible Addition of Thiolate Anions to Cyanamide. *J. Chem. Soc. Perkin Trans. 2* **1997**, 1555–1564. https://doi.org/10.1039/a700205j.
15. Gilbert, H. F.; Jencks, W. P. Mechanisms for Enforced General Acid Catalysis of the Addition of Thiol Anions to Acetaldehyde. *J. Am. Chem. Soc.* **1977**, 7931–7947. https://doi.org/10.1021/ja00466a029.
16. Miller, R. M.; Paavilainen, V. O.; Krishnan, S.; Serafimova, I. M.; Taunton, J. Electrophilic Fragment-Based Design of Reversible Covalent Kinase Inhibitors. *J. Am. Chem. Soc.* **2013**, *135* (14), 5298–5301.
17. Krishnan, S.; Miller, R. M.; Tian, B.; Mullins, R. D.; Jacobson, M. P.; Taunton, J. Design of Reversible, Cysteine-Targeted Michael Acceptors Guided by Kinetic and Computational Analysis. *J. Am. Chem. Soc.* **2014**, *136* (36), 12624–12630.
18. Bradshaw, J. M.; McFarland, J. M.; Paavilainen, V. O.; Bisconte, A.; Tam, D.; Phan, V. T.; Romanov, S.; Finkle, D.; Shu, J.; Patel, V.; Ton, T.; Li, X.; Loughhead, D. G.; Nunn, P. A.; Karr, D. E.; Gerritsen, M. E.; Funk, J. O.; Owens, T. D.; Verner, E.; Brameld, K. A.; Hill, R. J.; Goldstein, D. M.; Taunton, J. Prolonged and Tunable Residence Time Using Reversible Covalent Kinase Inhibitors. *Nat. Chem. Biol.* **2015**, *11* (7), 525–531.

19. Schwartz, P. A.; Kuzmic, P.; Solowiej, J.; Bergqvist, S.; Bolanos, B.; Almaden, C.; Nagata, A.; Ryan, K.; Feng, J.; Dalvie, D.; Kath, J. C.; Xu, M.; Wani, R.; Murray, B. W. Covalent EGFR Inhibitor Analysis Reveals Importance of Reversible Interactions to Potency and Mechanisms of Drug Resistance. *Proc. Natl. Acad. Sci. U. S. A.* **2014**, *111* (1), 173–178.
20. Strelow, J. M. A Perspective on the Kinetics of Covalent and Irreversible Inhibition. *SLAS Discov.* **2017**, *22* (1), 3–20.
21. Bisswanger, H. *Enzyme Kinetics: Principles and Methods*; John Wiley & Sons, 2017.
22. Di Trani, J. M.; De Cesco, S.; O'Leary, R.; Plescia, J.; do Nascimento, C. J.; Moitessier, N.; Mittermaier, A. K. Rapid Measurement of Inhibitor Binding Kinetics by Isothermal Titration Calorimetry. *Nat. Commun.* **2018**, *9* (1), 893.
23. Singh, J.; Petter, R. C.; Baillie, T. A.; Whitty, A. The Resurgence of Covalent Drugs. *Nat. Rev. Drug Discov.* **2011**, *10* (4), 307–317.
24. Kuzmič, P. A two-point IC50 method for evaluating the biochemical potency of irreversible enzyme inhibitors. *BioRxiv* **2020**. https://doi.org/10.1101/2020.06.25.171207.
25. Thurlkill, R. L.; Grimsley, G. R.; Martin Scholtz, J.; Nick Pace, C. pK Values of the Ionizable Groups of Proteins. *Protein Sci.* **2006**, *15*, 1214–1218. https://doi.org/10.1110/ps.051840806.
26. Flanagan, M. E.; Abramite, J. A.; Anderson, D. P.; Aulabaugh, A.; Dahal, U. P.; Gilbert, A. M.; Li, C.; Montgomery, J.; Oppenheimer, S. R.; Ryder, T.; Schuff, B. P.; Uccello, D. P.; Walker, G. S.; Wu, Y.; Brown, M. F.; Chen, J. M.; Hayward, M. M.; Noe, M. C.; Obach, R. S.; Philippe, L.; Shanmugasundaram, V.; Shapiro, M. J.; Starr, J.; Stroh, J.; Che, Y. Chemical and Computational Methods for the Characterization of Covalent Reactive Groups for the Prospective Design of Irreversible Inhibitors. *J. Med. Chem.* **2014**, *57* (23), 10072–10079.
27. Schwöbel, J. A. H.; Wondrousch, D.; Koleva, Y. K.; Madden, J. C.; Cronin, M. T. D.; Schüürmann, G. Prediction of Michael-Type Acceptor Reactivity toward Glutathione. *Chem. Res. Toxicol.* **2010**, *23* (10), 1576–1585.
28. MacFaul, P. A.; Morley, A. D.; Crawford, J. J. A Simple In Vitro Assay for Assessing the Reactivity of Nitrile Containing Compounds. *Bioorg. Med. Chem. Lett.* **2009**, *19* (4), 1136–1138.
29. Cee, V. J.; Volak, L. P.; Chen, Y.; Bartberger, M. D.; Tegley, C.; Arvedson, T.; McCarter, J.; Tasker, A. S.; Fotsch, C. Systematic Study of the Glutathione (GSH) Reactivity of N-Arylacrylamides: 1. Effects of Aryl Substitution. *J. Med. Chem.* **2015**, *58* (23), 9171–9178.
30. Ward, R. A.; Anderton, M. J.; Ashton, S.; Bethel, P. A.; Box, M.; Butterworth, S.; Colclough, N.; Chorley, C. G.; Chuaqui, C.; Cross, D. A. E.; Dakin, L. A.; Debreczeni, J.É.; Eberlein, C.; Finlay, M. R. V.; Hill, G. B.; Grist, M.; Klinowska, T. C. M.; Lane, C.; Martin, S.; Orme, J. P.; Smith, P.; Wang, F.; Waring, M. J. Structure- and Reactivity-Based Development of Covalent Inhibitors of the Activating and Gatekeeper Mutant Forms of the Epidermal Growth Factor Receptor (EGFR). *J. Med. Chem.* **2013**, *56* (17), 7025–7048.
31. Lonsdale, R.; Burgess, J.; Colclough, N.; Davies, N. L.; Lenz, E. M.; Orton, A. L.; Ward, R. A. Expanding the Armory: Predicting and Tuning Covalent Warhead Reactivity. *J. Chem. Inf. Model.* **2017**, *57* (12), 3124–3137.
32. Smith, J. M.; Rowley, C. N. Automated Computational Screening of the Thiol Reactivity of Substituted Alkenes. *J. Comput. Aided Mol. Des.* **2015**, *29* (8), 725–735.
33. Parr, R. G.; Szentpály, L. V.; Liu, S. Electrophilicity Index. *J. Am. Chem. Soc.* **1999**, *121*, 1922–1924. https://doi.org/10.1021/ja983494x.
34. Foster, J. M.; Boys, S. F. Canonical Configurational Interaction Procedure. *Rev. Mod. Phys.* **1960**, *32*, 300–302. https://doi.org/10.1103/revmodphys.32.300.

35. Pipek, J.; Mezey, P. G. A Fast Intrinsic Localization Procedure Applicable for Ab Initio and Semiempirical Linear Combination of Atomic Orbital Wave Functions. *J. Chem. Phys.* **1989**, 4916–4926. https://doi.org/10.1063/1.456588.
36. Hirshfeld, F. L. Bonded-Atom Fragments for Describing Molecular Charge Densities. *Theor. Chim. Acta* **1977**, *44*, 129–138. https://doi.org/10.1007/bf00549096.
37. Reed, A. E.; Weinstock, R. B.; Weinhold, F. Natural-population analysis. *J. Chem. Phys.* **1985**, *83*, 735–746. https://doi.org/10.1063/1.449486.
38. Smith, J. M.; Jami Alahmadi, Y.; Rowley, C. N. Range-Separated DFT Functionals Are Necessary to Model Thio-Michael Additions. *J. Chem. Theory Comput.* **2013**, *9* (11), 4860–4865.
39. Awoonor-Williams, E.; Isley, W. C., 3rd; Dale, S. G.; Johnson, E. R.; Yu, H.; Becke, A. D.; Roux, B.; Rowley, C. N. Quantum Chemical Methods for Modeling Covalent Modification of Biological Thiols. *J. Comput. Chem.* **2020**, *41* (5), 427–438.
40. Baillie, T. A. Targeted Covalent Inhibitors for Drug Design. *Angew. Chem. Int. Ed. Engl.* **2016**, *55* (43), 13408–13421.
41. Cohen, M. S.; Zhang, C.; Shokat, K. M.; Taunton, J. Structural Bioinformatics-Based Design of Selective, Irreversible Kinase Inhibitors. *Science* **2005**, *308* (5726), 1318–1321.
42. Leproult, E.; Barluenga, S.; Moras, D.; Wurtz, J.-M.; Winssinger, N. Cysteine Mapping in Conformationally Distinct Kinase Nucleotide Binding Sites: Application to the Design of Selective Covalent Inhibitors. *J. Med. Chem.* **2011**, *54*, 1347–1355. https://doi.org/10.1021/jm101396q.
43. Liu, Q.; Sabnis, Y.; Zhao, Z.; Zhang, T.; Buhrlage, S. J.; Jones, L. H.; Gray, N. S. Developing Irreversible Inhibitors of the Protein Kinase Cysteinome. *Chem. Biol.* **2013**, *20* (2), 146–159.
44. Zhang, Y.; Zhang, D.; Tian, H.; Jiao, Y.; Shi, Z.; Ran, T.; Liu, H.; Lu, S.; Xu, A.; Qiao, X.; Pan, J.; Yin, L.; Zhou, W.; Lu, T.; Chen, Y. Identification of Covalent Binding Sites Targeting Cysteines Based on Computational Approaches. *Mol. Pharm.* **2016**, *13* (9), 3106–3118.
45. Zhao, Z.; Liu, Q.; Bliven, S.; Xie, L.; Bourne, P. E. Determining Cysteines Available for Covalent Inhibition across the Human Kinome. *J. Med. Chem.* **2017**, *60* (7), 2879–2889.
46. Chaikuad, A.; Koch, P.; Laufer, S. A.; Knapp, S. The Cysteinome of Protein Kinases as a Target in Drug Development. *Angew. Chem. Int. Ed. Engl.* **2018**, *57* (16), 4372–4385.
47. Awoonor-Williams, E.; Rowley, C. N. How Reactive Are Druggable Cysteines in Protein Kinases? *J. Chem. Inf. Model.* **2018**, *58* (9), 1935–1946.
48. Pettinger, J.; Carter, M.; Jones, K.; Cheeseman, M. D. Kinetic Optimization of Lysine-Targeting Covalent Inhibitors of HSP72. *J. Med. Chem.* **2019**, *62* (24), 11383–11398.
49. Mukherjee, H.; Grimster, N. P. Beyond Cysteine: Recent Developments in the Area of Targeted Covalent Inhibition. *Curr. Opin. Chem. Biol.* **2018**, *44*, 30–38.
50. Reijenga, J.; van Hoof, A.; van Loon, A.; Teunissen, B. Development of Methods for the Determination of pK_a Values. *Anal. Chem. Insights* **2013**, *8*, 53–71.
51. Awoonor-Williams, E.; Rowley, C. N. Evaluation of Methods for the Calculation of the pKa of Cysteine Residues in Proteins. *J. Chem. Theory Comput.* **2016**, *12*, 4662–4673. https://doi.org/10.1021/acs.jctc.6b00631.
52. Pahari, S.; Sun, L.; Basu, S.; Alexov, E. DelPhiPKa: Including Salt in the Calculations and Enabling Polar Residues to Titrate. *Proteins* **2018**, *86* (12), 1277–1283.
53. Olsson, M. H. M.; Søndergaard, C. R.; Rostkowski, M.; Jensen, J. H. PROPKA3: Consistent Treatment of Internal and Surface Residues in Empirical pKa Predictions. *J. Chem. Theory Comput.* **2011**, 7 (2), 525–537.
54. Anandakrishnan, R.; Aguilar, B.; Onufriev, A. V. H++ 3.0: Automating pK Prediction and the Preparation of Biomolecular Structures for Atomistic Molecular Modeling and Simulations. *Nucleic Acids Res.* **2012**, *40* (Web Server issue), W537–W541.

55. Georgescu, R. E.; Alexov, E. G.; Gunner, M. R. Combining Conformational Flexibility and Continuum Electrostatics for Calculating pK_as in Proteins. *Biophys. J.* **2002**, 1731–1748. https://doi.org/10.1016/s0006-3495(02)73940-4.
56. Bernèche, S.; Roux, B. The Ionization State and the Conformation of Glu-71 in the KcsA K(+) Channel. *Biophys. J.* **2002**, *82* (2), 772–780.
57. Wallace, J. A.; Shen, J. K. Predicting pKa Values with Continuous Constant pH Molecular Dynamics. *Methods Enzymol.* **2009**, 455–475. https://doi.org/10.1016/s0076-6879(09)66019-5.
58. Radak, B. K.; Chipot, C.; Suh, D.; Jo, S.; Jiang, W.; Phillips, J. C.; Schulten, K.; Roux, B. Constant-pH Molecular Dynamics Simulations for Large Biomolecular Systems. *J. Chem. Theory Comput.* **2017**, *13* (12), 5933–5944.
59. Cruzeiro, V. W. D.; Feliciano, G. T.; Roitberg, A. E. Exploring Coupled Redox and pH Processes with a Force-Field-Based Approach: Applications to Five Different Systems. *J. Am. Chem. Soc.* **2020**, *142* (8), 3823–3835.
60. Meng, Y.; Roitberg, A. E. Constant pH Replica Exchange Molecular Dynamics in Biomolecules Using a Discrete Protonation Model. *J. Chem. Theory Comput.* **2010**, *6* (4), 1401–1412.
61. Alexov, E.; Mehler, E. L.; Baker, N.; Baptista, A. M.; Huang, Y.; Milletti, F.; Nielsen, J. E.; Farrell, D.; Carstensen, T.; Olsson, M. H. M.; Shen, J. K.; Warwicker, J.; Williams, S.; Word, J. M. Progress in the Prediction of pKa Values in Proteins. *Proteins* **2011**, *79* (12), 3260–3275.
62. Awoonor-Williams, E.; Rowley, C. N. The Hydration Structure of Methylthiolate from QM/MM Molecular Dynamics. *J. Chem. Phys.* **2018**, *149* (4), 045103.
63. Liu, R.; Yue, Z.; Tsai, C.-C.; Shen, J. Assessing Lysine and Cysteine Reactivities for Designing Targeted Covalent Kinase Inhibitors. *J. Am. Chem. Soc.* **2019**, *141* (16), 6553–6560.
64. Chatterjee, P.; Botello-Smith, W. M.; Zhang, H.; Qian, L.; Alsamarah, A.; Kent, D.; Lacroix, J. J.; Baudry, M.; Luo, Y. Can Relative Binding Free Energy Predict Selectivity of Reversible Covalent Inhibitors? *J. Am. Chem. Soc.* **2017**, *139* (49), 17945–17952.
65. Zhang, H.; Jiang, W.; Chatterjee, P.; Luo, Y. Ranking Reversible Covalent Drugs: From Free Energy Perturbation to Fragment Docking. *J. Chem. Inf. Model.* **2019**, *59*, 2093–2102. https://doi.org/10.1021/acs.jcim.8b00959.
66. Backus, K. M.; Correia, B. E.; Lum, K. M.; Forli, S.; Horning, B. D.; González-Páez, G. E.; Chatterjee, S.; Lanning, B. R.; Teijaro, J. R.; Olson, A. J.; Wolan, D. W.; Cravatt, B. F. Proteome-Wide Covalent Ligand Discovery in Native Biological Systems. *Nature* **2016**, *534* (7608), 570–574.
67. Zaro, B. W.; Whitby, L. R.; Lum, K. M.; Cravatt, B. F. Metabolically Labile Fumarate Esters Impart Kinetic Selectivity to Irreversible Inhibitors. *J. Am. Chem. Soc.* **2016**, *138*, 15841–15844. https://doi.org/10.1021/jacs.6b10589.
68. Inakollu, V. S.; Geerke, D. P.; Rowley, C. N.; Yu, H. Polarisable Force Fields: What Do they Add in Biomolecular Simulations? *Curr. Opin. Struct. Biol.* **2020**, *61*, 182–190.
69. Aleksandrov, A.; Roux, B.; MacKerell, A. D., Jr. P Calculations with the Polarizable Drude Force Field and Poisson-Boltzmann Solvation Model. *J. Chem. Theory Comput.* **2020**, *16* (7), 4655–4668.
70. Zheng, Y.; Cui, Q. Microscopic Mechanisms that Govern the Titration Response and pK Values of Buried Residues in Staphylococcal Nuclease Mutants. *Proteins* **2017**, *85* (2), 268–281.
71. Vant, J. W.; Lahey, S.-L. J.; Jana, K.; Shekhar, M.; Sarkar, D.; Munk, B. H.; Kleinekathöfer, U.; Mittal, S.; Rowley, C.; Singharoy, A. Flexible Fitting of Small Molecules into Electron Microscopy Maps Using Molecular Dynamics Simulations with Neural Network Potentials. *J. Chem. Inf. Model.* **2020**, *60* (5), 2591–2604.

72. Lahey, S.-L.; Phúc, T. N. T.; Rowley, C. Benchmarking force field and the ANI neural network potentials for the torsional potential energy surface of biaryl drug fragments. *ChemRxiv* **2020**. https://doi.org/10.26434/chemrxiv.12780056.
73. Lahey, S.-L. J.; Rowley, C. N. Simulating Protein–ligand Binding with Neural Network Potentials. *Chem. Sci.* **2020**, 2362–2368. https://doi.org/10.1039/c9sc06017k.
74. Awoonor-Williams, E.; Walsh, A. G.; Rowley, C. N. Modeling Covalent-Modifier Drugs. *Biochim. Biophys. Acta Proteins Proteomics* **2017**, *1865* (11 Pt B), 1664–1675.

CHAPTER SEVEN

Computational support of structure-based targeted covalent inhibitor design

Robert Abel[a,*] and Richard Friesner[b]
[a]Schrodinger, Inc., New York, NY, United States
[b]Columbia University, New York, NY, United States
[*]Corresponding author: e-mail address: robert.abel@schrodinger.com

Contents

1. Introduction

Over the last decade, there has been a resurgence of interest in targeted covalent inhibitors (TCIs) to support the discovery of novel and more efficacious drug therapies.[1–17] Following the guidance of Bianco et al., we include here all covalent inhibitors which exhibit selective covalent binding to an intended target, irrespective of whether or not the covalent linkage formed between the ligand and protein is reversible or irreversible, and further irrespective of whether or not the protein side chain to which the covalent bond will be formed is involved in enzymatic catalysis or distal from the active site.[14] We believe this broad definition to be most appropriate since the wide variety of members of this broad class of TCIs share many of the same drug design challenges. Following the recent work of Bianco et al., Awnoor-Williams et al., and Lonsdale et al., the focus of this review will be to summarize the advantages and challenges of a TCI therapeutic modality, and to describe the many ways that state of the art computational modeling can support the design of next generation TCI drug therapies.[13,14,17]

Annual Reports in Medicinal Chemistry, Volume 56
ISSN 0065-7743
https://doi.org/10.1016/bs.armc.2021.01.001

As implied by name, TCIs, be they reversible or irreversible, distinguish themselves from more commonly developed non-covalent small molecules inhibitors, by their containing an appropriate electrophilic reactive moiety, often called a warhead, which will react to form a covalent bond with the side chain of an appropriate nucleophilic protein residue, most commonly a cysteine, lysine, serine, or tyrosine.[1,18] Such TCIs are most broadly divided into two important subclasses, reversible and irreversible, on the basis of whether or not the covalent linkage between the protein and the ligand reaches a proper thermodynamic equilibrium on the time scale of the experiment, i.e., is reversible, or instead persists indefinitely on the time scale of the experiment once the covalent bond between the protein and the ligand is formed, i.e., irreversible. The essential differences in these two processes are depicted in Fig. 1. An example of a reversible TCI recently entering clinical trials is the fibroblast growth factor receptor 4 inhibitor roblitinib, and perhaps the most commonly encountered irreversible TCI is Asprin, which irreversibly inhibits cyclooxygenase-1.[19,20]

If the TCI under consideration is reversible, then the ratio of the free ligand and covalent protein-ligand adduct should reach a proper thermodynamic equilibrium on the time scale of the experimental data collection, as described by the equilibrium constant, K_i.[1] However, if the TCI under consideration is irreversible, then the formation of covalent protein-ligand

A) Reversible Inhibition:

$$P + I \overset{K_i}{\leftrightharpoons} P\text{-}I$$

B) Irreversible Inhibition:

$$P + I \overset{K_i}{\leftrightharpoons} P \bullet I \overset{k_{inact}}{\longrightarrow} P\text{-}I$$

Fig. 1 Graphical depiction of the binding processes of reversible and irreversible TCIs. Panel (A) depicts the reversible covalent linkage formed between protein "P" and reversible TCI "I" which will equilibrate to a thermodynamic equilibrium described by equilibrium constant Ki. The separated protein and ligand is depicted as "P + I" and the covalent adduct is depicted as "P-I." Panel (B) depicts the two step binding process between protein "P" and irreversible TCI "I" where the inhibitor may repeatedly bind and unbind from the protein occupying a pre-reaction complex "P●I" before eventually irreversibly reacting to form the covalent adduct "P-I." The equilibrium constant of the first step of the binding process is K_i and the rate constant for the second step of the process is k_{inact}.

adduct will monotonically increase with respect to time until either all the TCI or free protein has been consumed by the irreversible step of the binding process.[1,16] For such inhibitors, the critical parameter for optimization is the efficiency of covalent bond formation, k_{inact}/K_i, and all other properties being equal, it is the optimization of this ratio of constants that should be expected to translate to a lower therapeutic dose in the clinic.[16] It should also be noted observed rate k_{inact} is distinct from the intrinsic reactivity of the compound I, and for a particular P-I pair includes contributions from the intrinsic reactivity, the position and orientation of the warhead in the pre-reaction protein-ligand complex P●I, the microenvironment created by the protein around the ligand warhead and engaged residue in the pre-reaction complex P●I, and pKa value of the protein residue to which the covalent linkage will be formed.[21]

Typical values of K_i and k_{inact} values for marketed irreversible TCI drugs range from K_i values of 1–10 nM and K_{inact} values of 10^{-2}–10^{-3} s^{-1}.[12,21] Although such K_i value ranges likely readily mirror our expectations of the typical binding potencies of noncovalent drug therapies, an intuition of the desirability of a particular range of k_{inact} values may be more opaque. The meaning of this quantity can be made clearer by rewriting the the k_{inact} rate constant as a capture period (t_c), where $t_c = 1/k_{inact}$, and one readily finds the capture periods of marketed irreversible TCI drugs to range from minutes to tens of minutes, which likely mirrors our expectations of the residency time of the pre-reaction P●I complex from residency time studies of noncovalent inhibitors.[21] It further makes intuitive once of the central challenges of TCI design—i.e., to optimize the TCI warhead moiety to be sufficiently reactive to form the intended covalent linkage with the targeted protein nucleophilic residue, but not sufficiently reactive so to as unselectively react with off-target nucleophiles.[22]

It should be emphasized TCIs compose a broad class of compounds, and an intuition regarding the diversity of reactive electrophilic reactive moieties, variety targeted protein residues, and breadth of protein targets amenable to a TCI modality can be obtained via inspection of the community resource, CovalentIntDB, which aggregates experimental covalent inhibitor data, including associated structural biology characterization.[18] This resource at the time of this writing includes experimental data for 4511 covalent inhibitors, 57 distinct warhead chemistries, 280 protein targets amenable to covalent inhibition, and 68 marketed TCI drug therapies. From this data, the breadth of the applicability of TCIs for drug discovery should be self-evident.

2. TCI advantages

The preceding notwithstanding, given the expected additional complexities associated with introducing a reactive electrophilic moiety into a selective inhibitor, one might legitimately ask the question posed by Gonzalez-Bello et al., is designing a TCI worth the effort?[6] So far, at least for certain particularly amenable targets, the answer to that question has been a resounding "yes," especially so for kinases.[12,15] The excitement around TCIs as a therapeutic modality has been driven by a series of advantages provided by TCIs over more traditional noncovalent small molecule drug therapies, including that TCIs often exhibiting

- outstanding biochemical and cellular potencies, even in the presence of high concentrations of competing endogenous ligand and/or when binding less classically druggable binding sites[1,16,17];
- outstanding selectivity if targeting a nucleophilic residue not conserved within the gene family[12,17];
- greater robustness to mutational resistance[8];
- possibly manifesting a long duration of action which may decouple PD from PK, in turn enabling sustained efficacy beyond metabolic clearance[12,17];
- a lower clinical dose clinical dose and less frequent administration[12];
- if irreversible, a built in biomarker of target engagement through monitoring for the presence of the covalent adduct[23,24]; and
- If reversible, improved transport properties and tissue distribution through reversible reaction with naturally occurring biological nucleophiles such as glutathione and serum albumin.[1]

However, the above substantial advantages of TCIs must be weighed against the follow drawbacks:

- highly reactive electrophilic warheads may react promiscuously, leading to off-target and/or immunogenetic toxicity[12,13,17,25];
- the long residency times expected to be manifest by TCIs may be counterproductive if the target requires a short residency time or partial inhibition[26,27];
- if target resynthesis, i.e., the turn-over rate, is fast, then covalent inhibitors may provide little or no benefit over a noncovalent inhibition strategy[12,17]; and
- Developing a TCI requires an appropriate nucleophilic residue be present in the binding site of interest, which may not be the case for many medicinally interesting protein targets.[12]

Among these various potential drawbacks, special attention should be given to the concerns that a TCI under development might react promiscuously, in turn leading to off-target toxicity or immunogenic toxicity induced by covalent adducts. Although this has been a historically persuasive argument against the pursuit of TCI therapies, it has been increasingly realized that the single most reliable predictor of idiosyncratic toxicity is the daily administered dose of a drug, and to the extent the very high potencies of covalent inhibitors can be translated to lower administered doses in clinic, it is plausible such therapies may be safer in practice than other modalities requiring a larger daily administered therapeutic dose.[8,25] This observation is indeed broadly consistent with the empirical fact that a great many covalent drug therapies have been extensively prescribed for decades with few or no reports of idiosyncratic toxicity.[8] A such, if the reactivity of the inhibitor is optimized to produce highly efficient and specific covalent bond formation to the on target protein through careful monitoring of the k_{inact}/K_i parameter if an irreversible TCI is being developed, or K_i parameter if a reversible inhibitor is being developed, and the project team is monitoring the intrinsic reactivity of the matter in appropriate glutathione counterscreens, the risk of off-target and immunogenic toxicity can likely be managed using similar approaches to those routinely employed in classical noncovalent inhibitor drug discovery.

3. Computational support of structure-based TCI design

Given the substantial advantages provided by TCIs and the increasing recognition the drawbacks associated with TCIs can be managed, it is natural for a computational chemist to ask where computational modeling could most benefit and accelerate TCI drug discovery. In general, TCI drug discovery follows one of two paths:

1. Adapt a noncovalent inhibitor already in-hand to be a TCI by introduction of an appropriate warhead to engage an appropriate proximal Cysteine, Lysine, Serine, Tyrosine, or other appropriate residue[13,28]; or
2. Elaborate or modify a covalent inhibitor fragment, natural product, or initial hit compound into a drug-like TCI suitable for advancement into clinic.[17,28]

Ideally, computational methods should be able to provide equal support for either of these design paradigms, and in turn accelerate project execution to identify property satisfactory chemical matter. To that end, we suggest in agreement with others,[14,17] at least four core structure-based computational modeling techniques that may be helpful to support TCI drug discovery project execution:

1. *Protein-residue pKa calculation* to estimate the propensity of a residue adjacent to the binding site of interest to act as a nucleophile and appropriately react with the with the ligand electrophile warhead;
2. *Docking* to support ligand binding mode determination and structure-based virtual screening;
3. *Free energy calculations* of protein-TCI binding to support multiproperty optimization of the TCI inhibitors; and
4. *QM calculations* to optimize ligand warhead reactivity to balance on-target covalent linkage formation efficiency with off-target linkage formation.

Protein residue pKa prediction. The ability to calculate the pKa values of protein residues has broad utility in structure-based modeling, especially so for protein structure preparation, and a variety of approaches have been developed to meet this challenge including Propka,[29] H++,[30] MCCE,[31] as well as more computationally MD-based approaches.[31,32] For TCI drug discovery, the ability to predict the p*Ka* values of protein cysteine, serine, lysine and tyrosine residues, in principle, should allow the project team to accurately assess the extent to which such residues may be predisposed to act as a nucleophile. Further, if a nucleophilic residue within or adjacent to the binding site of interest exhibits an unusually extreme p*Ka* value, it might afford the project team an opportunity to use a less reactive warhead, thereby improving the likelihood the covalent linkage will be formed with high specificity with the residue of interest. Important work establishing the plausibility of this approach has been spearheaded by Awoonor-Williams et al. utilizing continuum electrostatic and MD-based methods to evaluate cysteine reactivity.[33,34] Although the numerical accuracy reported in these studies was problematic, with errors typically >2 log units, as the computational methods for p*Ka* prediction further improve, we anticipate this to be a fruitful application area in the future.

Docking. The ability to computationally determine a putative binding mode of a small molecule with a target protein and further evaluate the likelihood it will bind with appreciable affinity to the receptor, i.e., small molecule docking, has become a workhorse modeling technique for small molecule drug discovery.[35] To meet this need, a growing number of docking programs now support binding mode prediction and virtual screening of TCIs, including GOLD,[36] AutoDock,[37] CovalentDock,[38] DOCKovalent,[38,39] ICM-Pro,[40] FITTED,[41] DOCKTITE,[42] CovDock,[43,44] SCAR,[45] EnzyDock[46] Reactive Docking[14] and DUckCov.[47] Two recently published studies utilizing large publicly available datasets to profile the relative accuracy of a number of these TCI docking methods vs found the performance of the top methods to be comparable, and on par with the accuracy one would expect to generally observe then docking noncovalent small molecules.[48,49] Consistent with these retrospective analyses, prospective successes utilizing these docking methods to support virtual screening campaigns to discover novel TCI hit compounds have been reported for AmpC β-Lactamase, p90 ribosomal protein S6 kinase-2, JAK3 kinase, KRas, cathepsin K, ubiquitin-like poxvirus proteinase I7L, and prolyl oligopeptidase, among other targets.[39,47,50–52] This growing body of prospective successes should, in our view, serve to embolden discovery project teams to readily attempt computational virtual screening of covalent inhibitor compound collections to support the identification of TCI hit molecules that might serve as starting points to initiate TCI hit-to-lead chemistry effort.

Free energy calculations. As was sagely noted by London et al. in 2014, "A key unsolved problem in the discovery of [TCIs] is how to identify a protein-binding scaffold that optimally orients the electrophile while minimizing the number of compounds that must be synthesized and tested."[39] Following theoretical innovations first introduced in 2017, free energy calculations have begun to emerge as a fully satisfactory solution to this long standing challenge.[53–59] To our knowledge, the first study successfully adapting free energy calculations to the challenge of optimizing TCIs is the work of Kuhn et al.[53] An important innovation reported in that work was the construction of a thermodynamic cycle that would allow the shifts in the binding affinities of reversible covalent inhibitors to be accurately evaluated without the need to explicitly compute the energetics associated with the formation of the covalent linkage between the protein and the ligand. This free energy approach was prospectively tested head-to-head in that work vs other commonly used approaches to support lead optimization, including docking calculations, the recommendations of an expert medicinal chemist, and an expert structure-based modeler permitted to use any modeling technique other than free energy calculations. The outcome of that testing found the free energy calculations lead to 8 times as many tight binding molecules identified in follow on chemistry vs all other tested approaches (Fig. 2). The free energy calculations themselves were also quantitatively accurate, exhibiting a mean unsigned error of 1.06 kcal/mol.[60]

Since that initial work, the thermodynamic cycle utilized for enabling relative binding affinity calculations of reversible covalent inhibitors was further refined by Chatterlee et al. to support modifications that might

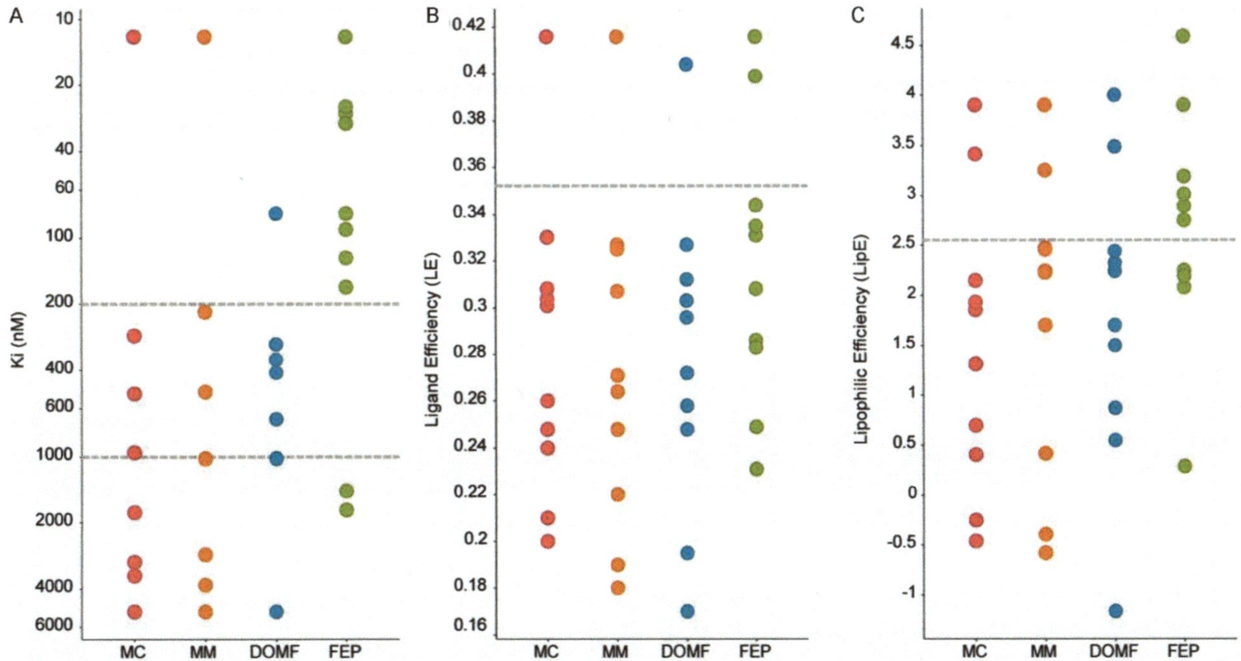

Fig. 2 Scatter plots displaying (A) the inhibitory constant Ki, (B) ligand efficiency and (C) lipophilic efficiency of synthesized covalent inhibitors grouped by selection method (MC = pink: Medicinal Chemist Sections, MM = orange: Conventional Structure-based Modeling Selections, DOMF = blue: Docking-based Selections, and FEP = green: Free Energy Calculation-based Selections). The dashed lines indicate the Ki, ligand efficiency, and lipophilic efficiency of reference compound 2 (see Fig. 1B, Ki = 200 nM, LE = 0.352, and LipE = 2.55). The second dashed line on scatter plot (A) shows the arbitrary threshold of 1 μM beyond which compounds are considered inactive. Free energy calculations were here found to identify eight times as many tighter binding than the starting compound vs any of the other selection methods tested. *Adapted from reference Kuhn B, Tichý M, Wang L, Robinson S, Martin RE, Kuglstatter A, et al. Prospective Evaluation of Free Energy Calculations for the Prioritization of Cathepsin L Inhibitors. J. Med. Chem. 2017;60: 2485–2497.*

substantially alter the intrinsic reactivity of the compounds.[55] Yu et al. have also further modified the method to be applicable to irreversible covalent inhibitors, enabling shifts in the k_{inact}/K_i ratio to be directly computed for congeneric compounds.[54] Zhang et al. have further extended the approach to facilitate calculation of the absolute binding affinities of reversible covalent inhibitors.[59] It should be noted that unlike the work of Kuhn et al., these more complex formulations require auxiliary QM calculations of the shifts in the reactivities of the ligands to be pursued side by side with the free energy calculations, and thus are substantially more complex to execute in practice. However, as will be described in the next section, if the warhead chemistries of TCIs are to be modified, explicitly QM-based modeling of the small molecule reactivity will likely be necessary.

QM calculations. A central goal of TCI drug discovery is to design matter that will selectively and efficiently form a covalent linkage with the target protein residue of interest. This requires that the reactivity of the electrophile warhead moiety of the ligand be carefully tuned to react effectively within the target protein binding site and minimally react with the many other nucleophiles present *in vivo*. QM calculations are ideally suited to address such questions of reactivity, and there is a growing body of literature suggesting that QM/MM calculations can be used to optimize warhead reactivity in the context of the binding site of the target protein.[55–58,61] Likewise QM calculations can be used to optimize the intrinsic reactivity of the ligand matter, which is an important optimization parameter to avoid unselective binding *in vivo*. Reaction with glutathione is also often used as an experimental proxy for the intrinsic reactivity of the warhead, and is expected to correlate with the degree of nonspecific reaction of the warhead *in vivo*.[22,62] A variety of methods have been proposed to approach this important experimental endpoint, including quantum transition state search,[62–64] proton affinity analysis,[65,66] and reaction energy analysis.[66] However, reliably achieving quantitative accuracy has been difficult even when using DFT modern functionals due to the complexity of the mechanisms manifest by some warheads.[67] Likewise, even if the mechanism of reaction is well modeled by a given level of theory, drug-like TCIs will often manifest extensive conformational flexibility, and moieties far from warhead may contribute substantially to the observed rate glutathione half life for reasons entirely unrelated to shifts in the intrinsic reactivity of the compound.[68] The preceding notwithstanding, we anticipate ongoing improvements to quantum transition state search methods and DFT functionals, will in the near future bring a greater level of reliability to the very important application area.[63]

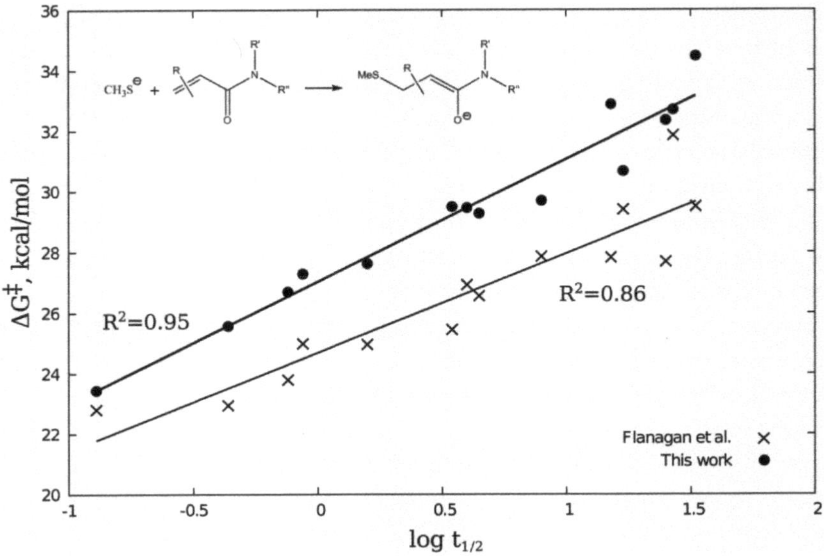

Fig. 3 Correlation between $\Delta G_{\ddagger}$ and log $t_{1/2}$ for the first 14 Michael additions from table 2 from the work of Flanagan et al.[62] for which a precise measurement of $t_{1/2}$ was available. The units of $t_{1/2}$ are hours. *Adapted from reference Jacobson LD, Bochevarov AD, Watson MA, Hughes TF, Rinaldo D, Ehrlich S, et al. Automated Transition State Search and its Application to Diverse Types of Organic Reactions. J. Chem. Theory Comput. 2017;13: 5780–5797.*

An example of such progress is the reported use of fully automated transition state search method, referred to as AutoTS, to successfully model glutathione reaction rates for a series of covalent inhibitors reported by Flanagan et al., as depicted in Fig. 3.[62,63]

4. Conclusions

TCIs compose a broad class of compounds which can manifest a wide range of advantages for pharmaceutical drug discovery, including improved potency, selectivity, robustness to mutational resistance, improved PD, lower clinical dose, and facile biomarker strategies in the clinic. The single major historical objection to the development of TCIs has been a concern that the reactivity of such compounds might lead to off-target and immunogenic toxicity. However, a number of recent developments, especially including the successes of kinase TCI drug therapies in the clinic, have led to this concern being reassessed and a new wave of excitement cresting around this therapeutic modality. Structure-based modeling is well positioned to support TCI drug design through (1) protein residue p*K*a calculation to identify protein nucleophiles appropriate for targeting with a TCI warhead, (2) docking calculations to support identification of TCI hit molecule of suitable quality to initiate hit-to-lead efforts, (3) free energy calculations to directly optimize the Ki values of reversible TCIs and k_{inact}/K_i values of irreversible TCIs, and (4) QM calculations to optimize the reactivity of the warhead necessary to efficiently react with the target protein without promoting promiscuous reactivity with other nucleophiles present *in vivo*. The multiple published prospective successes of docking and free energy calculations to advance hit discovery and hit-to-lead suggest project teams working to develop next generation TCI drug therapies would benefit from broad utilization of these techniques.

References

1. Tuley, A.; Fast, W. The Taxonomy of Covalent Inhibitors. *Biochemistry* **2018**, *57*, 3326–3337.
2. De Cesco, S.; Kurian, J.; Dufresne, C.; Mittermaier, A. K.; Moitessier, N. Covalent Inhibitors Design and Discovery. *Eur. J. Med. Chem.* **2017**, *138*, 96–114.
3. Lu, S.; Zhang, J. Designed Covalent Allosteric Modulators: An Emerging Paradigm in Drug Discovery. *Drug Discov. Today* **2017**, *22*, 447–453.
4. Visscher, M.; Arkin, M. R.; Dansen, T. B. Covalent Targeting of Acquired Cysteines in Cancer. *Curr. Opin. Chem. Biol.* **2016**, *30*, 61–67.
5. Bandyopadhyay, A.; Gao, J. Targeting Biomolecules with Reversible Covalent Chemistry. *Curr. Opin. Chem. Biol.* **2016**, *34*, 110–116.
6. González-Bello, C. Designing Irreversible Inhibitors—Worth the Effort? *ChemMedChem* **2016**, *11*, 22–30.
7. Baillie, T. A. Targeted Covalent Inhibitors for Drug Design. *Angew. Chem. Int. Ed. Eng.* **2016**, *55*, 13408–13421.
8. Bauer, R. A. Covalent Inhibitors in Drug Discovery: From Accidental Discoveries to Avoided Liabilities and Designed Therapies. *Drug Discov. Today* **2015**, *20*, 1061–1073.
9. Mah, R.; Thomas, J. R.; Shafer, C. M. Drug Discovery Considerations in the Development of Covalent Inhibitors. *Bioorg. Med. Chem. Lett.* **2014**, *24*, 33–39.
10. Kalgutkar, A. S.; Dalvie, D. K. Drug Discovery for a New Generation of Covalent Drugs. *Expert Opin. Drug Discovery* **2012**, 7, 561–581.
11. Singh, J.; Petter, R. C.; Baillie, T. A.; Whitty, A. The Resurgence of Covalent Drugs. *Nat. Rev. Drug Discov.* **2011**, *10*, 307–317.
12. Abdeldayem, A.; Raouf, Y. S.; Constantinescu, S. N.; Moriggl, R.; Gunning, P. T. Advances in Covalent Kinase Inhibitors. *Chem. Soc. Rev.* **2020**, *49*, 2617–2687.
13. Awoonor-Williams, E.; Walsh, A. G.; Rowley, C. N. Modeling covalent-modifier drugs. *Biochim. Biophys. Acta, Proteins Proteomics* **2017**, *1865*, 1664–1675.
14. Bianco, G.; Goodsell, D. S.; Forli, S. Selective and Effective: Current Progress in Computational Structure-Based Drug Discovery of Targeted Covalent Inhibitors. *Trends Pharmacol. Sci.* **2020**, *41*, 1038–1049.
15. Ghosh, A. K.; Samanta, I.; Mondal, A.; Liu, W. R. Covalent Inhibition in Drug Discovery. *ChemMedChem* **2019**, *14*, 889–906.

16. Strelow, J. M. A Perspective on the Kinetics of Covalent and Irreversible Inhibition. *SLAS Discov.* **2017**, *22*, 3–20.
17. Lonsdale, R.; Ward, R. A. Structure-Based Design of Targeted Covalent Inhibitors. *Chem. Soc. Rev.* **2018**, *47*, 3816–3830.
18. Du, H.; Gao, J.; Weng, G.; Ding, J.; Chai, X.; Pang, J.; et al. CovalentInDB: A Comprehensive Database Facilitating the Discovery of Covalent Inhibitors. *Nucleic Acids Res.* **2020**, *49*, D1122–D1129. https://doi.org/10.1093/nar/gkaa876.
19. Fairhurst, R. A.; Knoepfel, T.; Buschmann, N.; Leblanc, C.; Mah, R.; Todorov, M.; et al. Discovery of Roblitinib (FGF401) as a Reversible-Covalent Inhibitor of the Kinase Activity of Fibroblast Growth Factor Receptor 4. *J. Med. Chem.* **2020**, *63*, 12542–12573.
20. Tóth, L.; Muszbek, L.; Komáromi, I. Mechanism of the Irreversible Inhibition of Human Cyclooxygenase-1 by Aspirin as Predicted by QM/MM Calculations. *J. Mol. Graph. Model.* **2013**, *40*, 99–109.
21. Schwartz, P. A.; Kuzmic, P.; Solowiej, J.; Bergqvist, S.; Bolanos, B.; Almaden, C.; et al. Covalent EGFR Inhibitor Analysis Reveals Importance of Reversible Interactions to Potency and Mechanisms of Drug Resistance. *Proc. Natl. Acad. Sci. U. S. A.* **2014**, *111*, 173–178. https://doi.org/10.1073/pnas.1313733111.
22. Martin, J. S.; MacKenzie, C. J.; Fletcher, D.; Gilbert, I. H. Characterising Covalent Warhead Reactivity. *Bioorg. Med. Chem.* **2019**, *27*, 2066–2074.
23. Durham, T. B.; Blanco, M.-J. Target engagement in lead generation. *Bioorg. Med. Chem. Lett.* **2015**, *25*, 998–1008.
24. Advani, R. H.; Buggy, J. J.; Sharman, J. P.; Smith, S. M.; Boyd, T. E.; Grant, B.; et al. Bruton Tyrosine Kinase Inhibitor Ibrutinib (PCI-32765) Has Significant Activity in Patients with Relapsed/Refractory B-Cell Malignancies. *J. Clin. Oncol.* **2013**, *31*, 88–94.
25. Johnson, D. S.; Weerapana, E.; Cravatt, B. F. Strategies for Discovering and Derisking Covalent, Irreversible Enzyme Inhibitors. *Future Med. Chem.* **2010**, *2*, 949–964.
26. Lipton, S. A. Paradigm Shift in Neuroprotection by NMDA Receptor Blockade: Memantine and beyond. *Nat. Rev. Drug Discov.* **2006**, *5*, 160–170.
27. Ohlson, S. Designing Transient Binding Drugs: A New Concept for Drug Discovery. *Drug Discov. Today* **2008**, *13*, 433–439.
28. Kathman, S. G.; Statsyuk, A. V. Covalent Tethering of Fragments for Covalent Probe Discovery. *MedChemComm* **2016**, *7*, 576–585. https://doi.org/10.1039/c5md00518c.
29. Olsson, M. H. M.; Søndergaard, C. R.; Rostkowski, M.; Jensen, J. H. PROPKA3: Consistent Treatment of Internal and Surface Residues in Empirical pKa Predictions. *J. Chem. Theory Comput.* **2011**, *7*, 525–537.
30. Anandakrishnan, R.; Aguilar, B.; Onufriev, A. V. H++ 3.0: Automating pK Prediction and the Preparation of Biomolecular Structures for Atomistic Molecular Modeling and Simulations. *Nucleic Acids Res.* **2012**, *40*, W537–W541.
31. Song, Y.; Mao, J.; Gunner, M. R. MCCE2: Improving Protein pKa Calculations with Extensive Side Chain Rotamer Sampling. *J. Comput. Chem.* **2009**, *30*, 2231–2247.
32. Donnini, S.; Tegeler, F.; Groenhof, G.; Grubmüller, H. Correction to Constant pH Molecular Dynamics in Explicit Solvent with λ-Dynamics. *J. Chem. Theory Comput.* **2013**, *9*, 3261.
33. Awoonor-Williams, E.; Rowley, C. N. How Reactive Are Druggable Cysteines in Protein Kinases? *J. Chem. Inf. Model.* **2018**, *58*, 1935–1946.
34. Awoonor-Williams, E.; Rowley, C. N. Evaluation of Methods for the Calculation of the pKa of Cysteine Residues in Proteins. *J. Chem. Theory Comput.* **2016**, *12*, 4662–4673.
35. Wang, X.; Song, K.; Li, L.; Chen, L. Structure-Based Drug Design Strategies and Challenges. *Curr. Top. Med. Chem.* **2018**, *18*, 998–1006. https://doi.org/10.2174/1568026618666180813152921.

36. Jones, G.; Willett, P.; Glen, R. C.; Leach, A. R.; Taylor, R. Development and validation of a genetic algorithm for flexible docking 1 1Edited by F. E. Cohen. *J. Mol. Biol.* **1997**, *267* (3), 727–748. https://doi.org/10.1006/jmbi.1996.0897.
37. Bianco, G.; Forli, S.; Goodsell, D. S.; Olson, A. J. Covalent Docking Using Autodock: Two-Point Attractor and Flexible Side Chain Methods. *Protein Sci.* **2016**, *25*, 295–301. https://doi.org/10.1002/pro.2733.
38. Ouyang, X.; Zhou, S.; Su, C. T. T.; Ge, Z.; Li, R.; Kwoh, C. K. CovalentDock: Automated Covalent Docking With Parameterized Covalent Linkage Energy Estimation and Molecular Geometry Constraints. *J. Comb. Chem.* **2013**, *34*, 326–336. https://doi.org/10.1002/jcc.23136.
39. London, N.; Miller, R. M.; Krishnan, S.; Uchida, K.; Irwin, J. J.; Eidam, O.; et al. Covalent Docking of Large Libraries for the Discovery of Chemical Probes. *Nat. Chem. Biol.* **2014**, *10*, 1066–1072. https://doi.org/10.1038/nchembio.1666.
40. Abagyan, R.; Totrov, M.; Kuznetsov, D. ICM?A New Method for Protein Modeling and Design: Applications to Docking and Structure Prediction From the Distorted Native Conformation. *J. Comb. Chem.* **1994**, *15*, 488–506. https://doi.org/10.1002/jcc.540150503.
41. Corbeil, C. R.; Englebienne, P.; Moitessier, N. Docking Ligands into Flexible and Solvated Macromolecules. 1. Development and Validation of FITTED 1.0. *J. Chem. Inf. Model.* **2007**, *47*, 435–449. https://doi.org/10.1021/ci6002637.
42. Scholz, C.; Knorr, S.; Hamacher, K.; Schmidt, B. DOCKTITE—A Highly Versatile Step-by-Step Workflow for Covalent Docking and Virtual Screening in the Molecular Operating Environment. *J. Chem. Inf. Model.* **2015**, *55*, 398–406. https://doi.org/10.1021/ci500681r.
43. Zhu, K.; Borrelli, K. W.; Greenwood, J. R.; Day, T.; Abel, R.; Farid, R. S.; et al. Docking Covalent Inhibitors: A Parameter Free Approach To Pose Prediction and Scoring. *J. Chem. Inf. Model.* **2014**, *54*, 1932–1940. https://doi.org/10.1021/ci500118s.
44. Warshaviak, D. T.; Golan, G.; Borrelli, K. W.; Zhu, K.; Kalid, O. Structure-Based Virtual Screening Approach for Discovery of Covalently Bound Ligands. *J. Chem. Inf. Model.* **2014**, *54*, 1941–1950. https://doi.org/10.1021/ci500175r.
45. Ai, Y.; Yu, L.; Tan, X.; Chai, X.; Liu, S. Discovery of Covalent Ligands via Noncovalent Docking by Dissecting Covalent Docking Based on a "Steric-Clashes Alleviating Receptor (SCAR)" Strategy. *J. Chem. Inf. Model.* **2016**, *56*, 1563–1575. https://doi.org/10.1021/acs.jcim.6b00334.
46. Das, S.; Shimshi, M.; Raz, K.; Eliaz, N. N.; Mhashal, A. R.; Ansbacher, T.; et al. EnzyDock: Protein–Ligand Docking of Multiple Reactive States along a Reaction Coordinate in Enzymes. *J. Chem. Theory Comput.* **2019**, *15*, 5116–5134. https://doi.org/10.1021/acs.jctc.9b00366.
47. Rachman, M.; Scarpino, A.; Bajusz, D.; Pálfy, G.; Vida, I.; Perczel, A.; et al. DUckCov: A Dynamic Undocking-Based Virtual Screening Protocol for Covalent Binders. *ChemMedChem* **2019**, *14*, 1011–1021.
48. Scarpino, A.; Ferenczy, G. G.; Keserű, G. M. Comparative Evaluation of Covalent Docking Tools. *J. Chem. Inf. Model.* **2018**, *58*, 1441–1458.
49. Wen, C.; Yan, X.; Gu, Q.; Du, J.; Wu, D.; Lu, Y.; et al. Systematic Studies on the Protocol and Criteria for Selecting a Covalent Docking Tool. *Molecules* **2019**, *24*. https://doi.org/10.3390/molecules24112183.
50. Katritch, V.; Byrd, C. M.; Tseitin, V.; Dai, D.; Raush, E.; Totrov, M.; et al. Discovery of Small Molecule Inhibitors of Ubiquitin-like Poxvirus Proteinase I7L Using Homology Modeling and Covalent Docking Approaches. *J. Comput. Aided Mol. Des.* **2007**, *21*, 549–558.
51. Schröder, J.; Klinger, A.; Oellien, F.; Marhöfer, R. J.; Duszenko, M.; Selzer, P. M. Docking-Based Virtual Screening of Covalently Binding Ligands: An Orthogonal Lead Discovery Approach. *J. Med. Chem.* **2013**, *56*, 1478–1490.

52. De Cesco, S.; Deslandes, S.; Therrien, E.; Levan, D.; Cueto, M.; Schmidt, R.; et al. Virtual Screening and Computational Optimization for the Discovery of Covalent Prolyl Oligopeptidase Inhibitors with Activity in Human Cells. *J. Med. Chem.* **2012**, *55*, 6306–6315.
53. Kuhn, B.; Tichý, M.; Wang, L.; Robinson, S.; Martin, R. E.; Kuglstatter, A.; et al. Prospective Evaluation of Free Energy Calculations for the Prioritization of Cathepsin L Inhibitors. *J. Med. Chem.* **2017**, *60*, 2485–2497.
54. Yu, H. S.; Gao, C.; Lupyan, D.; Wu, Y.; Kimura, T.; Wu, C.; et al. Toward Atomistic Modeling of Irreversible Covalent Inhibitor Binding Kinetics. *J. Chem. Inf. Model.* **2019**, *59*, 3955–3967.
55. Chatterjee, P.; Botello-Smith, W. M.; Zhang, H.; Qian, L.; Alsamarah, A.; Kent, D.; et al. Can Relative Binding Free Energy Predict Selectivity of Reversible Covalent Inhibitors? *J. Am. Chem. Soc.* **2017**, *139*, 17945–17952.
56. da Costa, C. H. S.; Bonatto, V.; Dos Santos, A. M.; Lameira, J.; Leitão, A.; Montanari, C. A. Evaluating QM/MM Free Energy Surfaces for Ranking Cysteine Protease Covalent Inhibitors. *J. Chem. Inf. Model.* **2020**, *60*, 880–889.
57. Lameira, J.; Bonatto, V.; Cianni, L.; Dos Reis, R. F.; Leitão, A.; Montanari, C. A. Predicting the Affinity of Halogenated Reversible Covalent Inhibitors through Relative Binding Free Energy. *Phys. Chem. Chem. Phys.* **2019**, *21*, 24723–24730.
58. Mihalovits, L. M.; Ferenczy, G. G.; Keserű, G. M. Affinity and Selectivity Assessment of Covalent Inhibitors by Free Energy Calculations. *J. Chem. Inf. Model.* **2020**, *60*, 6579–6594.
59. Zhang, H.; Jiang, W.; Chatterjee, P.; Luo, Y. Ranking Reversible Covalent Drugs: From Free Energy Perturbation to Fragment Docking. *J. Chem. Inf. Model.* **2019**, *59*, 2093–2102.
60. Abel, R.; Bhat, S. Free Energy Calculation Guided Virtual Screening of Synthetically Feasible Ligand R-Group and Scaffold Modifications: An Emerging Paradigm for Lead Optimization. *Annu. Rep. Med. Chem.* **2017**, *50*, 237–262. https://doi.org/10.1016/bs.armc.2017.08.007.
61. Schirmeister, T.; Kesselring, J.; Jung, S.; Schneider, T. H.; Weickert, A.; Becker, J.; et al. Quantum Chemical-Based Protocol for the Rational Design of Covalent Inhibitors. *J. Am. Chem. Soc.* **2016**, *138*, 8332–8335.
62. Flanagan, M. E.; Abramite, J. A.; Anderson, D. P.; Aulabaugh, A.; Dahal, U. P.; Gilbert, A. M.; et al. Chemical and Computational Methods for the Characterization of Covalent Reactive Groups for the Prospective Design of Irreversible Inhibitors. *J. Med. Chem.* **2014**, *57*, 10072–10079.
63. Jacobson, L. D.; Bochevarov, A. D.; Watson, M. A.; Hughes, T. F.; Rinaldo, D.; Ehrlich, S.; et al. Automated Transition State Search and its Application to Diverse Types of Organic Reactions. *J. Chem. Theory Comput.* **2017**, *13*, 5780–5797.
64. Mulliner, D.; Wondrousch, D.; Schüürmann, G. Predicting Michael-Acceptor Reactivity and Toxicity through Quantum Chemical Transition-State Calculations. *Org. Biomol. Chem.* **2011**, *9*, 8400–8412.
65. Krishnan, S.; Miller, R. M.; Tian, B.; Mullins, R. D.; Jacobson, M. P.; Taunton, J. Design of Reversible, Cysteine-Targeted Michael Acceptors Guided by Kinetic and Computational Analysis. *J. Am. Chem. Soc.* **2014**, *136*, 12624–12630.
66. Lonsdale, R.; Burgess, J.; Colclough, N.; Davies, N. L.; Lenz, E. M.; Orton, A. L.; et al. Expanding the Armory: Predicting and Tuning Covalent Warhead Reactivity. *J. Chem. Inf. Model.* **2017**, *57*, 3124–3137.
67. Awoonor-Williams, E.; Isley, W. C., 3rd; Dale, S. G.; Johnson, E. R.; Yu, H.; Becke, A. D.; et al. Quantum Chemical Methods for Modeling Covalent Modification of Biological Thiols. *J. Comput. Chem.* **2020**, *41*, 427–438.
68. Voice, A.; Tresadern, G.; van Vlijmen, H.; Mulholland, A. Limitations of Ligand-Only Approaches for Predicting the Reactivity of Covalent Inhibitors. *J. Chem. Inf. Model.* **2019**, *59*, 4220–4227.

CHAPTER EIGHT

Covalent fragment screening

Amit Shraga†, Efrat Resnick†, Ronen Gabizon, and Nir London*
Department of Chemical and Structural Biology, The Weizmann Institute of Science, Rehovot, Israel
*Corresponding author: e-mail address: nir.london@weizmann.ac.il

Contents

Most human proteins lack selective small molecule binders, and in fact some protein classes are typically considered as challenging targets for small molecule ligand discovery. These include, for example, transcription factors, GTPases, and scaffolding proteins, which are frequently termed "undruggable."[1] Such targets are often challenging as they lack a large hydrophobic pocket evolved to bind a ligand, and therefore have limited "ligandability" which is considered necessary for a successful High Throughput Screening (HTS) campaign.[2] Fragment-based drug discovery (FBDD) has proven to be a powerful approach for lead identification against such targets, as well as for more traditional targets.

Fragments are loosely defined as small molecules with molecular weight (MW) of less than 300 Da. An early analysis of fragments that are suitable for screening for efficient lead discovery, defined the "*rule of three*": a set of physio-chemical properties which include MW <300 Da, number of H-bond donors ≤3, number of H-bond acceptors ≤3 and ClogP ≤3.[3]

Fragments at this size, compared to screening compounds in traditional HTS, minimize unfavorable interactions or steric hindrance between the ligand and the protein target.[4,5] In addition, since chemical space grows

† Contributed equally.

Annual Reports in Medicinal Chemistry, Volume 56
ISSN 0065-7743
https://doi.org/10.1016/bs.armc.2021.04.001

exponentially with the number of heavy-atoms, fragment based libraries allow more efficient exploration of their respective chemical space while using smaller size libraries with higher hit rates.[6,7] Lastly, such fragments have physio-chemical properties that are appealing for a hit-to-lead campaign, while still maintaining Lipinski's "*rule of five*"[8,9] for oral bioavailability.

Indeed, to date, the food and drug administration (FDA) has approved four drugs (vemurafenib, venetoclax, erdafitinib and pexidartinib), with over 40 additional compounds in ongoing clinical studies, all originating from FBDD.[10]

However, FBDD is not without difficulties. Owing to their small size, a typical fragment binder will have a limited number of interactions with its target, resulting in weak binding affinity. This weak affinity requires primary fragment screens to use very high concentrations of the fragments, sometimes limited by solubility. This weak binding and corresponding high concentration of the fragments also necessitates the use of sensitive biophysical assays,[5] such as NMR, X-ray crystallography, SPR and native mass spectrometry, each with its own advantages and challenges.

1. Covalent fragment screening

One approach to circumvent the low affinity of initial fragment hits is to screen fragments harboring an electrophilic moiety—covalent fragments. The covalent bond formed between a fragment and its target protein, typically enhances the potency of the fragment, thereby alleviating the requirement for screening at high concentrations. Second, the irreversible bond enables direct identification of labeling events using intact protein liquid chromatography/mass spectrometry (LC/MS). This solves two of the issues often encountered in fragment screens. An added benefit of covalent binders (not limited to fragments) is that their proteomic selectivity can be assessed via chemoproteomic approaches.[11,12] In the context of covalent fragments, this allows optimization of selectivity to be included early in the optimization campaign.

Using intact protein MS screening, additional information on the potential selectivity (or intrinsic reactivity) of covalent fragments can be gleaned directly from the primary screen. While a single labeling event could suggest specific labeling of the target, multiple labeling may represent lack of specificity of a covalent fragment. Moreover, screening by MS allows multiplexing—a mixture of fragments could be incubated with the target protein, and as long as their masses could be differentiated by the LC/MS analysis, distinct and separate binding events can be identified.

The pioneering method for covalent fragment screening was disulfide tethering.[13–15] Originally developed to address the problem of identifying weak interactions between fragments and their target, not necessarily in the context of covalent inhibitors. Fragments containing a disulfide can interact with a cysteine residue, either native or engineered, located on the surface of a protein *in vitro*, forming a covalent bond. Early disulfide screens took advantage of multiplexing, increasing the throughput of the assay and reducing the number of measurements taken. Moreover, by allowing the fragments in the mixture to compete with each other over the binding to the protein, the false positive rate is reduced. As noted, disulfide tethering was used for site-directed fragment discovery, even when the native protein target does not have an available cysteine residue.[16–18] First, a cysteine residue is introduced to the protein target, followed by a tethering screen, identifying potential hits targeting nearby pockets. Such hits would then be reverted to their non-covalent form to target the same pocket in the native protein target.

While covalent fragment screening may enable the discovery of non-covalent starting points for hit optimization, increasingly it is used to discover starting points for targeted covalent inhibitors.[19] Inhibitors that are able to form a covalent bond with their protein target have several advantages compared to their non-covalent counterparts. These include increased potency, potentially higher selectivity by targeting non-conserved nucleophilic residues, longer duration of activity for proteins not showing rapid turnover, potentially prolonged systemic exposure and reduced potential for resistance, as well as the ability to target challenging systems such as protein-protein-interactions and allosteric sites.[20–25] For such inhibitors, disulfide tethering hits would need to be modified to a carbon based electrophile as disulfides are only suitable for *in vitro* application, thus requiring screening and optimizing alternative warheads.

The main challenge that should be addressed in screening covalent fragments, is decoupling of the intrinsic reactivity of the covalent warhead from the contribution of the recognition motif to the binding affinity. A widely used method to assess the intrinsic thiol reactivity of electrophiles is by a glutathione (GSH) consumption assay.[26–28] The reaction of an electrophile with GSH is followed by LC/MS or NMR spectroscopy and the reactivity is reported as GSH t½. However, a major limitation of this approach is its low throughput. Several methods to streamline this process were reported using thiol reactive fluorophores[29,30] (described below). We reported an alternative that uses DTNB (Ellman's reagent: 5,5′-dithiobis(2-nitrobenzoic acid)) instead of GSH,[31] and following the disappearance of the absorbance of reduced DTNB. Both methods allow the use of a simple plate reader for the readout and have high throughput, but also have limitations such as limited dynamic range. Computational predictions provide a complementary approach for reactivity assessment, starting from simple substructure-based algorithms[32] through density functional theory (DFT)-based calculations[26] to electrophilicity index based predictions,[33] decision tree based combination of methods[27] and machine learning.[34]

Another avenue to address the reactivity of covalent fragments is by introducing reversible covalent binding. Covalent inhibitors that can dissociate from the target residue can have reduced off-target binding compared to their irreversible counterparts, and enable fine-tuning of the residence time of compounds on the target protein. Cyanoacrylamides and boronic acids are particularly important examples of such inhibitors. Miller et al. screened a small library of fragments equipped with cyanoacrylamides to target non catalytic cysteines, found a reversible-covalent fragment hit targeting the MSK/RSK kinase family.[35] Following X-ray crystallography of the complex between RSK2 and the initial hit, the inhibitor was optimized leading to a nanomolar inhibitor of RSK2. Selectivity versus related kinases was further improved by modifying the beta substituent of the cyanoacrylamide. Modifying the electron withdrawing groups near the electrophile can modulate the reversibility of the bond formation.[36] Moreover, by modifying the substituent of the beta position of cyanoacrylamides, Bradshaw et al.[37] developed inhibitors for BTK with tunable residence times. In a separate work, Lanier et al.[38] used boronic acid fragment screening to identify an inhibitor for the phospholipase Autotaxin.

Traditionally, the most common target for covalent inhibitors were enzymes, in which nucleophilic residues in the active site (such as cysteine, serine, histidine, lysine and tyrosine) were targeted by an electrophile in the inhibitor.[39–41] However, lately nucleophilic residues without a catalytic role that are present at or near a suitable binding-site are targeted as well.[42–45] Many proteins lack suitable nucleophiles in the target site, thus potentially restricting the possible target space. Yet, new chemistries keep expanding the scope of covalent binding and may allow covalent fragment screening against a wider variety of protein nucleophiles.[46–48]

2. Covalent fragment libraries

In recent years, several publications have reported screening small to medium sized libraries consisting of up to ~1000 covalent fragments, with a variety of electrophiles and corresponding reactivity. In this section we will review these library designs and their proof-of-concept applications (summarized in Table 1).

In a ground-breaking proof-of-concept study, Nonoo et al.[49] used "Kinetic Template-Guided Tethering" for the first time to screen a small library of about 10 fragments containing an acrylamide or vinyl sulfonamide electrophile. They were able, even with this very small collection, to identify a thymidylate synthase binder, targeting its catalytic cysteine, with selectivity over the phosphatase CDC25A and the protease papain. They recognized that vinyl sulfonamides were too reactive in their assay and had to be omitted from the screen. While their library was indeed very small, such libraries could be synthetically created in a one step reaction from an amine fragment building block. Importantly they showed that mixtures of 6–8 fragments can be screened simultaneously.

Jöst et al.[50] have created a small combinatorial library comprised of six reactive groups (acrylamides, chloroacetamides, dimethylsulfoniumacetylamides, bromodihydroisoxazoles, 2-cyanoacetamides and imidazole-1-carboxamides) attached to 11 fragment building blocks, resulting in a small library of 72 covalent fragments. This library was biochemically screened in enzymatic activity assays against 11 enzymes containing either a catalytic serine or cysteine residue. They also assessed rates of GSH labeling for the library. Despite the chloroacetamides being the only electrophiles for which they could detect GSH labeling, they did not inhibit any of the enzymes, nor did the acrylamide fragments. The authors also tested a small selection of compounds in a cell viability assay and showed that almost all of them were non-toxic up to 100 μM.

Kathman et al.[5] tried to identify the optimal electrophile for covalent fragment screening. They evaluated the intrinsic reactivity of 12 model compounds containing aniline based acrylamides or vinylsulfoneamides, beta-substituted methylacrylates and methyl-vinylsulfones. Due to the mild reactivity and small variability of the methylacrylates, they focused on this class of electrophiles. They synthesized a library of 100 methylacrylate fragments and screened them in pools of 10, against four model enzymes: Papain (a cysteine protease), human rhinovirus 3C protease, the deubiquitinase USP8, and the E2 ubiquitin-conjugating enzyme UbcH7. They identified

Table 1 Covalent fragment screens.

Citation	Year	Library size	Warheads	Incubation conditions	Hit rates	Protein targets tested
Nonoo et al.[49]	2012	10	Acrylamides, vinyl sulfonamides	400 or 100 μM fragment mixtures with 10 μM protein, 25 °C, 30–270 min	1/10	Thymidylate synthase
Jöst et al.[50]	2014	72	Acrylamides, chloroacetamides, dimethylsulfoniumacetylamides, Bromodihydroisoxazoles, 2-cyanoacetamides, imidazole-1-carboxamides	25 or 50 μM per fragment	1/72	DEN, WNV, THR, ecMetAP, hsMetAP-1, MurA, MurB, MurC, MurD, MurE, MurF
Kathman et al.[5]	2014	100 or 200	α,β-unsaturated methyl esters, methyl vinylsulfones	100 μM fragment mixture, 10 μM protein, 1 or 4 h, 23 °C	3/100 or 0/100	Papain, HRV3C, USP8, UbcH7
McShan et al.[51]	2015			10 μM fragments, protein, 30 min	7/200	Rhodesain
Kathman et al.[52]	2015			100 μM fragment mixture, 10 μM protein, 1 or 4 h, 23 °C	2/100	Nedd4-1
Olp et al.[53]	2020			100 μM fragment mixture, 20 μM protein, 8 h, 25 °C	20/200	BRD4-BD2

Craven et al.[30]	2018	138	Acrylamides, chloroacetamides, epoxides, SNAr substrates	500 μM fragments, 10 μM protein 4 °C	1/138	CDK2
Craven et al.[54]	2020			500 μM fragments, 10 μM protein 4 °C, 24 h	2/138	$CDK2^{F80C\ C177A}$
Resnick et al.[31]	2019	993	Acrylamides, chloroacetamides	200 μM fragment mixture, 2 or 10 μM protein, 24 h, 4 °C	2–4% for NNMT, OTUB2, and NUDT7 and 0.2–0.9% for other proteins	QSOX1, PCAF, PBP^{R504C}, $K\text{-}Ras^{G12C}$, USP8, NNMT, OTUB2, NUDT7, NV3CP, BSA
Dubiella et al.[55]	2019			200 μM fragment mixture, 2 μM protein, 24 h, 4 °C	11%	Pin1
Douangamath et al.[56]	2020			5 μM fragment mixture, 2 μM protein 1.5 h, 25 °C	8.5%	SARS CoV-2 M^{pro}
Johansson et al.[57]	2019	104	α,β-unsaturated methyl esters	20 μM fragment mixture, 2 μM protein, 0.5, 4 or 24 h, 25 °C	1/104	HOIP

three hits against Papain (labeling >50%; 100 μM fragment, 10 μM protein, 1 h) and three hits against the 3C protease (labeling 20–35%) with no hits against the other two enzymes. The papain hits were selective and showed k_{inact}/K_I values of 0.4–1.2 M^{-1} s^{-1}. In a subsequent study[51] they attempted to target another cysteine protease, Rhodesain, now screening the same 100 methylacrylates as well as the 100 corresponding methyl-vinylsulfones. Here, they screened at 10 μM using a biochemical assay, and got a similar hit rate of 3%, where two of the fragments that were validated in a secondary assay were in fact the identical hits found for Papain.

Virtual screening using covalent docking algorithms can also be used to increase the efficiency of fragment screening. Chowdhury et al.[58] found that in the library tested against Papain, the potency was well correlated with the score given by covalent docking. They therefore performed virtual screening of 1648 vinyl sulfone fragments against Cathepsin L. Following docking, five high-scoring compounds were synthesized, of which one compound inhibited Cathepsin L. A short optimization led to a compound inhibiting Cathepsin L with k_{inact}/K_I values of 5 M^{-1} s^{-1}, without inhibiting other proteases such as Cathepsin B. A crystal structure of the hit with Cathepsin L revealed additional directions to grow the fragment and to improve its potency and selectivity.

Craven et al.[30] developed "quantitative irreversible tethering (qIT)" as a method to optimize the kinetic selectivity of covalent fragments. The qIT method employs a fluorogenic probe to measure the kinetics of labeling between the covalent fragment and the protein thiol compared to GSH, yielding a rate of enhancement factor (rate vs protein/rate vs GSH). Thus, this method accounts for intrinsic reactivity of the fragments. To demonstrate the method, the authors constructed a library of 138 covalent fragments, largely adhering to the "*rule of three*," with mostly acrylamides but also chloroacetamides, epoxides and SNAr substrates. The exact library composition was not disclosed. The kinase CDK2 contains a single solvent exposed cysteine, making it suitable for screening by this approach. The authors showed that the fluorogenic probe indeed only shows single labeling of CDK2. They proceeded to screen their library against CDK2 and identified an acrylamide hit that showed a fivefold rate enhancement (over GSH) toward Cys177, located in an allosteric pocket in CDK2. They validated covalent binding via intact protein mass spectrometry against WT and C177A mutant CDK2 and were able to determine a co-crystal structure of the covalently labeled CDK2.

Next, the same group extended the qIT method to also include kinetic analysis, allowing to extract K_I and K_{inact} from the qIT data via hyperbolic-regression analysis.[54] Optimizing K_{inact}/K_i rather than just % labeling avoids potential false-positive highly-reactive electrophilic hits. To validate the approach, they used an electrophile sensitive[59–61] mutant of CDK2 (F80C). A screen against their 138 member fragment library identified two acrylamide hits (with rate-enhancement over GSH of 81 and 64). The hits were validated via intact protein mass and crystallography. Guided by structures and rate enhancements over GSH, they were able to merge the hit fragment with a close analog and improve the selectivity over GSH. Ultimately they were able to make an improved inhibitor with $k_{inact}/K_I = 17.7\ M^{-1}\ s^{-1}$ demonstrating how taking into account intrinsic reactivity during development can efficiently drive optimization.

Focusing on α,β-unsaturated methylacrylates, Johansson et al.[57] assembled a 106 member library, based on fragments from the GSK compound collection that were linked to the electrophile. They used intact protein LC/MS to identify hits from the library, screened in pools of 4–5 fragments separated by at least 5 Da, and have a promising lead against the HOIP RBR E3 ubiquitin ligase (see below).

Addressing the limited size of covalent FBDD libraries, we assembled a 993-member library comprised of commercially available acrylamide (24%) or chloroacetamide (76%) fragments adhering to the "*rule of three*," with 95% containing ≤20 HAC.[31] This library was then characterized by a high-throughput assay to evaluate their intrinsic thiol reactivity, followed by a screening in a 4 or 5-member mixtures with 10 different proteins containing target cysteines, and fragment labeling was measured via LC/MS. This methodology systematically identified that chloroacetamides are more reactive compared to acrylamides, similar to previously published results carried on a much smaller number of fragments.[26] Moreover, only very few fragments showed hyper intrinsic reactivity, reassuring that overall, the library showed only mild reactivity. Screening the library via intact protein mass spectrometry resulted in hits targeting 7 out of the 10 proteins tested. Only 27 fragments showed nonspecific irreversible binding (promiscuity) by labeling ≥3 proteins more than 50%. Interestingly, despite showing similar reactivity to the rest of the library, roughly ~10% (94) of the fragments showed "double-labeling" to the same protein, on two cysteine residues. Analyzing promiscuity toward the same protein negatively correlated with intrinsic reactivity of covalent fragments. Lastly, hits could be subjected to high-throughput crystallography to determine their binding mode and allow further development via fragment growing or merging. This is exemplified with proteins lacking validated binders: OTUB2 and NUDT7.

In the majority of studies using covalent fragments libraries, most of the variation is introduced to the non-covalent recognition element of the molecule, rather than the covalent warhead. However, different protein targets harbor nucleophilic residues with vastly different levels of reactivity, as well as distinct geometry and non-covalent interactions immediately surrounding the nucleophilic residue. Therefore, different warheads may be optimal for each target. With this in mind, Petri et al.[62] decorated a single fragment (3,5-bis(trifluoromethyl)phenyl) with a variety of cysteine-reactive electrophiles. After confirming the reactivity toward cysteine and the variability in reactivity by reaction with GSH, the authors screened the library against several proteins such as MurA, K-Ras^{G12C} and HDAC8. They found that while general trends in reactivity can be observed, with relatively consistent differences observed between different warheads, protein-dependent preferences for specific warheads are observed and therefore screening various warheads can be beneficial.

In another effort that focused on the exploration of electrophilicity, Keeley et al.[63] have compiled a library of 84 electrophilic heterocycles and characterized their stability and reactivity toward glutathione. These fragments spanned a wide range of reactivity. An important advantage of these fragments is that they show very minimal structural differences from non-electrophilic counterparts, and can therefore be incorporated into existing inhibitors with minimal perturbation of the binding mode.

The difficulty to prepare large libraries of compounds can be a major hindrance to screening campaigns. This problem can be partially addressed by the use of plate-based high throughput chemistry using low-volumes. Thomas et al.[64] used this approach to identify binders for carbonic anhydrase. The authors crosslinked a succinimidyl ester diazirine building block with a library of commercially available amines with >80% yield in most cases, incubated the crude products with carbonic anhydrase and used UV irradiation to crosslink the molecules to the protein, followed by intact LCMS. The screening resulted in a 0.7% hit rate, with clear structure-activity relationship (SAR) observed favoring primary sulfonamides, in agreement with previous studies of carbonic anhydrase inhibitors. The initial hits displayed weak binding affinities (>100 μM), and were used as a basis for the design of a second more focused library, using the same technique. Since the compounds discovered using this technique are not electrophilic in and of themselves, they can function primarily as chemical probes for cell culture applications, or as a route for the discovery of non-covalent binders.

3. Commercial covalent fragment libraries

As a testament to the increasing popularity of covalent fragment screening, several commercial vendors now furnish libraries of covalent fragments for screening. These libraries can offer a convenient way for initial hit discovery or even possibly some SAR for chemical probe development. We identified six vendors: Enamine, ChemSpace, Life Chemicals, Reaxense, ChemDiv, ComInnex, who made the content of their covalent fragment library available and analyzed their properties (all analyses were performed using pipeline pilot version 9.1). All six libraries appear to widely adhere to the "*rule of three*" (Fig. 1A–E). With the ComInnex library aside, the rest of the libraries in addition have similar distribution for H-bond donors and acceptors, rotatable bonds and lipophilicity (AlogP; Fig. 1B–E). The fragments in the ComInnex library are significantly smaller compared to the other libraries (average of 133 Da compared to 200–265 Da), which results in the shift of additional properties toward the low spectrum of values.

Comparing the distribution of MW of the other libraries (Fig. 1A) we can see in addition to the overall shift to higher (Enamine, ChemDiv) or lower (Reaxense) MW, some libraries such as Reaxense and ChemDiv have a wide range of available MW while Life Chemicals has a narrow distribution. Having a fragment library with lower MW can allow better coverage of the available chemical space, but risks too little reversible interactions to detect binding, where high MW will require a bigger library and might limit further development of the hits.

When deciding on a suitable library there are additional aspects that should be considered. A large library can help guarantee that a screening campaign will result with an adequate number of hits. However, the absolute number of compounds in the library might be misleading. Compounds in a big library might be very similar to one another and thus cover less chemical space than a very diverse small library. To assess this we performed identical clustering analysis for all the libraries (Cluster threshold Tanimoto coefficient = 0.5; ECFP4 fingerprints). The analysis reveals that ChemSpace which has the largest library indeed has the most clusters but relatively less clusters than one would expect when compared to smaller libraries (Fig. 1F). In further examining the Chemspace clusters it is evident that there are five clusters containing at least 100 fragments (2.7% of the library) and 140

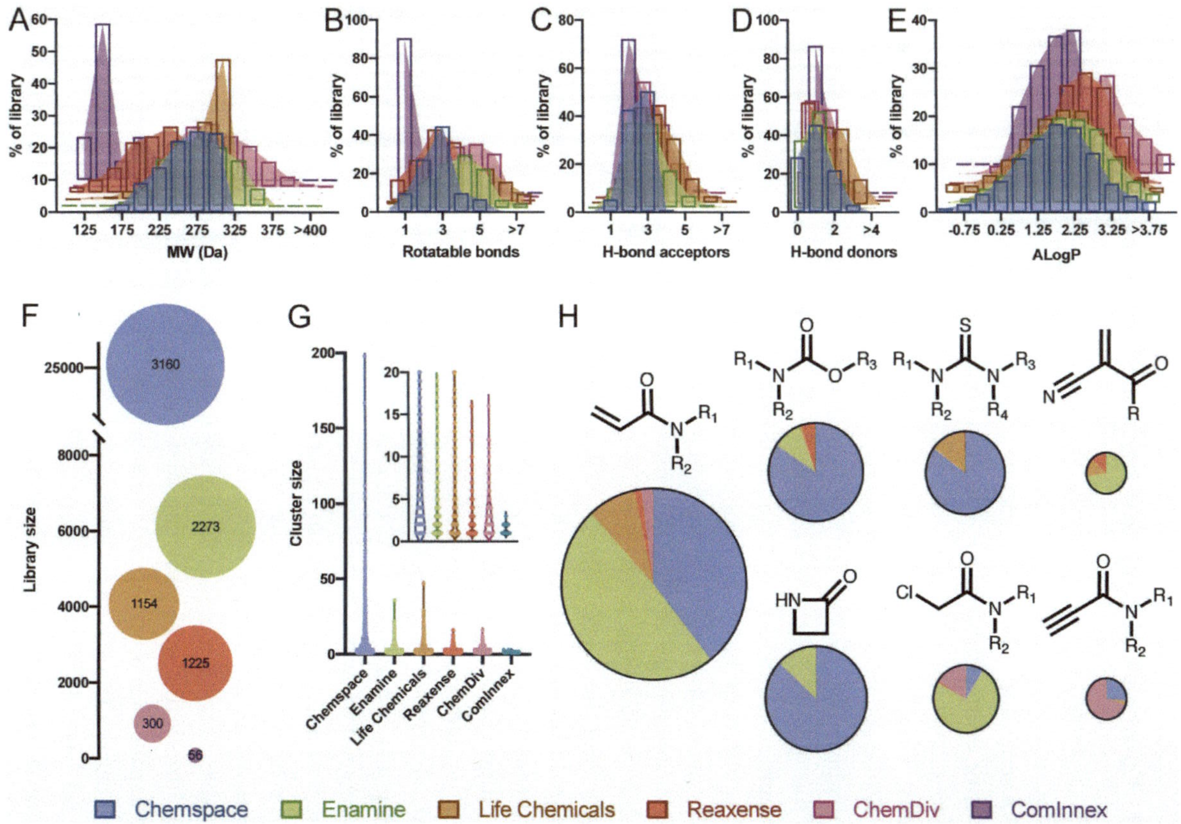

Fig. 1 Properties of commercially available covalent fragment libraries. Distribution of (A) molecular weight, (B) number of rotatable bonds, (C) number of H-bond acceptors, (D) number of H-bond donors and (E) lipophilicity (ALogP). (F) Number of clusters in each library (indicated by the number and size of each circle) compared to the library size (y-axis). (G) Distribution of cluster sizes in each library. (H) Most common warheads found in all the libraries. Circle size represents the overall amount of the specific warhead and the section describes the occurrences in the different libraries.

clusters each containing at least 30 fragments (25% of the library; Fig. 1G). Having many similar fragments might be useful for specific target exploration but will be problematic as a general library for initial fragment hit discovery. The other libraries occasionally have large clusters but overall their median size is about two fragments per cluster, indicating a potentially large chemical space coverage.

Lastly, for covalent fragment libraries the electrophile moiety is a crucial factor. Overall acrylamides are the most common electrophile found in the libraries and are present in all the libraries (Fig. 1H; except for Comlnnex which only contains activated heterocycles), with the majority being in the Enamine fragment library where in addition it is the most common warhead. Different warheads will present different ranges of reactivity and be able to target different side chains as discussed above, so in cases where there is a specific designated binding moiety this can in addition help choose an appropriate library. Most companies offer in addition to purchasing a complete library, the possibility to modify and customize in the final library.

4. Notable successful applications of covalent fragment screening

Perhaps the poster child of successful covalent fragment discovery is the oncogene K-Ras. This challenging target avoided drug discovery efforts for decades, until a few years ago the Shokat lab focused their efforts on a specific glycine to cysteine mutation of K-Ras, which is prevalent in lung cancer. Ostrem et al.[45] used the tethering approach to screen a library of 480 disulfide-fragments against K-Ras^{G12C} using intact protein mass spectrometry. Further screening of ~100 carbon based analogs (200 μM, 24 h) of the top tethering hit, they identified potent fragments binding K-Ras^{G12C} in a pocket formed under the switch-II loop, locking the protein in its inactive GDP bound state. These groundbreaking results initiated further efforts in developing K-Ras^{G12C} inhibitors with improved potency and pharmacological properties with several compounds including MRTX849,[23] ARS-3248 and AMG-510[65] (Fig. 2) all currently in clinical trials.

Kathman et al. screened their previously developed library[5] to identify an allosteric covalent inhibitor for the HECT Nedd4-1 E3 ubiquitin ligase. After treating the HECT domain of Nedd4-1 with mixtures of 10 covalent fragments (100 μM, 4 h), LC/MS identified two fragments that showed covalent labeling to a non-catalytic cysteine residue (as validated by counter screening against a Cys→Ala mutant). They were able to determine the

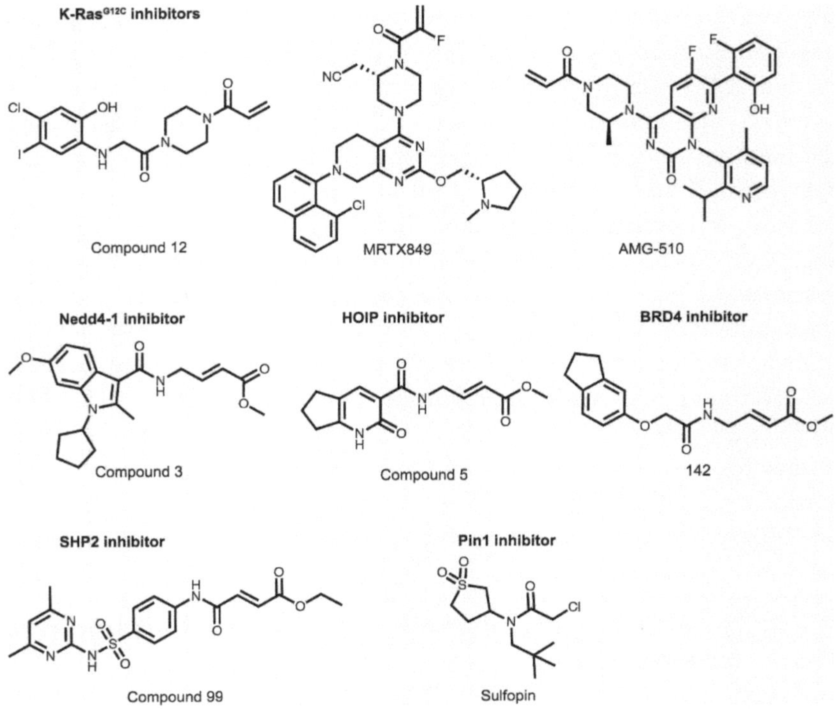

Fig. 2 Structures of notable successful applications developed from fragment screening.

crystal structure of the HECT domain with the fragment bound. Based on the structure, they explored additional analogs, resulting in a more potent inhibitor that displayed selectivity toward Nedd4 compared to other enzymes with reactive catalytic cysteine: deubiquitinase USP8, Human Rhinovirus 3C protease, E1 enzyme Ube1 and the E2 UbcH5a. They determined the K_{inact}/K_i of their lead fragment via fluorescence polarization to be $\sim 2\ M^{-1}\ s^{-1}$. Lastly, enzymatic activity assays revealed that the allosteric covalent binding switches Nedd4 from processive elongation of poly-ubiquitin chains, to a distributive activity.[52]

Using similar chemistry, Johansson et al. utilized a covalent FBDD approach to identify a lead fragment targeting a cysteine located in the active site of another E3 ubiquitin ligase: HOIP RBR, one of the members comprising the linear ubiquitin chain assembly complex (LUBAC).[57] One of their hits, Compound (5), demonstrated >85% labeling after 24 h incubation with recombinant HOIP (20 μM compound; 2 μM protein), with a

K_{inact}/K_i of ~1 M^{-1} s^{-1}. This hit was validated by using a Cys→Ser mutant which the compound was not able to label even at a 100-fold excess. They also used LC/MS to counter screen for covalent binding to off-targets including other RBR E3 ligases (HOIL-1L, HHARI, TRIAD3A, RNF144A RBR domains), HECT E3 ligases (NEDD4L, E6AP), E2 enzymes (UbcH5c, UbcH7), E1, and the DUBs (OTULIN, CYLD-1). The labeled target was shown to be inactive in a biochemical assay, and an improved analog of the initial hit was able to show inhibition in cellular assays. The latter was also shown to engage the target in a competitive chemoproteomic experiment. Finally, the binding mode of their covalent fragments was elucidated via protein crystallography.

Olp et al. used covalent fragment screening to identify a novel binding site in the BD2 domain of BRD4 which potentially can enable selective binding to this domain over the three other BET bromodomain family members, which lack selective binders.[53] They screened 200 acrylate methyl ester fragments in mixtures of 10 via LC/MS. 10% of the fragments showed >20% labeling with recombinant BRD4-BD2. Seven hits were chosen based on commercial availability, and showed prominent selectivity toward BD2 over two other BET bromodomains (BRD4-BD1 and, BRD3-BD2), as well as toward WT compared to point mutation of the cysteine in BRD4-BD2. Thermal shift assay (DSF) demonstrated that the covalent fragment indeed decreased the thermal stability of BRD4-BD2. MS/MS fragmentation experiment and protein NMR pinpointed the labeled residue to be Cys356, orthogonal to the enzyme binding site. Finally, using a PEG linker, the final hit was linked to a known BRD4 inhibitor, JQ1. Using a BRET assay the authors showed that the bifunctional, hybrid covalent inhibitor exhibited bivalent inhibition: it was able to target both BD1 (via JQ1) and BD2 (via the covalent fragment), and showed cellular activity for chromatin displacement of BRD4.

Recently, Kim et al.[66] used covalent FBDD to target a mutated cysteine residue, Y279C, in the tyrosine phosphatase SHP2 protein. This mutation is associated with Noonan syndrome, a developmental disorder. This is another nice example of targeting a non-conserved cysteine, or in this case, a mutated cysteine as a handle for a selective covalent inhibitor. The authors screened a subset of 200 electrophiles of the Life Chemicals commercial covalent fragment library (see above; including cyanoacrylamides, substituted acrylamides, chloroacetamides and maleimides) via an *in vitro* phosphatase assay, of the catalytic domain of SHP2-Y279C. Hits were then counter screened against the catalytic domain of WT-SHP2. Compound 99 showed potent and selective inhibition toward the mutant SHP2-Y279C with $IC_{50} \approx 6\ \mu M$, and no effect on the WT-SHP2 ($IC_{50} \gg 100\ \mu M$) as well as eight other tyrosine phosphatases and a Y279S-SHP2 mutant. LC/MS and washing experiments validated irreversible binding of the compound to SHP2-Y279C, with no labeling of the WT. Finally the compound was able to inhibit SHP2-Y279C in lysates.

A prominent anti-viral target for SARS-CoV-2[56] is the M^{pro} cysteine protease. At the onset of the COVID-19 pandemic, we collaborated with the Diamond Light Source to combine electrophilic and crystallographic fragment screening to find potential starting points for M^{pro} inhibitors. The library described in Resnick et al.[31] was used, with 8.5% hit rate of more than 30% labeling, some of which shared common motifs (*N*-chloroacetyl-N′-sulfonamido-piperazine or *N*-chloroacetylaniline) that potentially could serve as starting points for further lead development. Next, the lesser reactive fragment hits (68) were crystallized, together with other non-covalent fragments provided by collaborators, totaling 1176 unique fragments. The resulting 48 structures of covalent fragment binders provide insights on the binding mode, which gave rise to synthesis of a second generation series containing a *N*-chloroacetyl piperidinyl-4-carboxamide motif and tested again for protein labeling via MS. Interestingly, an unexpected covalent binding came from a non-covalent library termed FragLites,[67] and it appears that bromoalkynes could in fact bind covalently with activated cysteines, though their intrinsic reactivity is lower compared to acrylamides.[68] Potentially, by merging hits their potency could improve by synergistically binding several sub-pockets and might provide a treatment for SARS-CoV-2 in the future.

We applied covalent fragment screening to find a covalent inhibitor for Pin1.[55] Pin1 is a phosphorylation dependent proline isomerase that mediates many oncogenic processes, and controls the stability and function of numerous proteins. Its active-site proved challenging due to a shallow and positively charged pocket, evolved to bind phosphopeptides. We screened our previously reported library of chloroacetamides and acrylamides[31] in pools of five via intact protein LC/MS. 111 fragments showed >50% labeling (2 μM protein, 200 μM compound; 24 h at 4 °C), and 48 demonstrated >75%. Clear SAR emerged from the screen with an enrichment of sulfolane containing chloroacetamides. A focused library of analogs containing the sulfolane motif showed improved labeling with the top ten binders labeling 35–65% of the protein under more stringent conditions (equimolar concentration, 1 h at 25 °C). Sulfopin, showed the best trade-off of potency with markedly reduced intrinsic thiol reactivity. Two orthogonal chemoproteomic experiments identified Pin1 as the sole target for Sulfopin in cells. A crystal structure elucidated the preference for the sulfolane motif which buries two hydrogen bonds, and makes additional hydrophobic interactions. Sulfopin was not generally cytotoxic, and showed no toxicity when administered to mice. Using a desthiobiotin derivative, we were able to show its target engagement in cells and *in vivo*. RNA-seq suggested that Sulfopin mediates down regulation of MYC target genes, and thus we assessed it in zebrafish and mice models of MYC driven cancers such as Neuroblastoma and Pancreatic Ductal Adenocarcinoma. Sulfopin showed significant tumor reduction effects at doses of 40 mg/kg. This campaign which went all the way from a covalent fragment screen to *in vivo* proof of efficacy, highlights the potential of covalent fragment screening as an efficient tool for chemical probe and drug discovery.

5. Chemoproteomic fragment screens

An exciting development for covalent fragments is the ability to characterize their proteomic interaction landscape. In this chemoproteomic approach instead of screening many fragments against one defined target, a small number of fragments are "screened" against the entire proteome. This approach allows to discover unique targets, and is starting to map the proteome-wide covalent protein ligandability potential.

A proteome-wide study[12] attempted to identify possible cysteine residues that could be targeted by a covalent ligand. They used isoTOP-ABPP[69] for each of a library of ~60 covalent electrophilic fragments with either a chloroacetamide or acrylamide warhead. Overall, they identified 758 unique "ligandable" cysteines in 637 distinct proteins, including proteins previously considered "undruggable."[69] Only 14% of the proteins labeled in this assay appeared in the DrugBank at the time. This approach also allowed to assess the promiscuity of the different fragments which varied in their ability to label proteomic cysteines, with selectivity ranging from 0.1% to 15% of liganded cysteines. Overall, chloroacetamide fragments showed greater reactivity compared to acrylamide fragments, which were similar to cyanoacrylamide fragments. Interestingly, despite the relatively high concentration used in the screen (500 μM), most fragments did not react across the entire cysteinome. One hit was further developed to create a selective Caspase 8 covalent inhibitor.

Fragments that were identified to react with a very wide range of proteins were termed "scout fragments" and were used in subsequent studies to investigate targeted protein degradation,[70,71] cysteine reactivity changes upon T-cell activation,[72] and recently as a tool to facilitate covalent inhibitor optimization against specific targets.[73]

Recent technical improvements to the proteomic analysis pipeline allowed to significantly increase the throughput of covalent chemoproteomics.[74] The SLC-ABPP (streamline cysteine activity-based protein profiling) method uses sample multiplexing and 5-fold reduction in sample input, which allowed reducing by 42-fold the screening time per compound and were able to profile >8000 reactive cysteines in 3 h. The method was additionally used in a high throughput screen of 285 electrophiles across three cell lines, identifying >20,000 unique cysteine sites in >6000 proteins.

A covalent bond with the target protein can also be mediated by other means and not necessarily by an electrophile. Parker et al.[75] created a 450-member fragment library (~250 Da) linking a recognition motif with two functional moieties; an alkyne and a photoactivatable diazerene. After incubation in cells, exposure to UV light induces the formation of a covalent bond with a protein to which the fragment was hitherto reversibly bound. Using click-chemistry, the fragment bound proteins were enriched and analyzed via MS. This analysis enabled identification of functional hits. Following up on one of their hits enabled the development of a selective inhibitor targeting a poorly characterized membrane protein PGRMC2, promoting adipocyte differentiation.

An extension for this approach[76] introduced "enantio-probes" which are enantiomeric probe pairs. These allow the identification of enantioselective protein-ligand interactions and filter some of the non-selective binding artifacts. The researchers evaluated eight pairs of "enantio-probes" (incubated for 30 min, 200 μM), UV light exposure induced formation of covalent bond and samples were clicked with azide probe before analysis via isoTOP-labeling, which detected 176 stereoselective ligandable proteins targets, binding in functionally relevant binding sites. Interestingly, there was no overall targeting preference of either *R* or *S* enantiomers, and 80% of the liganded proteins showed selectivity toward only one pair of "enantio-probes."

6. Conclusions and outlook

The renewed interest in targeted covalent inhibitors necessitates new technologies for their discovery. Covalent fragment screening has established itself as an effective and viable approach to discover starting points for such covalent inhibitors, in particular against targets where no preliminary chemical matter is available.

Methods for high-throughput characterization of intrinsic thiol reactivity and accumulated screening data converged to a productive reactivity window in which the fragments are sufficiently mild to show selectivity, but sufficiently reactive to allow irreversible binding driven by low-affinity reversible recognition. Chloroacetamides and methylacrylate esters seem to reside in this reactivity window and produced several successful inhibitors (Fig. 2). Despite their presence in screening libraries, almost no acrylamides are reported as hits in screens against recombinant proteins, which may suggest that this electrophile is not sufficiently reactive to drive covalent bond formation in this low affinity regime. It is interesting to note that acrylamide fragments were identified as hits in unbiased proteomic screens,[77–79] which demonstrates that they are not inherently unsuitable, but require a much larger search space (i.e., the entire proteome). On the other end of the reactivity spectrum, vinyl sulfonamides were found to be too reactive.[5,49]

With regards to library size, while screens of 100–1000 compounds were able to identify progressable hits, it stands to reason that larger libraries of 1000–10,000 fragments will be more productive and perhaps the standard in the near future. First, this mirrors similar sizes of current non-covalent fragment libraries.[80,81] Second, libraries of these sizes are already commercially available (Fig. 1), and finally due to their irreversible binding mode and MS based readout, covalent fragment screening in particular lands itself to screening directly from crude reactions.[64,82] Screening from crude reaction mixtures would allow the effective and economic generation of larger covalent fragment libraries.

While most of the effort so far focused on cysteine targeting covalent fragments, the same methods and principles apply to chemistries that target other amino-acids. The productive reactivity window (and perhaps corresponding size of the reversible recognition element) may differ to address the fact that cysteine is very scarce in the proteome while nucleophiles such as lysine and tyrosine are abundant.

Successful applications such as we described above, and in particular potential drugs such as the K-Ras^{G12C} binders (Fig. 2) that are currently in the clinic, and originated from covalent fragment screening just a few years ago, demonstrate the utility of covalent fragment screening. We look forward to wider adaptation of this technology by the field and to the new generation of covalent probes and drug candidates that will be discovered through it.

References

1. Crews, C. M. Targeting the Undruggable Proteome: The Small Molecules of my Dreams. *Chem. Biol.* **2010**, *17*, 551–555.
2. Surade, S.; Blundell, T. L. Structural Biology and Drug Discovery of Difficult Targets: The Limits of Ligandability. *Chem. Biol.* **2012**, *19*, 42–50.
3. Congreve, M.; Carr, R.; Murray, C.; Jhoti, H. A 'Rule of Three' for Fragment-Based Lead Discovery? *Drug Discov. Today* **2003**, *8*, 876–877.
4. Scott, D. E.; Coyne, A. G.; Hudson, S. A.; Abell, C. Fragment-Based Approaches in Drug Discovery and Chemical Biology. *Biochemistry* **2012**, *51*, 4990–5003.
5. Kathman, S. G.; Xu, Z.; Statsyuk, A. V. A Fragment-Based Method to Discover Irreversible Covalent Inhibitors of Cysteine Proteases. *J. Med. Chem.* **2014**, *57*, 4969–4974.
6. Hall, R. J.; Mortenson, P. N.; Murray, C. W. Efficient Exploration of Chemical Space by Fragment-Based Screening. *Prog. Biophys. Mol. Biol.* **2014**, *116*, 82–91.
7. Hann, M. M.; Leach, A. R.; Harper, G. Molecular Complexity and Its Impact on the Probability of Finding Leads for Drug Discovery. *J. Chem. Inf. Comput. Sci.* **2001**, *41*, 856–864.
8. Heynen-Genel, S.; et al. Screening for Selective Ligands for GPR55—Antagonists. In *Probe Reports from the NIH Molecular Libraries Program*; National Center for Biotechnology Information: US, 2010.
9. Lipinski, C. A.; Lombardo, F.; Dominy, B. W.; Feeney, P. J. Experimental and Computational Approaches to Estimate Solubility and Permeability in Drug Discovery and Development Settings. *Adv. Drug Deliv. Rev.* **2001**, *46*, 3–26.
10. Osborne, J.; Panova, S.; Rapti, M.; Urushima, T.; Jhoti, H. Fragments: Where Are We Now? *Biochem. Soc. Trans.* **2020**, *48*, 271–280.
11. Browne, C. M.; et al. A Chemoproteomic Strategy for Direct and Proteome-Wide Covalent Inhibitor Target-Site Identification. *J. Am. Chem. Soc.* **2019**, *141*, 191–203.
12. Backus, K. M.; et al. Proteome-Wide Covalent Ligand Discovery in Native Biological Systems. *Nature* **2016**, *534*, 570–574.
13. Erlanson, D. A.; et al. Site-Directed Ligand Discovery. *Proc. Natl. Acad. Sci. U. S. A.* **2000**, *97*, 9367–9372.
14. Hardy, J. A.; Lam, J.; Nguyen, J. T.; O'Brien, T.; Wells, J. A. Discovery of an Allosteric Site in the Caspases. *Proc. Natl. Acad. Sci. U. S. A.* **2004**, *101*, 12461–12466.
15. Erlanson, D. A.; Wells, J. A.; Braisted, A. C. Tethering: Fragment-Based Drug Discovery. *Annu. Rev. Biophys. Biomol. Struct.* **2004**, *33*, 199–223.
16. Buck, E.; Wells, J. A. Disulfide Trapping to Localize Small-Molecule Agonists and Antagonists for a G Protein-Coupled Receptor. *Proc. Natl. Acad. Sci. U. S. A.* **2005**, *102*, 2719–2724.
17. Sadowsky, J. D.; et al. Turning a Protein Kinase On or Off From a Single Allosteric Site Via Disulfide Trapping. *Proc. Natl. Acad. Sci. U. S. A.* **2011**, *108*, 6056–6061.
18. Sijbesma, E.; et al. Site-Directed Fragment-Based Screening for the Discovery of Protein–Protein Interaction Stabilizers. *J. Am. Chem. Soc.* **2019**, *141*, 3524–3531.
19. Lu, W.; et al. Fragment-Based Covalent Ligand Discovery. *RSC Chem. Biol.* **2021**, *2*, 354–367.
20. Way, J. C. Covalent Modification as a Strategy to Block Protein-Protein Interactions With Small-Molecule Drugs. *Curr. Opin. Chem. Biol.* **2000**, *4*, 40–46.
21. Noe, M. C.; Gilbert, A. M. Targeted Covalent Enzyme Inhibitors. In *Annual Reports in Medicinal Chemistry*; vol. 47; Elsevier, 2012; pp. 413–439.
22. Sameshima, T.; et al. Discovery of an Irreversible and Cell-Active BCL6 Inhibitor Selectively Targeting Cys53 Located at the Protein-Protein Interaction Interface. *Biochemistry* **2018**, *57*, 1369–1379.

23. Fell, J. B.; et al. Identification of the Clinical Development Candidate MRTX849, a Covalent KRASG12C Inhibitor for the Treatment of Cancer. *J. Med. Chem.* **2020**, *63*, 6679–6693.
24. Singh, J.; Petter, R. C.; Baillie, T. A.; Whitty, A. The Resurgence of Covalent Drugs. *Nat. Rev. Drug Discov.* **2011**, *10*, 307–317.
25. Vita, E. D. 10 Years into the Resurgence of Covalent Drugs. *Future Med. Chem.* **2021**, *13*, 193–210.
26. Flanagan, M. E.; et al. Chemical and Computational Methods for the Characterization of Covalent Reactive Groups for the Prospective Design of Irreversible Inhibitors. *J. Med. Chem.* **2014**, *57*, 10072–10079.
27. Lonsdale, R.; et al. Expanding the Armory: Predicting and Tuning Covalent Warhead Reactivity. *J. Chem. Inf. Model.* **2017**, *57*, 3124–3137.
28. Cheng, H.; et al. Discovery of 1-${(3R, 4R)-3-[({5-Chloro-2-[(1-methyl-1 H-pyrazol-4-yl) amino]-7 H-pyrrolo [2, 3-d] pyrimidin-4-yl} oxy) methyl]-4-methoxypyrrolidin-1-yl}$ prop-2-en-1-one (PF-06459988), a Potent, WT Sparing, Irreversible Inhibitor of T790M-Containing EGFR Mutants. *J. Med. Chem.* **2016**, *59*, 2005–2024.
29. Epps, D. E.; Taylor, B. M. A Competitive Fluorescence Assay to Measure the Reactivity of Compounds. *Anal. Biochem.* **2001**, *295*, 101–106.
30. Craven, G. B.; et al. High-Throughput Kinetic Analysis for Target-Directed Covalent Ligand Discovery. *Angew. Chem. Int. Ed. Engl.* **2018**, *57*, 5257–5261.
31. Resnick, E.; et al. Rapid Covalent-Probe Discovery by Electrophile-Fragment Screening. *J. Am. Chem. Soc.* **2019**, *141*, 8951–8968.
32. Metz, J. T.; Huth, J. R.; Hajduk, P. J. Enhancement of Chemical Rules for Predicting Compound Reactivity Towards Protein Thiol Groups. *J. Comput. Aided Mol. Des.* **2007**, *21*, 139–144.
33. Palazzesi, F.; Grundl, M. A.; Pautsch, A.; Weber, A.; Tautermann, C. S. A Fast ab Initio Predictor Tool for Covalent Reactivity Estimation of Acrylamides. *J. Chem. Inf. Model.* **2019**, *59*, 3565–3571.
34. Palazzesi, F.; et al. BIreactive: A Machine-Learning Model to Estimate Covalent Warhead Reactivity. *J. Chem. Inf. Model.* **2020**, *60*, 2915–2923.
35. Miller, R. M.; Paavilainen, V. O.; Krishnan, S.; Serafimova, I. M.; Taunton, J. Electrophilic Fragment-Based Design of Reversible Covalent Kinase Inhibitors. *J. Am. Chem. Soc.* **2013**, *135*, 5298–5301.
36. Krishnan, S.; et al. Design of Reversible, Cysteine-Targeted Michael Acceptors Guided by Kinetic and Computational Analysis. *J. Am. Chem. Soc.* **2014**, *136*, 12624–12630.
37. Bradshaw, J. M.; et al. Prolonged and Tunable Residence Time Using Reversible Covalent Kinase Inhibitors. *Nat. Chem. Biol.* **2015**, *11*, 525–531.
38. Lanier, M.; et al. Repurposing Suzuki Coupling Reagents as a Directed Fragment Library Targeting Serine Hydrolases and Related Enzymes. *J. Med. Chem.* **2017**, *60*, 5209–5215.
39. Walsh, C. T. Suicide Substrates, Mechanism-Based Enzyme Inactivators: Recent Developments. *Annu. Rev. Biochem.* **1984**, *53*, 493–535.
40. Palmer, J. T.; Rasnick, D.; Klaus, J. L.; Brömme, D. Vinyl Sulfones as Mechanism-Based Cysteine Protease Inhibitors. *J. Med. Chem.* **1995**, *38*, 3193–3196.
41. Leung, D.; Hardouin, C.; Boger, D. L.; Cravatt, B. F. Discovering Potent and Selective Reversible Inhibitors of Enzymes in Complex Proteomes. *Nat. Biotechnol.* **2003**, *21*, 687–691.
42. Cohen, M. S.; Zhang, C.; Shokat, K. M.; Taunton, J. Structural Bioinformatics-Based Design of Selective, Irreversible Kinase Inhibitors. *Science* **2005**, *308*, 1318–1321.
43. Honigberg, L. A.; et al. The Bruton Tyrosine Kinase Inhibitor PCI-32765 Blocks B-Cell Activation and is Efficacious in Models of Autoimmune Disease and B-Cell Malignancy. *Proc. Natl. Acad. Sci. U. S. A.* **2010**, *107*, 13075–13080.

44. Xia, G.; et al. A Chemical Tuned Strategy to Develop Novel Irreversible EGFR-TK Inhibitors With Improved Safety and Pharmacokinetic Profiles. *J. Med. Chem.* **2014**, *57*, 9889–9900.
45. Ostrem, J. M.; Peters, U.; Sos, M. L.; Wells, J. A.; Shokat, K. M. K-Ras(G12C) Inhibitors Allosterically Control GTP Affinity and Effector Interactions. *Nature* **2013**, *503*, 548–551.
46. Gehringer, M.; Laufer, S. A. Emerging and Re-Emerging Warheads for Targeted Covalent Inhibitors: Applications in Medicinal Chemistry and Chemical Biology. *J. Med. Chem.* **2019**, *62*, 5673–5724.
47. Zanon, P. R. A.; et al. *Profiling the Proteome-Wide Selectivity of Diverse Electrophiles*; *ChemRxiv*, 2021. https://doi.org/10.26434/chemrxiv.14186561.v1.
48. Shannon, D. A.; Weerapana, E. Covalent Protein Modification: The Current Landscape of Residue-Specific Electrophiles. *Curr. Opin. Chem. Biol.* **2015**, *24*, 18–26.
49. Nonoo, R. H.; Armstrong, A.; Mann, D. J. Kinetic Template-Guided Tethering of Fragments. *ChemMedChem* **2012**, 7, 2082–2086.
50. Jöst, C.; Nitsche, C.; Scholz, T.; Roux, L.; Klein, C. D. Promiscuity and Selectivity in Covalent Enzyme Inhibition: A Systematic Study of Electrophilic Fragments. *J. Med. Chem.* **2014**, *57*, 7590–7599.
51. McShan, D.; et al. Identification of Non-Peptidic Cysteine Reactive Fragments as Inhibitors of Cysteine Protease Rhodesain. *Bioorg. Med. Chem. Lett.* **2015**, *25*, 4509–4512.
52. Kathman, S. G.; et al. A Small Molecule That Switches a Ubiquitin Ligase From a Processive to a Distributive Enzymatic Mechanism. *J. Am. Chem. Soc.* **2015**, *137*, 12442–12445.
53. Olp, M. D.; et al. Covalent-Fragment Screening of BRD4 Identifies a Ligandable Site Orthogonal to the Acetyl-Lysine Binding Sites. *ACS Chem. Biol.* **2020**, *15*, 1036–1049.
54. Craven, G. B.; et al. Multiparameter Kinetic Analysis for Covalent Fragment Optimization by Using Quantitative Irreversible Tethering (qIT). *Chembiochem* **2020**, *21*, 3417–3422.
55. Dubiella, C.; Pinch, B. J.; Koikawa, K.; Zaidman, D.; Poon, E.; Manz, T. D.; et al. Sulfopin is a Covalent Inhibitor of Pin1 That Blocks Myc-Driven Tumors in Vivo. *Nature Chemical Biology* **2021**. https://doi.org/10.1038/s41589-021-00786-7. May.
56. Douangamath, A.; et al. Crystallographic and Electrophilic Fragment Screening of the SARS-CoV-2 Main Protease. *Nat. Commun.* **2020**, *11*, 5047.
57. Johansson, H.; et al. Fragment-Based Covalent Ligand Screening Enables Rapid Discovery of Inhibitors for the RBR E3 Ubiquitin Ligase HOIP. *J. Am. Chem. Soc.* **2019**, *141*, 2703–2712.
58. Chowdhury, S. R.; et al. Discovery of Covalent Enzyme Inhibitors Using Virtual Docking of Covalent Fragments. *Bioorg. Med. Chem. Lett.* **2019**, *29*, 36–39.
59. Garske, A. L.; Peters, U.; Cortesi, A. T.; Perez, J. L.; Shokat, K. M. Chemical Genetic Strategy for Targeting Protein Kinases Based on Covalent Complementarity. *Proc. Natl. Acad. Sci. U. S. A.* **2011**, *108*, 15046–15052.
60. Koch, A.; Rode, H. B.; Richters, A.; Rauh, D.; Hauf, S. A Chemical Genetic Approach for Covalent Inhibition of Analogue-Sensitive Aurora Kinase. *ACS Chem. Biol.* **2012**, 7, 723–731.
61. Kung, A.; et al. A Chemical-Genetic Approach to Generate Selective Covalent Inhibitors of Protein Kinases. *ACS Chem. Biol.* **2017**, *12*, 1499–1503.
62. Petri, L.; et al. Assessment of Tractable Cysteines for Covalent Targeting by Screening Covalent Fragments. *Chembiochem* **2021**, *22*, 743–753.
63. Keeley, A.; Ábrányi-Balogh, P.; Keserű, G. M. Design and Characterization of a Heterocyclic Electrophilic Fragment Library for the Discovery of Cysteine-Targeted Covalent Inhibitors. *Medchemcomm* **2019**, *10*, 263–267.

64. Thomas, R. P.; et al. *A Direct-to-Biology High-Throughput Chemistry Approach to Reactive Fragment Screening*; *ChemRxiv*, 2021. https://doi.org/10.26434/chemrxiv.14261780.v1.
65. Canon, J.; et al. The Clinical KRAS(G12C) Inhibitor AMG 510 Drives Anti-Tumour Immunity. *Nature* **2019**, *575*, 217–223.
66. Kim, J. Y.; Plaman, B. A.; Bishop, A. C. Targeting a Pathogenic Cysteine Mutation: Discovery of a Specific Inhibitor of Y279C SHP2. *Biochemistry* **2020**, *59*, 3498–3507.
67. Wood, D. J.; et al. FragLites-Minimal, Halogenated Fragments Displaying Pharmacophore Doublets. An Efficient Approach to Druggability Assessment and Hit Generation. *J. Med. Chem.* **2019**, *62*, 3741–3752.
68. Mons, E.; et al. The Alkyne Moiety as a Latent Electrophile in Irreversible Covalent Small Molecule Inhibitors of Cathepsin K. *J. Am. Chem. Soc.* **2019**, *141*, 3507–3514.
69. Weerapana, E.; et al. Quantitative Reactivity Profiling Predicts Functional Cysteines in Proteomes. *Nature* **2010**, *468*, 790–795.
70. Zhang, X.; Crowley, V. M.; Wucherpfennig, T. G.; Dix, M. M.; Cravatt, B. F. Electrophilic PROTACs that Degrade Nuclear Proteins by Engaging DCAF16. *Nat. Chem. Biol.* **2019**, *15*, 737–746.
71. Zhang, X.; et al. DCAF11 Supports Targeted Protein Degradation by Electrophilic Proteolysis-Targeting Chimeras. *J. Am. Chem. Soc.* **2021**, *143*, 5141–5149.
72. Vinogradova, E. V.; et al. An Activity-Guided Map of Electrophile-Cysteine Interactions in Primary Human T Cells. *Cell* **2020**, *182*, 1009–1026.e29.
73. Crowley, V. M.; Thielert, M.; Cravatt, B. F. Functionalized Scout Fragments for Site-Specific Covalent Ligand Discovery and Optimization. *ACS Cent. Sci.* **2021**, 7 (4), 613–623. https://doi.org/10.1021/acscentsci.0c01336.
74. Kuljanin, M.; et al. Reimagining High-Throughput Profiling of Reactive Cysteines for Cell-Based Screening of Large Electrophile Libraries. *Nat. Biotechnol.* **2021**, *39*, 630–641. https://doi.org/10.1038/s41587-020-00778-3.
75. Parker, C. G.; et al. Ligand and Target Discovery by Fragment-Based Screening in Human Cells. *Cell* **2017**, *168*, 527–541.e29.
76. Wang, Y.; et al. Expedited Mapping of the Ligandable Proteome Using Fully Functionalized Enantiomeric Probe Pairs. *Nat. Chem.* **2019**, *11*, 1113–1123.
77. Bateman, L. A.; et al. Chemoproteomics-Enabled Covalent Ligand Screen Reveals a Cysteine Hotspot in Reticulon 4 That Impairs ER Morphology and Cancer Pathogenicity. *Chem. Commun.* **2017**, *53*, 7234–7237.
78. Counihan, J. L.; Wiggenhorn, A. L.; Anderson, K. E.; Nomura, D. K. Chemoproteomics-Enabled Covalent Ligand Screening Reveals ALDH3A1 as a Lung Cancer Therapy Target. *ACS Chem. Biol.* **2018**, *13*, 1970–1977.
79. Chung, C. Y.-S.; et al. Covalent Targeting of the Vacuolar H+-ATPase Activates Autophagy Via mTORC1 Inhibition. *Nat. Chem. Biol.* **2019**, *15*, 776–785.
80. Murray, C. W.; Rees, D. C. The Rise of Fragment-Based Drug Discovery. *Nat. Chem.* **2009**, *1*, 187–192.
81. Konteatis, Z. What Makes a Good Fragment in Fragment-Based Drug Discovery? *Expert Opin. Drug Discovery* **2021**, 1–4. https://doi.org/10.1080/17460441.2021.1905629.
82. Sutanto, F.; et al. Multicomponent Reaction-Derived Covalent Inhibitor Space. *Sci. Adv.* **2021**, 7 (6), eabd9307. https://doi.org/10.1126/sciadv.abd9307.

CHAPTER NINE

The future of covalent inhibition

György M. Keserű[a,*] and Daniel A. Erlanson[b,*]
[a]Research Center for Natural Sciences (RCNS), Budapest, Hungary
[b]Frontier Medicines Corporation, San Francisco, CA, United States
*Corresponding authors: e-mail address: keseru.gyorgy@ttk.mta.hu; daniel.erlanson@frontiermeds.com

Contents

Most FDA-approved drugs bind reversibly, with non-covalent interactions to their corresponding macromolecular targets. New drug applications during the last decade, however, show a rapid increase of covalent inhibitors. Although compounds forming covalent bonds with their drug targets were known, most of these were discovered serendipitously. Covalent mechanisms of action were underrepresented for multiple reasons. First, electrophilic compounds were routinely removed from screening cascades due to their inherent risk of immunogenicity and idiosyncratic toxicity. Second, this mechanism is simply not feasible for many of the potential targets due to the lack of tractable nucleophilic residues at the binding site. Third, neither the theoretical background, nor the design principles, nor suitable experimental models were widely available for implementing this approach in drug discovery settings, while straightforward, reversible approaches were widely available. During the last decade, however, active research has provided guidelines for the discovery and development of compounds that covalently label their targets. Since the most important contributions are systematically covered in the previous chapters of this book, here we focus on emerging trends and perspectives of compounds acting by covalent mechanisms of action.

Annual Reports in Medicinal Chemistry, Volume 56
ISSN 0065-7743
https://doi.org/10.1016/bs.armc.2020.10.003

Future directions of covalent inhibitors can be considered in terms of both compounds and targets. From the ligand point of view, there are considerable efforts to discover new specific warheads and chemistries available for covalently labeling a variety of protein nucleophiles. In addition to the developments in labeling chemistry, established drug discovery technologies could also benefit from covalent targeting. Fragment-based approaches using covalent and photoaffinity-tagged fragments are emerging in target validation and hit discovery programs. Heterobifunctional small molecules used in the proteolysis targeting chimera (PROTAC) methodology might be equipped with electrophilic warheads that covalently bind unwanted proteins to enhance their removal. Electrophilic compounds can be used in proteomics technologies and other cell-based approaches that might enhance target finding and validation by identifying ligandable proteins for further hit discovery efforts.

From the target perspective, there is widening scope in both the residues and target classes considered. The serines, threonines, and cysteines targeted most frequently in the last decade are now joined by other potential nucleophiles such as lysines, tyrosines and histidines. Although the first successful attempts of rationally designed covalent inhibitors were focused on protein kinases, more challenging targets are now being explored by a covalent mechanism of action. These targets include proteins with shallow binding sites involved in protein-protein interactions (PPIs), intrinsically disordered proteins (IDPs), proteins with allosteric and/or cryptic binding sites, and further challenging drug targets such as the oncology signaling integrator KRAS or the COVID-19 related main protease of SARS-CoV-2 virus. In the next sections, we review these emerging areas of covalent approaches and provide case studies for the most promising trends.

1. New electrophilic warheads

Mechanism-based inhibitors developed against the catalytic serine and cysteine residues of certain proteases might be considered as the first examples of targeted covalent compounds. Early irreversible covalent approaches typically targeted cysteine residues due to the low occurrence of cysteine in the human proteome (2.3%).[1] In addition, low conservation of non-catalytic cysteines alleviates selectivity issues and might help avoid the development of resistance to kinase inhibitors.[2] The cysteine thiol is highly reactive due to its high electron density and polarizability. Therefore, this functional group can be attacked with low reactivity ligands, as preferred to minimize off-target interactions.[3] Moreover, cysteines play a significant role in a variety of functions including nucleophilic and redox catalysis, metal binding and allosteric regulation. These functional cysteines are found on diverse proteins such as proteases, oxidoreductases and kinases and therefore cysteine-targeted electrophiles can be utilized to affect the function of a wide range of proteins.[4,5] As summarized in chapter "Beyond cysteine" by Jones, other residues, such as serine, threonine, tyrosine and lysine, can also be engaged. On one hand, extending the scope of potential protein nucleophiles widens the range of the potential targets addressable by covalent inhibitors. On the other hand, labeling other and typically less nucleophilic residues often requires new labeling chemistries and particularly new warheads. There are a number of warheads available for covalent labeling of serine and threonine residues. In addition to the nitriles, epoxides, and boronic acids covered in earlier chapters, these includes 1,2,5-thiadiazolidin-3-ones, benzoxazin-4-ones, γ-lactams, γ-lactones, γ-sultones,[6] 3-oxo-β-sultams,[7] heterocyclic ureas and cyclic boronates,[8] β-sultames, phosphonates, phosphonyl fluorides and isocoumarin derivatives,[9] and isoxazolidinones.[10] The aromatic hydroxyl group of tyrosine residues can be labeled by substituted imines, diazocarboxamides and diazonium salts[11] and fluorosialyl fluorides.[9] Somewhat selective labeling of lysine sidechains can be performed by difluoroalkenes,[12] alkyl sulfonamides,[9] specific furane derivatives[13] and isocyanates.[9]

Other sidechains are typically labeled less specifically; however, electrophiles that react against several different protein nucleophiles might be useful tools in chemogenomic and chemoproteomic approaches. Sulfur containing electrophiles such as fluorosulfates,[14] sulfamoyl fluorides and sulfonyl fluorides[15] and sulfonate esters[11] were demonstrated to react with multiple residues including cysteines, serines, tyrosines, lysines, histidines and even aspartates and glutamates. Although chemical biology approaches benefit significantly from these probes, specific labeling that is more challenging for non-cysteine residues requires further investigation. Developing new warheads therefore should be one of the key objectives of future research.

2. Covalent fragment approaches

Fragment screening uses relatively small libraries of less complex and highly soluble molecules (fragments) and demonstrates multiple advantages over large scale screens of druglike molecules.[16,17] Although fragments cover a larger fraction of the available chemical space and provide a higher hit rate for structurally diverse starting points, typical fragment hits have low affinity that makes their detection, validation and optimization challenging. Since electrophiles form covalent bonds with the target, fragments equipped with such functionalities might alleviate these problems. Covalent fragments label the target chemically and so might have higher affinity that improves the chance of detecting the binding event. Furthermore, their fixed binding mode would make their optimization less ambiguous. These advantages of covalent fragments apply in many drug discovery applications from target identification and target validation studies to covalent inhibitor programs.[18] The improved target engagement achieved has prompted significant research in this direction.[19–23] In addition to target identification, screening covalent fragments could assess the ligandability and chemical tractability of new drug targets.[24] This approach might be particularly useful against unexplored and transient binding sites typically located in allosteric or cryptic pockets. In optimization settings, covalent fragment hits provide a viable starting point for classical fragment-based drug discovery strategies including fragment growing or merging with hits identified from non-covalent fragment libraries screened in parallel.

In addition to these benefits, screening covalent fragments has inherent challenges. One of the key issues is clearly related to the balance between the available non-covalent interactions and the reactivity of the warhead. Selecting appropriate warheads is not a trivial task since warhead reactivity measured in surrogate assays can change significantly at the binding site.[25] On the other hand, the intrinsic reactivity does not necessarily correlate with the labeling efficiency.[26] Depending on the nucleophilicity of the available residues, covalent fragments might label multiple sites. Promiscuity might be avoided by filtering out high reactivity fragments or using engineered proteins in covalent fragment screens.[27] Considering structural and pharmacophore complexity, non-covalent interactions might be influenced significantly upon the formation of the covalent bond with the targeted residue. Furthermore, electronic crosstalk between the reaction center of the warhead and the non-covalent fragment core would complicate the identification of optimal warhead types. These circumstances make the interpretation of structure-activity relationships (SAR) difficult for covalent fragments and suggest that confirmed covalent hits from mass spectrometry should be followed up by structural studies.

Significant improvements in the design principles of covalent fragment libraries together with methodological developments observed in protein- and cell-based assays will make covalent fragments increasingly popular in drug discovery settings. Drug discovery applications of covalent fragment libraries have been recently reviewed[19] and further applications are expected in the future.

3. Covalent PROTAC methods

An emerging class of drugs seeks not merely to inhibit target proteins, but to degrade them. In so-called PROTACs this is done by using bivalent ligands, in which one portion binds to the target of interest, while the other recruits an E3 ligase. The ligase can cause ubiquitylation of the target protein, ultimately leading to degradation in the proteasome. As compared to traditional inhibitors, PROTACS show multiple advantages that include sub-stoichiometric target occupancy,[28,29] enhanced specificity,[30,31] sustained pharmacodynamic effects,[32] and powerful management of drug resistance by degrading the target.[33] Since covalent ligands show strong target affinity and high target occupancy, linking the E3 ligase recognition element with a covalent inhibitor of the target seems to be a rational strategy.

There is no consensus yet on the efficacy of covalent PROTACs compared to non-covalent PROTACS in the literature. Although the seminal paper[34] introducing the technology used a covalent PROTAC, it was characterized only in vitro. Recently, a number of covalent PROTACs with cellular efficacy have been published,[26,35–37] however, some conflicting results were also reported. Despite the high target occupancy and exceptional clinical efficacy of the covalent Bruton's tyrosine kinase (BTK) inhibitor ibrutinib, PROTACs based on this ligand were unable to induce protein degradation.[38] The authors concluded that irreversible binding of covalent PROTACs might contribute to the loss of the sub-stoichiometric occupancy.

One straightforward option for BTK was changing the irreversible warheads to reversible covalent functionalities. Comparative analysis of reversible non-covalent, reversible covalent, and irreversible covalent BTK-targeting PROTACs revealed that a reversible covalent PROTAC might have improved intracellular accumulation and target engagement.[39] If confirmed on a wider set of targets, this strategy might be used more generally and provide improved efficacy for PROTAC degraders. Interestingly, however, another BTK-targeted PROTAC equipped with an ibrutinib analogue degraded the target efficiently.[40] This finding is at odds with the limited catalytic efficiency of irreversible covalent PROTACs. The community will need to clarify the scope and limitations of this approach (see also below).

In addition to covalent approaches applied for liganding proteins targeted for degradation, covalent approaches have recently been applied for the E3 ligase recognizing part of PROTACs. The key motivation of these studies was broadening the available set of E3 recruiter ligands. Screening covalent ligands by activity-based protein profiling (ABPP) identified covalent hits reacting with the E3 ubiquitin ligase RNF4 without influencing its activity. After further potency optimization, the compound was linked with an inhibitor of the BET family of bromodomains and the resulted bifunctional degrader removed BRD4 in a proteasome- and RNF4-dependent manner.[41] Efforts identifying specific covalent inhibitors of E3 ligases[42] will increase the number of available ligands for PROTAC development and the methodology might contribute to developing new PROTACs based on additional ubiquitin ligases.

4. Covalent proteomics and phenotypic approaches

Most of this chapter—and indeed most of this book—has focused on target-based lead discovery, in which covalent inhibitors are generated for isolated protein targets. However, in recent years chemoproteomic approaches have become sufficiently advanced to enable covalent lead discovery in more complicated systems, such as whole cells.

Some of the earliest work along these lines, dating back more than 20 years, was conducted by Ben Cravatt and colleagues at the Scripps Research Institute. Starting with molecules containing warheads directed toward serine hydrolases, such as fluorophosphonates, they screened cells and cell lysates in an approach called activity-based protein profiling (ABPP).[43] Initially this consisted of fluorophore-labeled electrophiles and gel-based separation of labeled proteins, but as mass spectrometry technology has advanced this technique has come to the fore. The technique is the subject of hundreds of papers and the interested reader is directed to recent reviews.[44,45]

Many chemoproteomics efforts have centered on examining selectivity of known covalent molecules.[46] Given the potential for off-target toxicity, this is an important area of research. Indeed, ABPP profiling of the investigational drug BIA 10-2474, a fatty acid amide hydrolase inhibitor, revealed inhibition of several other lipolytic enzymes, including one linked to the neurotoxicity of organophosphates.[47] Unfortunately, this study was conducted after a tragic clinical trial in which one subject died and four others were hospitalized. It is likely that covalent drugs will increasingly be profiled prospectively to look for dangerous off-targets, though it is important to note that a covalent drug could also bind non-covalently to targets, and this would likely not be picked up in an ABPP screen.

In addition to assessing the selectivity of existing molecules, an exciting area of research consists of using phenotypic screens of covalent molecules to identify new targets. This "chemical genetics" approach is facilitated by the fact that the molecules are covalent, facilitating target identification using mass spectrometry. In an early application, Cravatt and colleagues screened a library of just 50 fragments against cells and cell lysates and identified several chloroacetamide-containing molecules that modified the active-site cysteine of caspase-8.[21] Interestingly though, these were selective for the pre-processed (or zymogen) form of the enzyme. More recently, Roberto Zoncu, Daniel Nomura, and colleagues conducted a cell-based screen of 217 potentially covalent molecules to look for those that could activate autophagy. One of the hits that did so was found to bind to the ATP6V1A subunit of the lysosomal v-ATPase.[48]

Previously this chapter highlighted the role of covalent inhibitors in developing PROTACs, and here too cell-based screens have a role. The Cravatt group coupled three particularly reactive "scout fragments" to a ligand for FKBP12 and conducted cellular assays to determine whether these bifunctional ligands would cause degradation of FKBP12. One chloroacetamide-based molecule caused degradation of FKBP12, and further work determined that the degradation was caused by engaging the E3 ligase DCAF16.[49] Independently, Dan Nomura's group at UC Berkeley identified chloroacetamide-containing, as well as natural-product based, molecules that can recruit the E3 ligase RNF4.[41,50] The Nomura lab has also made PROTACs targeting the E3 ligase KEAP1 using a known covalent binder.[51] These examples add to the growing list of E3 ligases that can be coopted for targeted protein degradation, although the researchers note that the covalent ligands can engage proteins other than the E3 ligases identified.

Although these examples are interesting, most of the reports to date have consisted of initial hits, with little follow-up work or medicinal chemistry optimization. Moreover, many of the hits have been chloroacetamides, a warhead likely too promiscuous and unstable to be used in drugs. In some cases it may be possible to "swap" this moiety with a more attractive warhead, and while this can be challenging due to different geometries, recent reports suggest that a chlorofluoroacetamide has improved stability and specificity.[52,53] Also, these are early days. The ability to identify covalent ligands against a wide variety of targets—and rapidly characterize them—will undoubtedly open new areas of biology and drug discovery.

5. Novel target types

As of mid-2020 six cysteine-reactive kinase inhibitors have been clinically approved, starting with ibrutinib in 2013, and the topic has been reviewed extensively, including earlier in this book.[54,55] But of course there are many other drug targets beyond kinases, as examined below

5.1 KRAS^{G12C}

The RAS family of proteins has been known to be involved in cancer for decades. However, due to the lack of previously targetable small-molecule binding sites, identifying potent inhibitors of these proteins has been hugely challenging.

The glycine to cysteine mutation at the twelfth amino acid in KRAS is present in nearly 6.7% of lung adenocarcinomas, 2.6% of colorectal adenocarcinomas, and 1.7% pancreatic ductal adenocarcinomas.[56] The mutant cysteine residue is located next to the nucleotide binding pocket and—analogously to work on kinases that makes use of purine mimetics—Nathanael Gray and colleagues synthesized a guanosine analog containing a chloroacetamide warhead.[57] Although this molecule could covalently modify KRAS^{G12C}, the diphosphate linker between the warhead and the nucleoside was problematic for cell permeability. Moreover, the extremely high affinity of KRAS for GDP and GTP, combined with the high intracellular concentrations of these nucleotides, further complicate the potential for direct competitors.

Kevan Shokat and colleagues also recognized the potential of the cysteine residue as a handle for covalent inhibition and took a de novo approach on KRAS, using disulfide tethering.[58] In this approach, a library of disulfide-containing fragments is screened against a cysteine-containing protein under partially reducing conditions. If the fragment binds in close proximity to the cysteine it can form a disulfide bond which can be stabilized against reduction, allowing identification of the bound fragment by mass spectrometry. The disulfide moiety is sterically and electronically separated by a few bonds from the variable portion, providing a clean signal for binding without

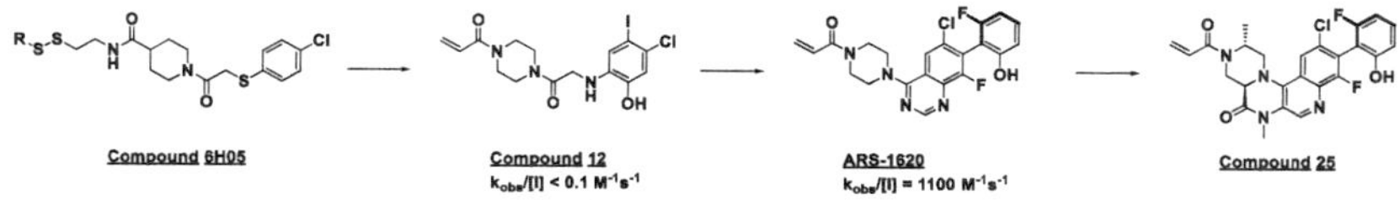

Fig. 1 Optimization of the KRAS^{G12C} inhibitor ARS-1620.

complications from differences in reactivity between different fragments. After screening 480 fragments, the researchers identified hits such as 6H05. Replacing the disulfide with an acrylamide and further SAR led ultimately to Compound 12, which reacts covalently with KRAS^{G12C} and blocks binding of the effector protein RAF (Fig. 1).[27]

This series of compounds was licensed to the biotechnology company Araxes, which further developed them to more potent molecules such as ARS-1620.[59] This compound shows potent biochemical and cell-based activity as well as tumor regression in mouse xenograft models. Araxes partnered with Janssen on further development, and JNJ-74699157 entered clinical trials in 2019, although in 2020 patient recruitment was halted with only 10 of 140 projected subjects enrolled.[60] Nonetheless, the multiple patent applications published by Araxes proved useful for several other companies. Indeed, by the end of 2019 more than three dozen patent applications had been filed from various companies claiming covalent KRAS^{G12C} inhibitors,[61,62] many of them with clear origins in this chemical matter. For example, AstraZeneca recently published a series of inhibitors (such as compound 25) in which conformationally locking ARS-1620 improved potency.[63]

An independent series of KRAS^{G12C} inhibitors was discovered by researchers from Carmot Therapeutics and Amgen, using a fragment-based approach called Chemotype Evolution, which involves iterative synthesis and screening of thousands of molecules.[64] An initial weak hit (compound 2) was improved to compound 1 (Fig. 2). Although this molecule demonstrated more rapid modification of the protein than ARS-1620, the pharmacokinetic properties were not suitable for advancement. However, crystallography revealed that the tetrahydroisoquinoline moiety was binding in a newly-formed subpocket, and filling this with a pyridyl moiety led to AMG 510, or sotorasib, the first KRAS^{G12C} inhibitor to be evaluated in clinical trials.[65,66] The second agent to begin clinical development, MRTX849 (Fig. 2), is also structurally related to these molecules, although the reported origins are from a covalent screening effort at Array BioPharma.[67–69] Both AMG 510 and MRTX849 have demonstrated

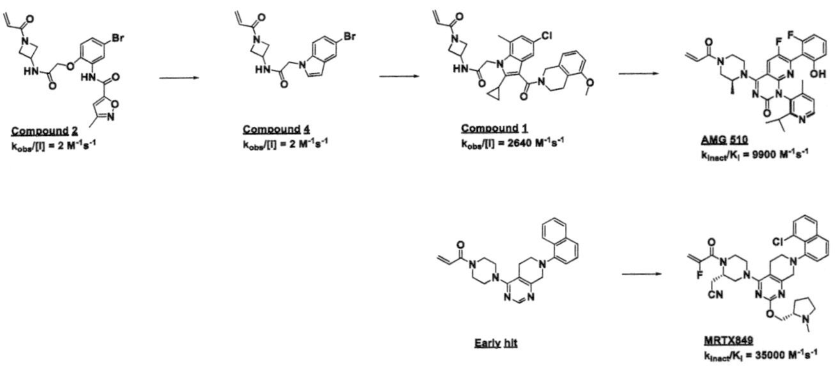

Fig. 2 Optimization of covalent KRAS^{G12C} inhibitors AMG510 and MRTX849.

encouraging activity in non-small-cell lung cancer, and it will be exciting to watch their progress.[70]

In contrast to covalent kinase inhibitor drugs, all of which have high affinities for their target proteins, the KRAS^{G12C} inhibitors rely on a high inactivation rate with a relatively low inherent affinity. Indeed, the K_I for MRTX849 was determined to be 3.7 μM, two to three orders of magnitude weaker than many drugs.[69] Nonetheless, the selectivity of these molecules for KRAS^{G12C} is remarkable as assessed by proteomic profiling. The alkyl acrylamide warhead is not particularly reactive, but when the molecules bind to the protein, a catalytic lysine residue forms a hydrogen bond to the acrylamide carbonyl oxygen, thereby increasing its reactivity.[71]

Recently two groups have published molecules that use a covalent ligand for KRAS^{G12C} to cause its degradation. In one, a molecule based on ARS-1620 and carrying a ligand for the E3 ligase cereblon was able to degrade a fusion protein of green fluorescent protein and KRAS^{G12C}, but not endogenous KRAS^{G12C}.[72] However, a separate report in which MRTX849 was linked to a VHL ligand was successful at causing degradation of KRAS^{G12C}.[37] Obviously much remains to be determined as to the differences between these results, as well as whether KRAS degradation will have different biological effects from inhibition.

All of these molecules target a single mutant form of KRAS. However, while the G12C mutation is clinically relevant, it is less common than G12D in several forms of cancer. To try to expand to this mutation, Shokat and colleagues replaced the acrylamide moiety with eight different warheads.[73] Unfortunately, although slight modification of the G12D and G12S mutants could be observed, the level of modification was considerably lower than the G12C mutant. Moving beyond cysteine residues remains an active challenge.

5.2 SARS-CoV-2

At the time of writing, SARS-CoV-2, the virus responsible for COVID-19, has infected more than 35 million people, of which more than a million have died. This is greater than the number of people killed annually by HIV or malaria. In response, researchers around the world have turned their attention to fighting the virus. Vaccines and antibodies are advancing most rapidly in the clinic, but small molecule drug discovery efforts are also underway. One of the key viral targets is the main protease, or M^{pro}, which cleaves the newly translated viral polyproteins into individual functional proteins. This cysteine protease contains a hyperreactive cysteine residue, C145, and thus has been the focus of multiple covalent efforts.

Repurposing existing drugs would potentially provide the fastest path to the clinic, and multiple groups have explored this approach. Research led by Leike Zhang and Haitao Yang revealed that the anti-cancer drug carmofur can react covalently with M^{pro} at the catalytic cysteine, generating an alkyl carbamothioate, and confirmed this crystallographically.[74] Carmofur is a prodrug of the antimetabolite 5-fluorouracil, itself a covalent inhibitor of thymidylate synthase, though it is also a potent inhibitor of acid ceramidase, like M^{pro} another cysteine hydrolase.[75] However, the double digit micromolar cellular potency for inhibiting SARS-CoV-2 viral replication in cells makes therapeutic applications unlikely. The same is true for covalent drugs such as disulfiram.[76] Both of these drugs likely owe their activity against M^{pro} to nonspecific reactions with the highly reactive active-site cysteine. A more promising line of research involves covalent drugs developed for other proteases. Boceprevir, for example, is a serine protease inhibitor that was approved by the FDA as a treatment for hepatitis C. Newer, more effective drugs caused this drug to be pulled from the market, but boceprevir and several other related α-ketoamides have been reported as low micromolar M^{pro} inhibitors.[77,78]

Although repurposing existing drugs could provide a stopgap measure, directed drug discovery is more likely to produce potent, efficacious drugs, as has been the case with HIV and hepatitis C. The α-ketoamide warhead has been extensively used for both serine and cysteine hydrolases. Previous structure-based design work on the main protease of SARS-CoV and MERS-CoV has led to peptidomimetic-based α-ketoamides with sub-micromolar activity against SARS-CoV-2 M^{pro}.[79] Some effort has been expended on improving the pharmacokinetic properties of these molecules, and although oral bioavailability remains a challenge for these peptidomimetics, inhalation may be an option. A similar approach has led to mid-nanomolar potency aldehyde-based inhibitors of M^{pro}.[80] In general, aldehyde-containing molecules have suboptimal pharmacokinetic properties, but the researchers found reasonable half-lives in rats and dogs, and preliminary toxicology studies were also encouraging. Very recently, researchers from Pfizer have reported low nanomolar hydroxymethylketone-based inhibitors of M^{pro} and advanced a prodrug of one of them, PF-07304814, into clinical trials.[81,82]

As noted above, most protease inhibitors have been developed using structure-based design starting with the natural substrate and introducing an electrophilic warhead. While this can produce potent inhibitors, the peptide origins result in molecules with multiple hydrogen bond donors and high polar surface areas, challenging oral delivery. Intravenous drugs such as PF-07304814 may be useful in the hospital setting, but for a drug to be widely useful it needs to be able to be formulated as an oral treatment. Thus, in addition to peptidomimetic approaches, there has been a focus on fragment-based screening efforts. The hypothesis behind this approach is that starting from small fragments, the overall physicochemical properties can be more tightly controlled, leading to superior drugs.[17] Historically fragment-based screening has focused on non-covalent molecules, but in recent years attention has also turned to covalent fragments.[18]

A large multinational fragment screen was conducted against M^{pro} at the start of the pandemic, in February and March 2020. One component included a mass-spectrometry screen of 1000 electrophilic fragments.[83] This library had previously been screened against at least 10 other targets at 200 μM for 24 h,[20] but doing so with M^{pro} resulted in so many hits that the researchers needed to reduce the concentration of each fragment to 5 μM and reduce the time to just 90 min. Even this stringent approach led to an 8.5% hit rate, emphasizing the reactivity of the active-site cysteine. A high-throughput crystallography campaign of the hits produced crystal structures of 48 of these bound to the active-site cysteine. The majority of these came from chloroacetamides, which as noted above tend not to be pursued due to their relatively high chemical reactivity. However, two amino acid derived bromoalkynes were also found to bind to the active-site cysteine; these electrophiles are generally inert.[84] Additionally, two heterocyclic electrophiles also modified the protein; these particular molecules are reasonably stable to glutathione.[85] Gratifyingly, all of the crystal structures were released to the public shortly after they were solved and even before publication of the paper. Currently an open-source consortium called the COVID Moonshot is coordinating researchers worldwide attempting to advance molecules coming out of this screen.[86] Interestingly though, as of August 2020 much of the effort had turned to non-covalent inhibitors.[87]

6. Conclusion

It is probably fair to say that every major pharmaceutical company is pursuing some level of covalent drug discovery. In many cases this involves rationally appending warheads onto known high-affinity binders, as for covalent kinase inhibitors and peptide-based inhibitors. In other cases, screening libraries of covalent molecules against individual targets produces hits that can be optimized, as for $KRAS^{G12C}$ and fragment-based efforts on SARS-CoV-2. And in still other cases, the field of covalent inhibition has come full circle. The earliest covalent drugs, such as aspirin, penicillin, and mustard agents, were discovered serendipitously. Indeed, the biological targets were often not known until decades after the drugs came into widespread use. Phenotypic screens today can identify interesting biological effects, and chemoproteomic methods can readily identify the targets.

After a long period in which covalent drugs were seen as too risky, the early 21st century has seen a resurgence of interest in covalent drugs. Multiple new molecules are rapidly advancing through the clinic, and fundamental new research is expanding the range of what is possible. But there are plenty of opportunities for further research. As noted above, targeting residues beyond cysteine and catalytic serine or threonine remains underdeveloped. New warhead development and characterization remains an urgent task. Moreover, as seen for BIA 10-2474, covalent modifiers can have serious off-target effects—though the same can be said for non-covalent drugs. Hopefully this book demonstrates that covalent drugs have already played a major role in expanding humanity's pharmacopeia. As we tackle increasingly challenging targets, it is likely that covalent drugs will play an ever larger role.

References

1. Hallenbeck, K. K.; Turner, D. M.; Renslo, A. R.; Arkin, M. R. *Curr. Top. Med. Chem.* **2017**, *17*, 4–15.
2. Zhao, Z.; Bourne, P. E. *Drug Discovery Today* **2018**, *23*, 727–735.
3. Lagoutte, R.; Patouret, R.; Winssinger, N. *Curr. Opin. Chem. Biol.* **2017**, *39*, 54–63.
4. Pace, N. J.; Weerapana, E. *ACS Chem. Biol.* **2013**, *8*, 283–296.
5. Weerapana, E.; Wang, C.; Simon, G. M.; Richter, F.; Khare, S.; Dillon, M. B.; Bachovchin, D. A.; Mowen, K.; Baker, D.; Cravatt, B. F. *Nature* **2010**, *468*, 790–795.
6. Powers, J. C.; Asgian, J. L.; Ekici, Ö. D.; James, K. E. *Chem. Rev.* **2002**, *102*, 4639–4750.

7. Carvalho, L. A. R.; Almeida, V. T.; Brito, J. A.; Lum, K. M.; Oliveira, T. F.; Guedes, R. C.; Gonçalves, L. M.; Lucas, S. D.; Cravatt, B. F.; Archer, M.; Moreira, R. *ACS Chem. Biol.* **2020**, *15*, 878–883.
8. Tooke, C. L.; Hinchliffe, P.; KRAJNC, A.; Mulholland, A. J.; Brem, J.; Schofield, C. J.; Spencer, J. *RSC Med. Chem.* **2020**, *11*, 491–496.
9. Shannon, D. A.; Weerapana, E. *Curr. Opin. Chem. Biol.* **2015**, *24*, 18–26.
10. Gersch, M.; Kreuzer, J.; Sieber, S. A. *Nat. Prod. Rep.* **2012**, *29*, 659–682.
11. Tamura, T.; Hamachi, I. *J. Am. Chem. Soc* **2019**, *141*, 2782–2799.
12. Pettinger, J.; Jones, K.; Cheeseman, M. D. *Angew. Chem. Int. Ed.* **2017**, *56*, 15200–15209.
13. Drahl, C.; Cravatt, B. F.; Sorensen, E. J. *Angew. Chem. Int. Ed.* **2005**, *44*, 5788–5809.
14. Gehringer, M.; Laufer, S. A. *J. Med. Chem.* **2019**, *62*, 5673–5724.
15. Cheng, M.; Guo, C.; Gross, M. L. *Angew. Chem. Int. Ed.* **2019**, *59*, 5880–5889.
16. Hajduk, P.; Greer, J. *Nat. Rev. Drug Discov.* **2007**, *6*, 211–219.
17. Erlanson, D. A.; Fesik, S. W.; Hubbard, R. E.; Jahnke, W.; Jhoti, H. Twenty Years on: The Impact of Fragments on Drug Discovery. *Nat. Rev. Drug Discov.* **2016**, *15*, 605–619.
18. Keeley, A.; Petri, L.; Ábrányi-Balogh, P.; Keserű, G. M. *Drug Discovery Today* **2020**, *25*, 983–996.
19. Campos, S.; Dalton, S. E. *ChemBioChem* **2020**, *21*, 1080–1100.
20. Resnick, E.; Bradley, A.; Gan, J.; Douangamath, A.; Krojer, T.; Sethi, R.; et al. Rapid Covalent-Probe Discovery by Electrophile-Fragment Screening. *J. Am. Chem. Soc.* **2019**, *141*, 8951–8968.
21. Backus, K. M.; Correia, B. E.; Lum, K. M.; Forli, S.; Horning, B. D.; González-Páez, G. E.; Chatterjee, S.; Lanning, B. R.; Teijaro, J. R.; Olson, A. J.; Wolan, D. W.; Cravatt, B. F. *Nature* **2016**, *534*, 570–574.
22. Kathman, S. G.; Statsyuk, A. V. *Med. Chem. Commun.* **2016**, 7, 576–585.
23. Parker, C. G.; Galmozzi, A.; Wang, Y.; Correia, B. E.; Sasaki, K.; Joslyn, C. M.; Kim, A. S.; Cavallaro, C. L.; Lawrence, R. M.; Johnson, S. R.; Narvaiza, I.; Saez, E.; Cravatt, B. F. *Cell* **2017**, *168*, 527–541.e29.
24. Zhang, T.; Hatcher, J. M.; Teng, M.; Gray, N. S.; Kostic, M. *Cell Chem. Biol.* **2019**, *26*, 1486–1500.
25. Ábrányi-Balogh, P.; Petri, L.; Imre, T.; Szijj, P.; Scarpino, A.; Hrast, M.; Mitrović, A.; Fonovič, U. P.; Németh, K.; Barreteau, H.; Roper, D. I.; Horváti, K.; Ferenczy, G. G.; Kos, J.; Ilaš, J.; Gobec, S.; Keserű, G. M. *Eur. J. Med. Chem.* **2018**, *160*, 94–107.
26. Jöst, C.; Nitsche, C.; Scholz, T.; Roux, L.; Klein, C. D. *J. Med. Chem.* **2014**, *57*, 7590–7599.
27. Ostrem, J. M.; Peters, U.; Sos, M. L.; Wells, J. A.; Shokat, K. M. *Nature* **2013**, *503*, 548–551.
28. Bondeson, D. P.; Mares, A.; Smith, I. E.; Ko, E.; Campos, S.; Miah, A. H.; Mulholland, K. E.; Routly, N.; Buckley, D. L.; Gustafson, J. L.; Zinn, N.; Grandi, P.; Shimamura, S.; Bergamini, G.; Faelth-Savitski, M.; Bantscheff, M.; Cox, C.; Gordon, D. A.; Willard, R. R.; Flanagan, J. J.; Casillas, L. N.; Votta, B. J.; den Besten, W.; Famm, K.; Kruidenier, L.; Carter, P. S.; Harling, J. D.; Churcher, I.; Crews, C. M. *Nat. Chem. Biol.* **2015**, *11*, 611–617.
29. Burslem, G. M.; Smith, B. E.; Lai, A. C.; Jaime-Figueroa, S.; McQuaid, D. C.; Bondeson, D. P.; Toure, M.; Dong, H.; Qian, Y.; Wang, J.; Crew, A. P.; Hines, J.; Crews, C. M. *Cell Chem. Biol.* **2018**, *25*, 67–77.e63.
30. Gadd, M. S.; Testa, A.; Lucas, X.; Chan, K. H.; Chen, W.; Lamont, D. J.; Zengerle, M.; Ciulli, A. *Nat. Chem. Biol.* **2017**, *13*, 514–521.
31. Smith, B. E.; Wang, S. L.; Jaime-Figueroa, S.; Harbin, A.; Wang, J.; Hamman, B. D.; Crews, C. M. *Nat. Commun.* **2019**, *10*, 131.

32. Mares, A.; Miah, A. H.; Smith, I.; Rackham, M.; Thawani, A. R.; Cryan, J.; Haile, P. A.; Votta, B. J.; Beal, A. M.; Capriotti, C.; Reilly, M. A.; Fisher, D. T.; Zinn, N.; Bantscheff, M.; MacDonald, T. T.; Vossenkamper, A.; Dace, P.; Churcher, I.; Benowitz, A. B.; Watt, G.; Denyer, J.; Scott-Stevens, P.; Harling, J. D. *Commun. Biol.* **2020**, *3*, 140.
33. Bai, L.; Zhou, H.; Xu, R.; Zhao, Y.; Chinnaswamy, K.; McEachern, D.; Chen, J.; Yang, C. Y.; Liu, Z.; Wang, M.; Liu, L.; Jiang, H.; Wen, B.; Kumar, P.; Meagher, J. L.; Sun, D.; Stuckey, J. A.; Wang, S. *Cancer Cell* **2019**, *36*, 498–511.e417.
34. Sakamoto, K. M.; Kim, K. B.; Kumagai, A.; Mercurio, F.; Crews, C. M.; Deshaies, R. J. *Proc. Natl. Acad. Sci. U. S. A.* **2001**, *98*, 8554–8559.
35. Lebraud, H.; Wright, D. J.; Johnson, C. N.; Heightman, T. D. *ACS Cent. Sci.* **2016**, *2*, 927–934.
36. Peng, L.; Zhang, Z.; Lei, C.; Li, S.; Zhang, Z.; Ren, X.; Chang, Y.; Zhang, Y.; Xu, Y.; Ding, K. *ACS Med. Chem. Lett.* **2019**, *10*, 767–772.
37. Bond, M. J.; Chu, L.; Nalawansha, D. A.; Li, K.; Crews, C. M. *ACS Cent. Sci.* **2020**, *6*, 1367–1375.
38. Tinworth, C. P.; Lithgow, H.; Dittus, L.; Bassi, Z. I.; Hughes, S. E.; Muelbaier, M.; Dai, H.; Smith, I.; Kerr, W. J.; Burley, G. A.; Bantscheff, M.; Harling, J. D. *ACS Chem. Biol.* **2019**, *14*, 342–347.
39. Gabizon, R.; Shraga, A.; Gehrtz, P.; Livnah, E.; Shorer, Y.; Gurwicz, N.; Avram, L.; Unger, T.; Aharoni, H.; Albeck, S.; Brandis, A.; Shulman, Z.; Katz, B. Z.; Herishanu, Y.; London, N. *J. Am. Chem. Soc.* **2020**, *142*, 11734–11742.
40. Xue, G.; Chen, J.; Liu, L.; Zhou, D.; Zuo, Y.; Fu, T.; Pan, Z. *Chem. Commun.* **2020**, *56*, 1521–1524.
41. Ward, C. C.; Kleinman, J. I.; Brittain, S. M.; Lee, P. S.; Chung, C.; Kim, K.; Petri, Y.; Thomas, J. R.; Tallarico, J. A.; McKenna, J. M.; Schirle, M.; Nomura, D. K. *ACS Chem. Biol.* **2019**, *14*, 2430–2440.
42. Johansson, H.; Isabella Tsai, Y. C.; Fantom, K.; Chung, C. W.; Kümper, S.; Martino, L.; Thomas, D. A.; Eberl, H. C.; Muelbaier, M.; House, D.; Rittinger, K. *J. Am. Chem. Soc.* **2019**, *141*, 2703–2712.
43. Liu, Y.; Patricelli, M. P.; Cravatt, B. F. *Proc. Natl. Acad. Sci. U. S. A.* **1999**, *96* (26), 14694–14699.
44. Benns, H. J.; Wincott, C. J.; Tate, E. W.; Child, M. A. *Curr. Opin. Chem. Biol.* **2020**, *60*, 20–29.
45. Deng, H.; Lei, Q.; Wu, Y.; He, Y.; Li, W. *Eur. J. Med. Chem.* **2020**, *191*, 112151.
46. Lanning, B. R.; Whitby, L. R.; Dix, M. M.; Douhan, J.; Gilbert, A. M.; Hett, E. C.; Johnson, T. O.; Joslyn, C.; Kath, J. C.; Niessen, S.; Roberts, L. R.; Schnute, M. E.; Wang, C.; Hulce, J. J.; Wei, B.; Whiteley, L. O.; Hayward, M. M.; Cravatt, B. F. *Nat. Chem. Biol.* **2014**, *10*, 760–767.
47. van Esbroeck, A.; Janssen, A.; Cognetta, A. B., 3rd; Ogasawara, D.; Shpak, G.; van der Kroeg, M.; Kantae, V.; Baggelaar, M. P.; de Vrij, F.; Deng, H.; Allarà, M.; Fezza, F.; Lin, Z.; van der Wel, T.; Soethoudt, M.; Mock, E. D.; den Dulk, H.; Baak, I. L.; Florea, B. I.; Hendriks, G.; van der Stelt, M. *Science* **2017**, *356*, 1084–1087.
48. Chung, C. Y.; Shin, H. R.; Berdan, C. A.; Ford, B.; Ward, C. C.; Olzmann, J. A.; Zoncu, R.; Nomura, D. K. *Nat. Chem. Biol.* **2019**, *15*, 776–785.
49. Zhang, X.; Crowley, V. M.; Wucherpfennig, T. G.; Dix, M. M.; Cravatt, B. F. *Nat. Chem. Biol.* **2019**, *15*, 737–746.
50. Spradlin, J. N.; Hu, X.; Ward, C. C.; Brittain, S. M.; Jones, M. D.; Ou, L.; To, M.; Proudfoot, A.; Ornelas, E.; Woldegiorgis, M.; Olzmann, J. A.; Bussiere, D. E.; Thomas, J. R.; Tallarico, J. A.; McKenna, J. M.; Schirle, M.; Maimone, T. J.; Nomura, D. K. *Nat. Chem. Biol.* **2019**, *15*, 747–755.

51. Tong, B.; Luo, M.; Xie, Y.; Spradlin, J. N.; Tallarico, J. A.; McKenna, J. M.; Schirle, M.; Maimone, T. J.; Nomura, D. K. *Sci. Rep.* **2020**, *10*, 15543.
52. Shindo, N.; Fuchida, H.; Sato, M.; Watari, K.; Shibata, T.; Kuwata, K.; Miura, C.; Okamoto, K.; Hatsuyama, Y.; Tokunaga, K.; Sakamoto, S.; Morimoto, S.; Abe, Y.; Shiroishi, M.; Caaveiro, J.; Ueda, T.; Tamura, T.; Matsunaga, N.; Nakao, T.; Koyanagi, S.; Ohdo, S.; Yamaguchi, Y.; Hamachi, I.; Ono, M.; Ojida, A. *Nat. Chem. Biol.* **2019**, *15*, 250–258.
53. Sato, M.; Fuchida, H.; Shindo, N.; Kuwata, K.; Tokunaga, K.; Xiao-Lin, G.; Inamori, R.; Hosokawa, K.; Watari, K.; Shibata, T.; Matsunaga, N.; Koyanagi, S.; Ohdo, S.; Ono, M.; Ojida, A. *ACS Med. Chem. Lett.* **2020**, *11*, 1137–1144.
54. Gehringer, M. *Top. Med. Chem.* **2020**, 1–52.
55. Abdeldayem, A.; Raouf, Y. S.; Constantinescu, S. N.; Moriggl, R.; Gunning, P. T. *Chem. Soc. Rev.* **2020**, *49*, 2617–2687.
56. Li, S.; Balmain, A.; Counter, C. M. *Nat. Rev. Cancer* **2018**, *18* (12), 767–777.
57. Lim, S. M.; Westover, K. D.; Ficarro, S. B.; Harrison, R. A.; Choi, H. G.; Pacold, M. E.; Carrasco, M.; Hunter, J.; Kim, N. D.; Xie, T.; Sim, T.; Jänne, P. A.; Meyerson, M.; Marto, J. A.; Engen, J. R.; Gray, N. S. *Angew. Chem. Int. Ed. Engl.* **2014**, *53*, 199–204.
58. Erlanson, D. A.; Braisted, A. C.; Raphael, D. R.; Randal, M.; Stroud, R. M.; Gordon, E. M.; Wells, J. A. Site-Directed Ligand Discovery. *Proc. Natl. Acad. Sci. U. S. A.* **2000**, *2000* (97), 9367–9372.
59. Janes, M. R.; Zhang, J.; Li, L. S.; Hansen, R.; Peters, U.; Guo, X.; Chen, Y.; Babbar, A.; Firdaus, S. J.; Darjania, L.; Feng, J.; Chen, J. H.; Li, S.; Li, S.; Long, Y. O.; Thach, C.; Liu, Y.; Zarieh, A.; Ely, T.; Kucharski, J. M.; Kessler, L. V.; Wu, T.; Yu, K.; Wang, Y.; Yao, Y.; Deng, X.; Zarrinkar, P. P.; Brehmer, D.; Dhanak, D.; Lorenzi, M. V.; Hu-Lowe, D.; Patricelli, M. P.; Ren, P.; Liu, Y. *Cell* **2018**, *172*, 578–589.e17.
60. https://clinicaltrials.gov/ct2/show/record/NCT04006301.
61. Kettle, J. G.; Cassar, D. J. *Expert Opin. Ther. Pat.* **2020**, *30*, 103–120.
62. Orgován, Z.; Keserű, G. M. *Cancer Metastasis Rev.* **2020**. https://doi.org/10.1007/s10555-020-09911-9.
63. Kettle, J. G.; Bagal, S. K.; Bickerton, S.; Bodnarchuk, M. S.; Breed, J.; Carbajo, R. J.; Cassar, D. J.; Chakraborty, A.; Cosulich, S.; Cumming, I.; Davies, M.; Eatherton, A.; Evans, L.; Feron, L.; Fillery, S.; Gleave, E. S.; Goldberg, F. W.; Harlfinger, S.; Hanson, L.; Howard, M.; Howells, R.; Jackson, A.; Kemmitt, P.; Kingston, J. K.; Lamont, S.; Lewis, H. J.; Li, S.; Liu, L.; Ogg, D.; Phillips, C.; Polanski, R.; Robb, G.; Robinson, D.; Ross, S.; Smith, J. M.; Tonge, M.; Whiteley, R.; Yang, J.; Zhang, L.; Zhao, X. *J. Med. Chem.* **2020**, *63*, 4468–4483.
64. Shin, Y.; Jeong, J. W.; Wurz, R. P.; Achanta, P.; Arvedson, T.; Bartberger, M. D.; Campuzano, I. D. G.; Fucini, R.; Hansen, S. K.; Ingersoll, J.; Iwig, J. S.; Lipford, J. R.; Ma, V.; Kopecky, D. J.; McCarter, J.; San Miguel, T.; Mohr, C.; Sabet, S.; Saiki, A. Y.; Sawayama, A.; Sethofer, S.; Tegley, C. M.; Volak, L. P.; Yang, K.; Lanman, B. A.; Erlanson, D. A.; Cee, V. J. *ACS Med. Chem. Lett.* **2019**, *10*, 1302–1308.
65. Lanman, B. A.; Allen, J. R.; Allen, J. G.; Amegadzie, A. K.; Ashton, K. S.; Booker, S. K.; Chen, J. J.; Chen, N.; Frohn, M. J.; Goodman, G.; Kopecky, D. J.; Liu, L.; Lopez, P.; Low, J. D.; Ma, V.; Minatti, A. E.; Nguyen, T. T.; Nishimura, N.; Pickrell, A. J.; Reed, A. B.; Shin, Y.; Siegmund, A. C.; Tamayo, N. A.; Tegley, C. M.; Walton, M. C.; Wang, H. L.; Wurz, R. P.; Xue, M.; Yang, K. C.; Achanta, P.; Bartberger, M. D.; Canon, J.; Hollis, L. S.; McCarter, J. D.; Mohr, C.; Rex, K.; Saiki, A. Y.; San Miguel, T.; Volak, L. P.; Wang, K. H.; Whittington, D. A.; Zech, S. G.; Lipford, J. R.; Cee, V. J. *J. Med. Chem.* **2020**, *63*, 52–65.
66. Canon, J.; Rex, K.; Saiki, A. Y.; Mohr, C.; Cooke, K.; Bagal, D.; Gaida, K.; Holt, T.; Knutson, C. G.; Koppada, N.; Lanman, B. A.; Werner, J.; Rapaport, A. S.; San

Miguel, T.; Ortiz, R.; Osgood, T.; Sun, J. R.; Zhu, X.; McCarter, J. D.; Volak, L. P.; Houk, B. E.; Fakih, M. G.; O'Neil, B. H.; Price, T. J.; Falchook, G. S.; Desai, J.; Kuo, J.; Govindan, R.; Hong, D. S.; Ouyang, W.; Henary, H.; Arvedson, T.; Cee, V. J.; Lipford, J. R. *Nature* **2019**, *575*, 217–223.
67. Fell, J. B.; Fischer, J. P.; Baer, B. R.; Ballard, J.; Blake, J. F.; Bouhana, K.; Brandhuber, B. J.; Briere, D. M.; Burgess, L. E.; Burkard, M. R.; Chiang, H.; Chicarelli, M. J.; Davidson, K.; Gaudino, J. J.; Hallin, J.; Hanson, L.; Hee, K.; Hicken, E. J.; Hinklin, R. J.; Marx, M. A.; Mejia, M. J.; Olson, P.; Savechenkov, P.; Sudhakar, N.; Tang, T. P.; Vigers, G. P.; Zecca, H.; Christensen, J. G. *ACS Med. Chem. Lett.* **2018**, *9*, 1230–1234.
68. Hallin, J.; Engstrom, L. D.; Hargis, L.; Calinisan, A.; Aranda, R.; Briere, D. M.; Sudhakar, N.; Bowcut, V.; Baer, B. R.; Ballard, J. A.; Burkard, M. R.; Fell, J. B.; Fischer, J. P.; Vigers, G. P.; Xue, Y.; Gatto, S.; Fernandez-Banet, J.; Pavlicek, A.; Velastagui, K.; Chao, R. C.; Barton, J.; Pierobon, M.; Baldelli, E.; Patricoin, E. F., 3rd; Cassidy, D. P.; Marx, M. A.; Rybkin, I. I.; Johnson, M. L.; Ou, S. I.; Lito, P.; Papadopoulos, K. P.; Jänne, P. A.; Olson, P.; Christensen, J. G. *Cancer Discov.* **2020**, *10*, 54–71.
69. Fell, J. B.; Fischer, J. P.; Baer, B. R.; Blake, J. F.; Bouhana, K.; Briere, D. M.; Brown, K. D.; Burgess, L. E.; Burns, A. C.; Burkard, M. R.; Chiang, H.; Chicarelli, M. J.; Cook, A. W.; Gaudino, J. J.; Hallin, J.; Hanson, L.; Hartley, D. P.; Hicken, E. J.; Hingorani, G. P.; Hinklin, R. J.; Mejia, M. J.; Olson, P.; Otten, J. N.; Rhodes, S. P.; Rodriguez, M. E.; Savechenkov, P.; Smith, D. J.; Sudhakar, N.; Sullivan, F. X.; Tang, T. P.; Vigers, G. P.; Wollenberg, L.; Christensen, J. G.; Marx, M. A. *J. Med. Chem.* **2020**, *63*, 6679–6693.
70. Hong, D. S.; Fakih, M. G.; Strickler, J. H.; Desai, J.; Durm, G. A.; Shapiro, G. I.; Falchook, G. S.; Price, T. J.; Sacher, A.; Denlinger, C. S.; Bang, Y. J.; Dy, G. K.; Krauss, J. C.; Kuboki, Y.; Kuo, J. C.; Coveler, A. L.; Park, K.; Kim, T. W.; Barlesi, F.; Munster, P. N.; Ramalingam, S. S.; Burns, T. F.; Meric-Bernstam, F.; Henary, H.; Ngang, J.; Ngarmchamnanrith, G.; Kim, J.; Houk, B. E.; Canon, J.; Lipford, J. R.; Friberg, G.; Lito, P.; Govindan, R.; Li, B. T. *N. Engl. J. Med.* **2020**, *383*, 1207–1217.
71. Hansen, R.; Peters, U.; Babbar, A.; Chen, Y.; Feng, J.; Janes, M. R.; Li, L. S.; Ren, P.; Liu, Y.; Zarrinkar, P. P. *Nat. Struct. Mol. Biol.* **2018**, *25*, 454–462.
72. Zeng, M.; Xiong, Y.; Safaee, N.; Nowak, R. P.; Donovan, K. A.; Yuan, C. J.; Nabet, B.; Gero, T. W.; Feru, F.; Li, L.; Gondi, S.; Ombelets, L. J.; Quan, C.; Jänne, P. A.; Kostic, M.; Scott, D. A.; Westover, K. D.; Fischer, E. S.; Gray, N. S. *Cell Chem. Biol.* **2020**, *27*, 19–31.e6.
73. McGregor, L. M.; Jenkins, M. L.; Kerwin, C.; Burke, J. E.; Shokat, K. M. *Biochemistry* **2017**, *56*, 3178–3183.
74. Jin, Z.; Zhao, Y.; Sun, Y.; Zhang, B.; Wang, H.; Wu, Y.; Zhu, Y.; Zhu, C.; Hu, T.; Du, X.; Duan, Y.; Yu, J.; Yang, X.; Yang, X.; Yang, K.; Liu, X.; Guddat, L. W.; Xiao, G.; Zhang, L.; Yang, H.; Rao, Z. *Nat. Struct. Mol. Biol.* **2020**, *27*, 529–532.
75. Realini, N.; Solorzano, C.; Pagliuca, C.; Pizzirani, D.; Armirotti, A.; Luciani, R.; Costi, M. P.; Bandiera, T.; Piomelli, D. *Sci. Rep.* **2013**, *3*, 1035.
76. Jin, Z.; Du, X.; Xu, Y.; Deng, Y.; Liu, M.; Zhao, Y.; Zhang, B.; Li, X.; Zhang, L.; Peng, C.; Duan, Y.; Yu, J.; Wang, L.; Yang, K.; Liu, F.; Jiang, R.; Yang, X.; You, T.; Liu, X.; Yang, X.; Bai, F.; Liu, H.; Liu, X.; Guddat, L. W.; Xu, W.; Xiao, G.; Qin, C.; Shi, Z.; Jiang, H.; Rao, Z.; Yang, H. *Nature* **2020**, *582*, 289–293.
77. Ma, C.; Sacco, M. D.; Hurst, B.; Townsend, J. A.; Hu, Y.; Szeto, T.; Zhang, X.; Tarbet, B.; Marty, M. T.; Chen, Y.; Wang, J. *Cell Res.* **2020**, *30*, 678–692.
78. Ullrich, S.; Nitsche, C. *Bioorg. Med. Chem. Lett.* **2020**, *30*, 127377.

79. Zhang, L.; Lin, D.; Sun, X.; Curth, U.; Drosten, C.; Sauerhering, L.; Becker, S.; Rox, K.; Hilgenfeld, R. *Science* **2020**, *368*, 409–412.
80. Dai, W.; Zhang, B.; Jiang, X. M.; Su, H.; Li, J.; Zhao, Y.; Xie, X.; Jin, Z.; Peng, J.; Liu, F.; Li, C.; Li, Y.; Bai, F.; Wang, H.; Cheng, X.; Cen, X.; Hu, S.; Yang, X.; Wang, J.; Liu, X.; Xiao, G.; Jiang, H.; Rao, Z.; Zhang, L. K.; Xu, Y.; Yang, H.; Liu, H. *Science* **2020**, *368*, 1331–1335.
81. https://chemrxiv.org/articles/preprint/The_Discovery_of_Ketone-Based_Covalent_Inhibitors_of_Coronavirus_3CL_Proteases_for_the_Potential_Therapeutic_Treatment_of_COVID-19/12631496/1.
82. https://clinicaltrials.gov/ct2/show/NCT04535167.
83. Douangamath, A.; Fearon, D.; Gehrtz, P.; Krojer, T.; Lukacik, P.; Owen, C. D.; Resnick, E.; Strain-Damerell, C.; Aimon, A.; Ábrányi-Balogh, P.; Brandão-Neto, J.; Carbery, A.; Davison, G.; Dias, A.; Downes, T. D.; Dunnett, L.; Fairhead, M.; Firth, J. D.; Jones, S. P.; Keeley, A.; Keserü, G. M.; Klein, H. F.; Martin, M. P.; Noble, M. E. M.; O'Brien, P.; Powell, A.; Reddi, R. N.; Skyner, R.; Snee, M.; Waring, M. J.; Wild, C.; London, N.; von Delft, F.; Walsh, M. A. *Nat. Comm.* **2020**, *11*, 5047.
84. Mons, E.; Jansen, I. D. C.; Loboda, J.; van Doodewaerd, B. R.; Hermans, J.; Verdoes, M.; van Boeckel, C. A. A.; van Veelen, P. A.; Turk, B.; Turk, D.; Ovaa, H. *J. Am. Chem. Soc.* **2019**, *141*, 3507–3514.
85. Keeley, A.; Ábrányi-Balogh, P.; Hrast, M.; Imre, T.; Ilaš, J.; Gobec, S.; Keserű, G. M. *Arch. Pharm.* **2018**, *351*, e1800184.
86. https://covid.postera.ai/covid.
87. F. von Delft, Drug Discovery Chemistry Virtual Meeting, 2020.

MIX

Papier | Fördert gute Waldnutzung

FSC® C083411

Kolibri 360 GmbH
Ferdinand-Jühlke-Straße 7
99095 Erfurt, Deutschland
produktsicherheit@kolibri360.de